Progress in Nuclear Energy New Series. Series IV.

Nuclear Power Safety

Georgia Institute of Technology Series
in Nuclear Engineering

Other Titles in Georgia Institute of Technology Series

Karam, R.A. & Morgan, K.Z. — *Environmental Impact of Nuclear Power Plants*

Kallflez, J.M. & Karam, R.A. — *Advanced Reactors: Physics, Design and Economics*

Karam, R.A. & Morgan, K.Z. — *Energy and the Environment Cost-Benefit Analysis*

**Georgia Institute of Technology Series
in Nuclear Engineering**

Nuclear Power Safety

Edited by
James H. Rust and Lynn E. Weaver

PERGAMON PRESS INC.

New York / Toronto / Oxford / Sydney / Frankfurt / Paris

Pergamon Press Offices:

U.S.A.	Pergamon Press Inc., Maxwell House, Fairview Park, Elmsford, New York 10523, U.S.A.
U.K.	Pergamon Press Ltd., Headington Hill Hall, Oxford OX3, OBW, England
CANADA	Pergamon of Canada, Ltd., 207 Queen's Quay West, Toronto 1, Canada
AUSTRALIA	Pergamon Press (Aust) Pty. Ltd., 19a Boundary Street, Rushcutters Bay, sN.S.W. 2011, Australia
FRANCE	Pergamon Press SARL, 24 rue des Ecoles, 75240 Paris, Cedex 05, France
WEST GERMANY	Pergamon Press GmbH, 6242 Kronberg/Taunus, Frankfurt-am-Main, West Germany

Copyright © 1976 Pergamon Press Inc.

Library of Congress Cataloging in Publication Data

Main entry under title:

Nuclear power safety.

   Includes index.
   1. Nuclear engineering -- Safety measures -- Addresses, essays, lectures. 2. Atomic power-plants -- Safety measures -- Addresses, essays, lectures. I. Rust, James H. II. Weaver Lynn E.
TK9152.N79   1976   621.48'35   76-44475
ISBN 0-08-021417-7
ISBN 0-08-021744-3 (pbk.)

All Rights Reserved. No part of this publication may be reproduced, stored in a retrieval system or transmitted in any form or by any means: electronic, electrostatic, magnetic tape, mechanical, photocopying, recording or otherwise, without permission in writing from the publishers.

Printed in the United States of America

CONTENTS

PREFACE

SAFETY ASPECTS OF CONTAINMENT SYSTEM DESIGN FOR NUCLEAR
   POWER PLANTS
      Harold Oslick, Ebasco Services, Inc.   1

POWER PLANT SITING
      Geoffrey G. Eichholz, Georgia Institute of Technology. 33

NUCLEAR REGULATORY COMMISSION REGULATIONS AND
   LICENSING
      Robert Minogue, U.S. Nuclear Regulatory Commission 57

INSPECTION AND QUALITY ASSURANCE
      Hugh Dance, U.S. Nuclear Regulatory Commission 87

RELEASE OF RADIOACTIVE MATERIALS FROM REACTORS
      Karl Z. Morgan, Georgia Institute of Technology. 101

WAYS OF REDUCING RADIATION EXPOSURE IN A FUTURE
   NUCLEAR POWER ECONOMY
      Karl Z. Morgan, Georgia Institute of Technology. 155

REACTOR CONTROL
      Lynn E. Weaver, Georgia Institute of Technology. 169

FLUID FLOW AND HEAT TRANSFER IN WATER-COOLED REACTORS
      Roger W. Carlson, Georgia Institute of Technology. 209

MANAGEMENT OF RADIOACTIVE WASTES
      J. O. Blomke, Oak Ridge National Laboratory. 255

GENERAL FEATURES OF EMERGENCY CORE COOLING SYSTEMS
      R. W. Shumway, Aerojet Nuclear Company 281

A REVIEW OF ACCIDENT RISKS IN LIGHT-WATER-COOLED
   NUCLEAR POWER PLANTS
      Lynn E. Weaver, Georgia Institute of Technology. 303

CURRENT PROBLEMS IN REACTOR SAFETY
      John T. Telford, U.S. Nuclear Regulatory Commission. 351

TRANSPORTATION OF IRRADIATED FUELS
      Jack Rollins, Nuclear Assurance Corporation. 359

Appendices
  A--ACCIDENT ANALYSIS 381
  B--ABBREVIATIONS 403
INDEX. 407

PREFACE

Nuclear Power Safety is receiving much attention in the popular news media and technical journals and is often a topic of discussion and debate at meetings of civic clubs, professional societies, etc. The main concerns of the public are focused on 1) radioactivity in routine effluents and its effect on human health and the environment, 2) the possibility of serious reactor accidents and their consequences, 3) transportation accidents involving radioactive waste, 4) the disposal of radioactive waste particularly high-level wastes and, 5) the possible theft of special nuclear materials and their fabrication into a weapon by terrorists. In response to these concerns the nuclear industry has adopted the Defense-In-Depth approach to nuclear power safety. This approach relies upon 1) the careful design, construction and operation of nuclear facilities so that malfunctions which could lead to major accidents will be highly improbable; 2) the incorporation of systems to prevent such malfunctions as do occur from turning into major accidents, e.g., automatic shutdown (SCRAM) and leak detection systems; and 3) built in systems to limit offsite consequences of postulated, major accidents, e.g., emergency core cooling systems.

The Defense-In-Depth concept can be viewed as having three echelons. The first echelon provides accident prevention through 1) sound system design that can be built and operated with stringent quality standards, 2) a high degree of freedom from faults and errors, 3) a high tolerance for malfunctions should they occur, 4) tested components and materials, and 5) redundancy of instrumentation and controls. The second echelon of defense assumes there will be human error or equipment failure and provides protection systems to maintain safe operation or shut the plant down safely when such accidents occur. Some examples are 1) redundant sources of in-plant electric power, 2) sensitive detection systems to warn of incipient failures of fuel cladding or coolant systems, and 3) systems for automatic shutdown of reactors on a signal from monitoring instruments. The third echelon of defense provides additional margins to protect the public should severe failure occur despite the first two echelons. Examples of the third echelon include 1) concrete containment buildings, typically 4 ft thick, reinforced with steel and 2) Emergency Core Cooling Systems (ECCS) to flood the reactor core with water if the coolant is lost through a breach of the core coolant system.

The material in this book is derived from the notes for a short course on Nuclear Power Safety offered at Georgia Tech and addresses itself to those issues of public concern listed above and the implementation of the Defense-In-Depth concept of nuclear power safety. It presents an overview of the subject and is written for those who are beginning to study the problems and their solutions associated with safety in the nuclear power industry. It is assumed that the reader has a basic understanding of scientific and engineering principles and some familarity with nuclear power reactors. Considering the wide attention being given to the subject of nuclear power safety and the lack of organized up-to-date information, this publication is timely.

    The Editors are grateful to the many contributors who gave of their time so that the information could be shared and to each of the secretaries in the office of the School of Nuclear Engineering at Georgia Tech for her work in typing the final material.

                                            James H. Rust and Lynn E. Weaver

July, 1976   School of Nuclear Engineering
Atlanta, Georgia   Georgia Institute of Technology

SAFETY ASPECTS OF CONTAINMENT SYSTEM DESIGN
FOR NUCLEAR POWER PLANTS

Harold Oslick
Chief Engineer, Nuclear Licensing
Ebasco Services, Inc.

New York, New York 10006

INTRODUCTION

In the design of containment systems for nuclear power plants it is necessary to think in terms of a systems concept and that the containment is not a structure, but it is a system. This system is engineered to protect the public from the consequences of any credible break in the reactor coolant system. This purpose is accomplished through the functioning of the numerous subsystems that comprise the total containment system. The principal functions which must be performed by the containment system are as follows:

a) withstand the pressure buildup within the containment structure from the postulated reactor coolant system breaks (hereafter referred to as Loss-of-Coolant Accident--LOCA), and maintain its structural integrity indefinitely after the LOCA;

b) operate in conjunction with the emergency core cooling systems (ECCS) to limit energy releases from the LOCA to prevent pressure buildup in the containment structure in excess of design limits;

c) limit offsite releases of radioactive materials during and following a LOCA to below federal established guidelines (10 CFR 100) (1),

d) reduce pressure and temperature in the containment structure following a LOCA to permit recovery; and

e) provide protection for the reactor coolant system (RCS) from external environmental influences.

The first four of these functions are all related to protection against a postulated LOCA. In order to understand how the containment system accomplishes these functions, it is necessary for us to briefly trace through the course of a LOCA. We will concentrate on light water reactors (LWR). High temperature gas-cooled reactors (HTGR) are beyond the scope of this presentation but some differences will be highlighted.

In a LOCA the initial phase consists of a rapid expulsion of fluid and energy from the break. This energy finds its way into the containment structure where it results in an increase in pressure and temperature. The accident sequence assumes a loss of offsite power and so the standby diesel generators start on the LOCA signal. The containment barrier to fission product escape is completed on either a LOCA or high radiation signal through functioning of the containment isolation system. The ECCS functions to cool the core and the containment cooling systems remove energy from the containment volume, thereby lowering the pressure and temperature. Post LOCA conditions also result in hydrogen generation through radiolysis of water and this must be dealt with through atmospheric control systems to prevent accumulation of an explosive mixture in the containment.

CONTAINMENT CONCEPTS

Until such time as the containment active subsystems can begin to function, pressure and temperature within the containment structure will increase. During this early phase, the only functioning energy absorption mechanism for which credit is taken is direct absorption onto surfaces in the structure, or some passive means of condensing steam. It is principally this early phase of the accident that has led to the development of the different containment concepts in use today.

Most pressurized water reactor (PWR) plants built today utilize a dry containment concept, wherein energy is dissipated via heat transfer to exposed surfaces. A typical dry containment arrangement is shown in Fig. 1.

In an effort to reduce containment size and cost, Westinghouse introduced the ice condenser version of a pressure suppression containment (2). In the pressure suppression concept, the steam from the break is channeled through a condensing medium, in this case ice, which results in a greatly reduced containment post LOCA pressure and therefore a reduced containment design pressure. This type of containment has two distinct steam volume compartments operating at different pressures. This can be seen in Fig. 2. Ice condenser containments have been used in a few pressurized water reactor installations and are planned for use on off-shore power systems where size and weight of containment structure are a major consideration.

To date, all boiling water reactor (BWR) plants have been designed with pressure suppression type of containments. However, instead of ice they pass the steam released from the LOCA through a pool of water. Again this type is composed of two distinct steam volumes. This BWR pressure suppression concept has evolved through three models, the earliest Mark I, known as the Drywell/Torus or light bulb type, the Mark II over and under type, and the current Mark III model. These are shown in Figs. 3, 4, and 5.

During the past couple of years, concern has been voiced over the appropriate design loads to be used for the design of the BWR suppression pool type of containment. Phenomena associated with the existence of a large pool of water are now considered in detail in the design. This includes seismic sloshing of the water under earthquake conditions, the effect of variation of depth of submersion of discharge vents due to seismic sloshing, and the swell and frothing effect of bubbling large quantities of a steam and air mixture through the water. These considerations have been studied both analytically

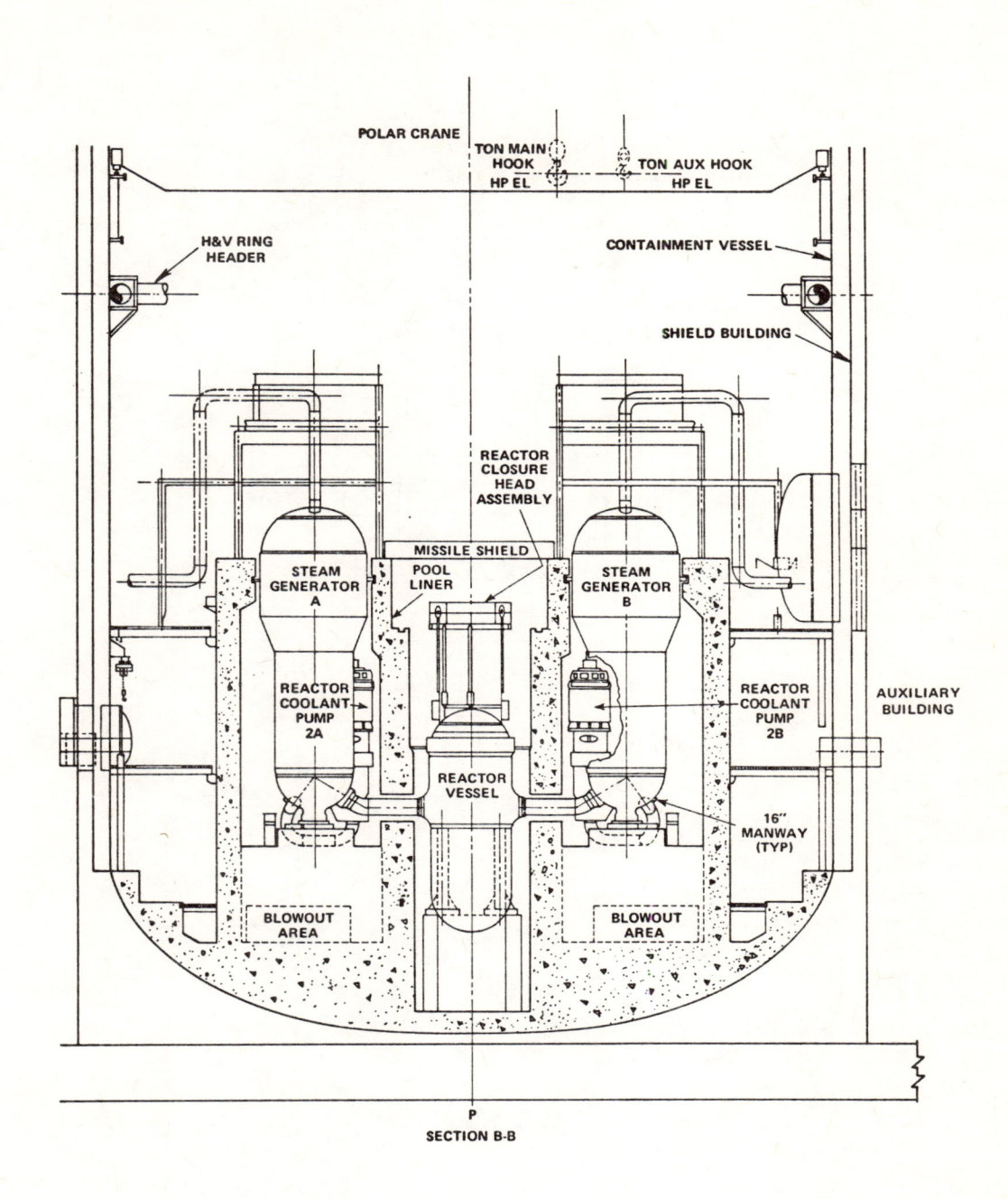

Fig. 1. Typical PWR Dry Containment

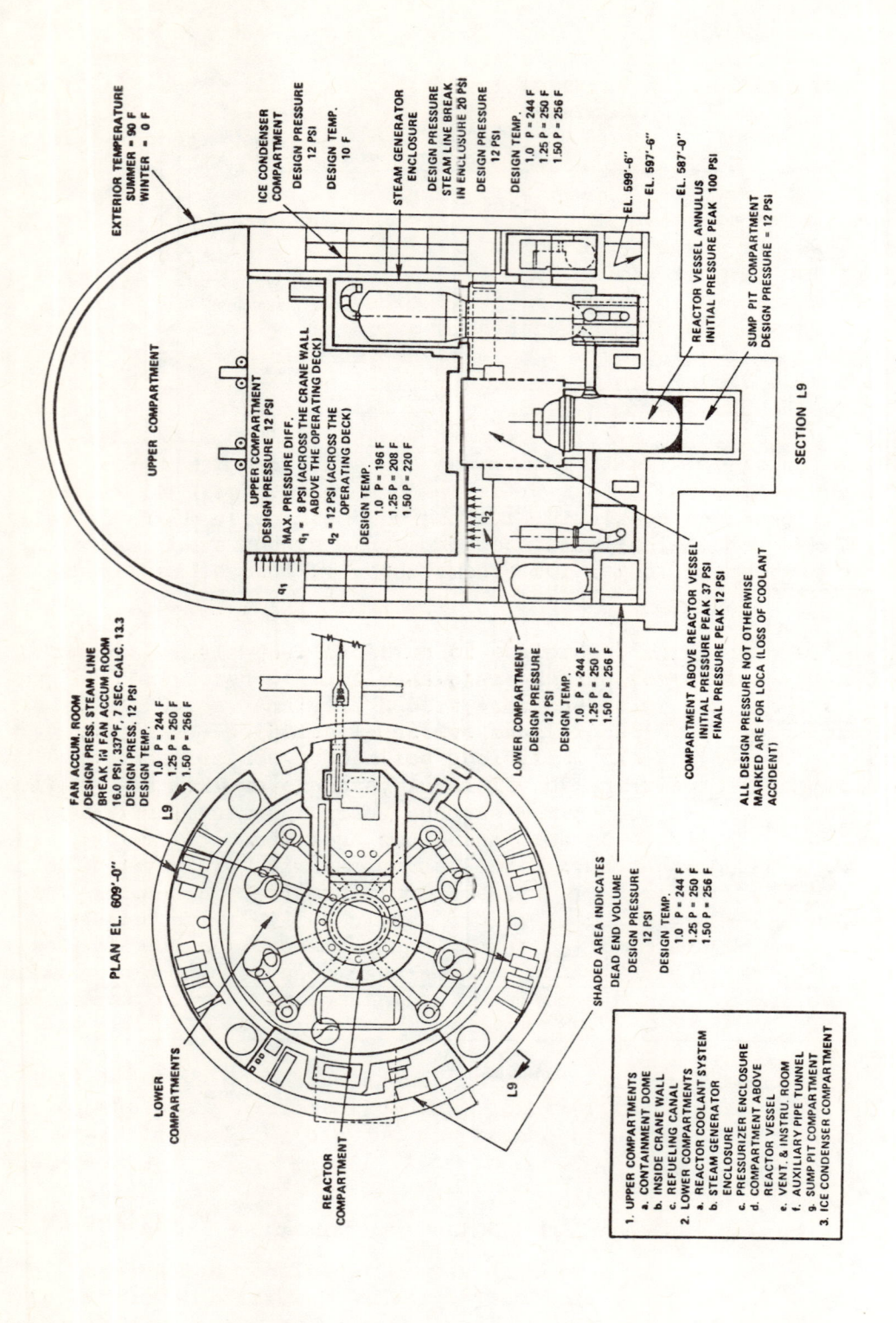

Fig. 2. PWR Ice Condenser Containment

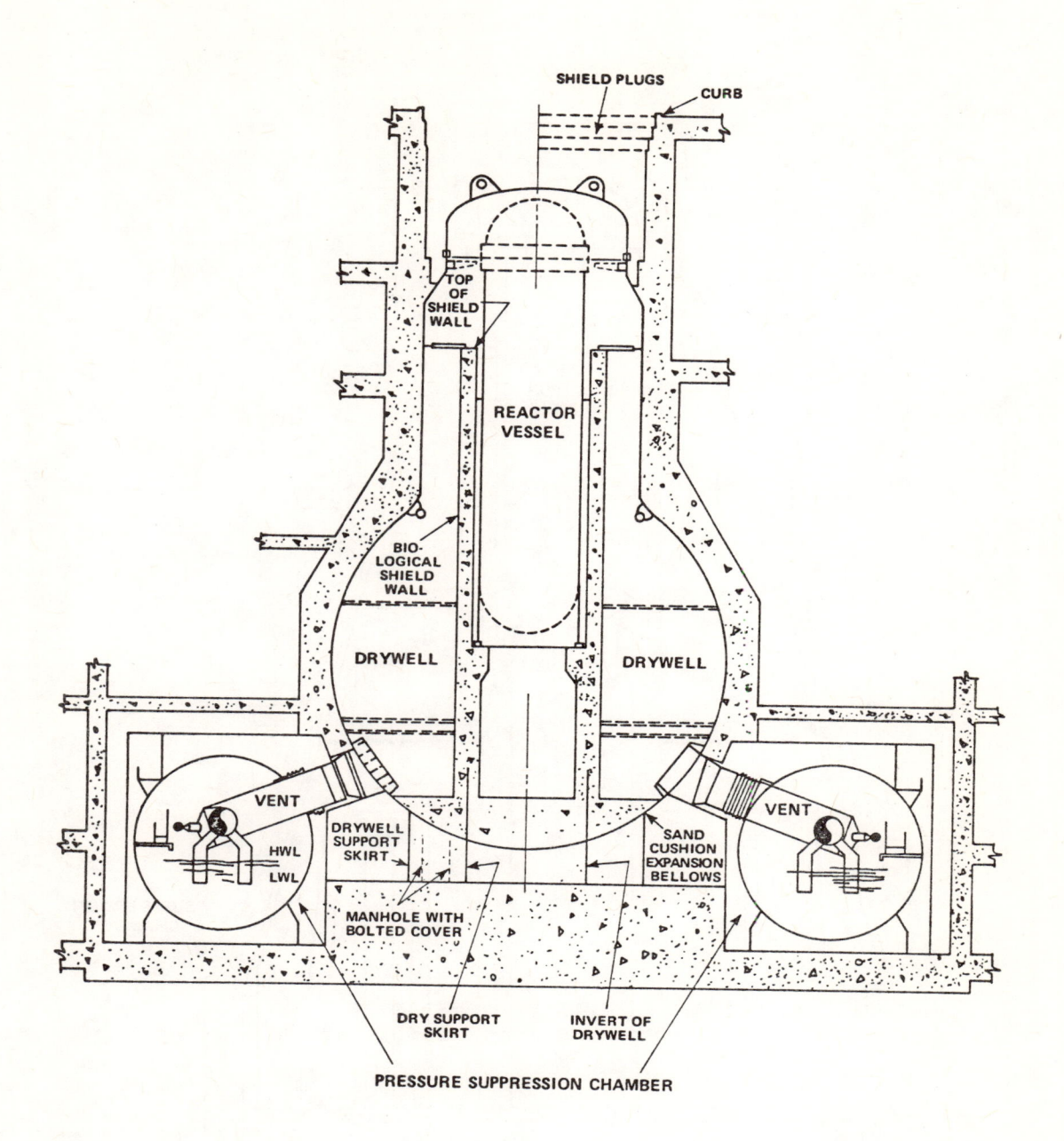

Fig. 3. BWR Light Bulb Torus Containment

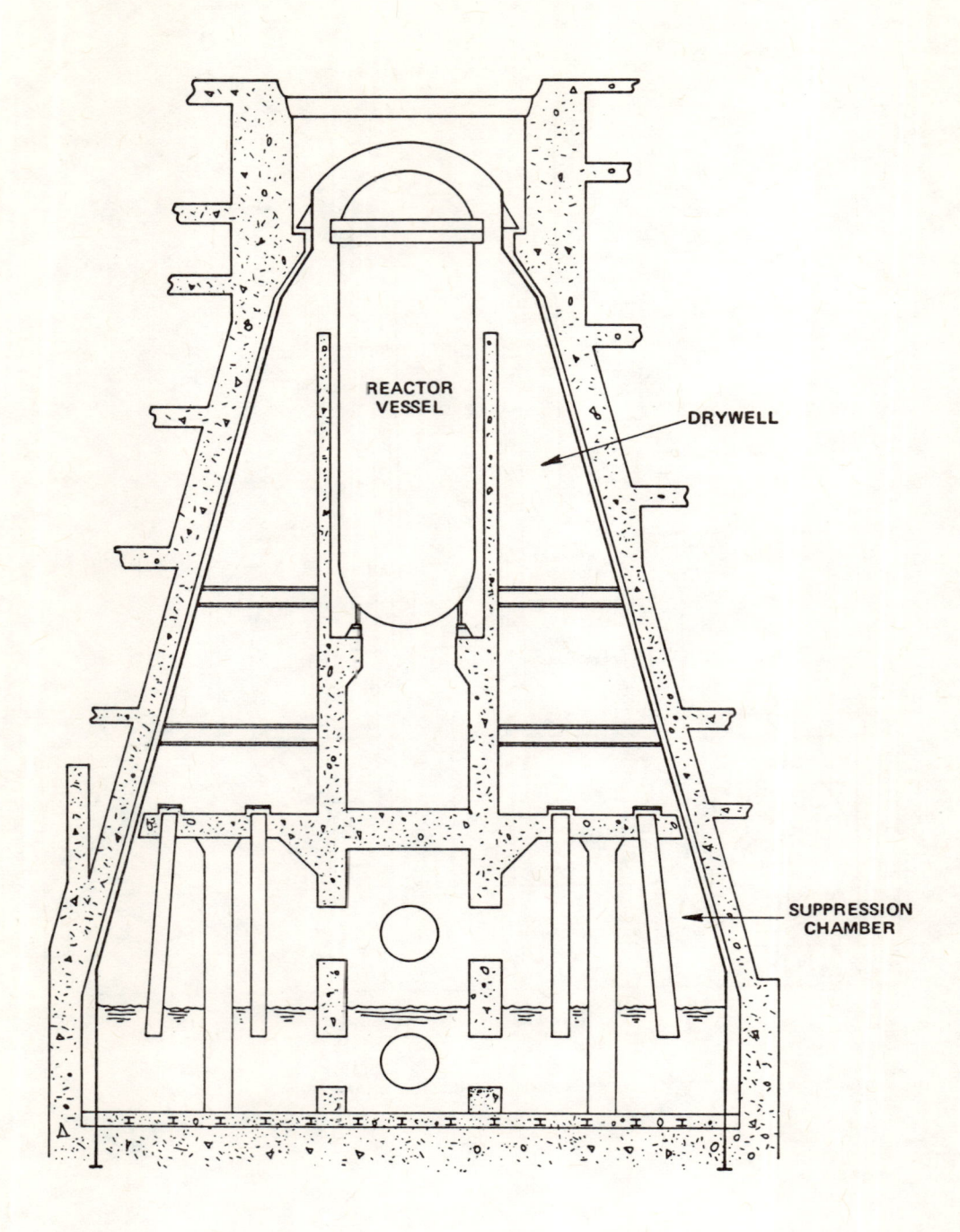

Fig. 4. BWR Mark II Containment

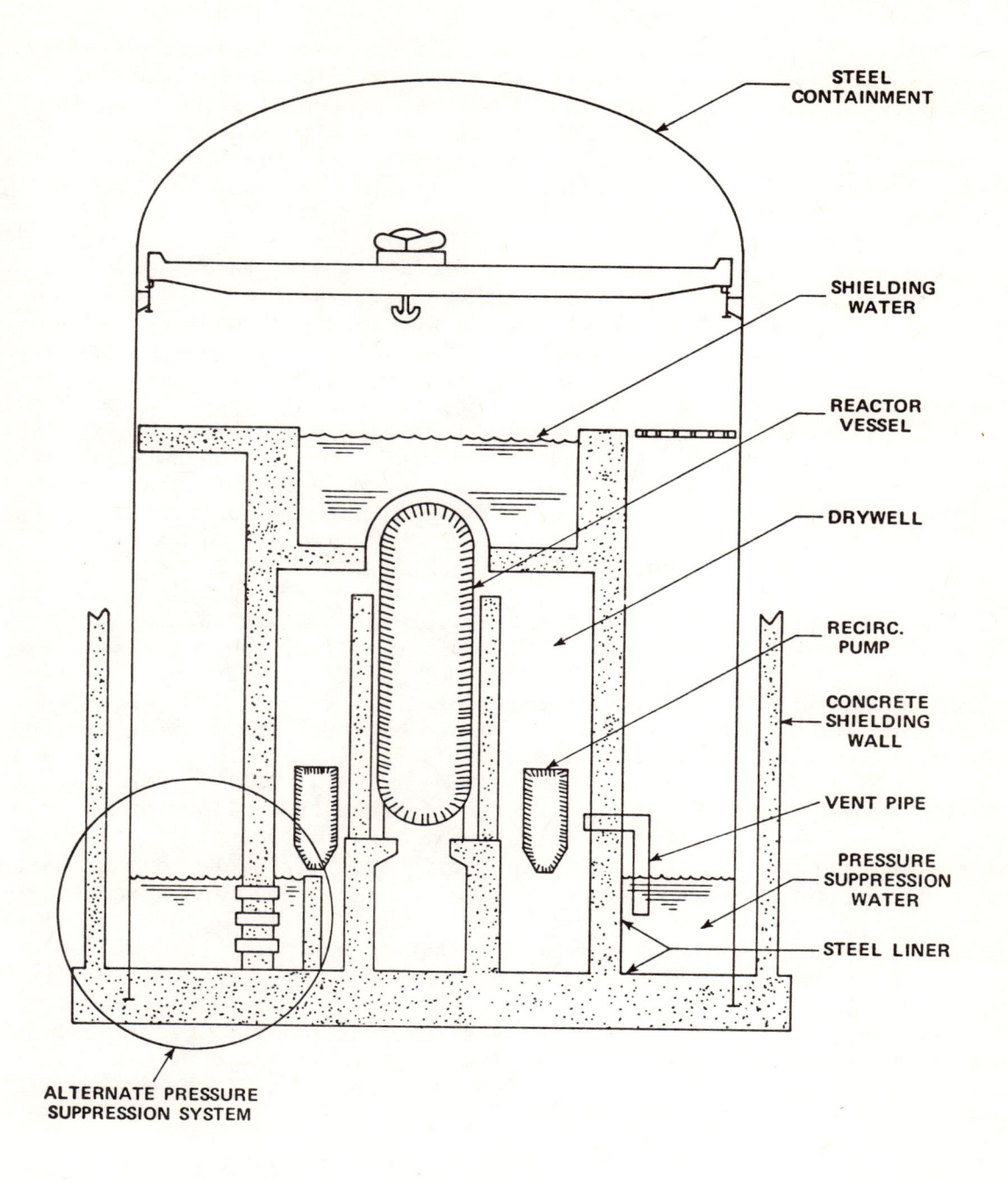

Fig. 5. BWR Mark III Containment

and experimentally in depth by the General Electric Company. Design loads for these phenomena are now available.

The reasons for the development of the different containment concepts have, for the most part, been related to economics. The prevailing thought was the lower the design pressure, the lower the evaluated plant cost would be. The same philosophy was applied to containment size. As a result, we had everyone striving for the smallest containment with the lowest design pressure. The results, in my opinion, were disasterous, as we begin to examine system related problems and interfaces.

With respect to containment size, one need only walk through some of the smaller ones to admit a mistake. The problems associated with maintenance accessibility and inservice inspection are a mechanic's nightmare. A good comparison in miniature can be found under the hood of your modern car. The effects of lower design pressure were more subtle and were not realized until the ECCS crisis. The development of more sophisticated computer codes showed us that ECCS performance was enhanced with increased containment back pressure, this being much more pronounced for the PWR than for the BWR systems. Systems design considerations now have to balance reduced design pressure versus degraded ECCS performance in developing an optimum containment design.

FISSION PRODUCT BARRIER

A basic assumption accompanying the postulated LOCA is the loss of off-site electrical power. Therefore, the functioning of any active system must be assumed to be delayed until diesel generator power is available. The only exception to the rule is the containment isolation system (CIS) which receives its actuating power either from storage batteries or instrument air reservoirs designed to safety criteria.

During normal plant operations, there are a number of functioning processes which require communication between the interior volume of the containment structure and systems and equipment located elsewhere in the plant complex. These include component or service water cooling, instrument air lines, sampling system lines, reactor coolant water purification and radioactive waste treatment lines, and some ventilation and purge lines. Were these process communication links to remain open following a LOCA, the fission product retention function of the containment system would be compromised. Since the LOCA postulation considers the utterly ridiculous assumption that large quantities of fission products somehow materialize within the containment boundary at the onset of a LOCA, we cannot await the availability of diesel power to isolate these links.

The CIS consists of redundant valves in series on all containment fluid penetrations. Wherever possible, one valve is inside and the other outside the containment structural boundary. Each of these valves is powered either from a compressed air reservoir located at the valve, or by a dc operated solenoid with spring loaded gate. The entire CIS, from signal sensor through the valves and power supplies, is designed so that no single failure will cause loss of the fission product retention function. The only process lines allowed to remain open following the LOCA are those essential to safety system operation. All of this latter group of process penetrations are designed to seismic category I and other appropriate safety standards and completely closed to the atmosphere outside of the containment, thus effectively preventing fission products leaking beyond their boundary. Some typical containment isolation arrangements are shown in Fig. 6 through 6E.

The containment pressure retention boundary and the valves of the containment isolation system form the primary fission product boundary. Leakage rates through this boundary vary from 0.1 to 0.5% of the enclosed volume per day, depending on site boundary distance, meteorology, population density, and the plant fission product treatment and removal system. The successful accomplishment of the fission product retention function consists not only of maintenance of the primary fission product barrier, but also depends on the interrelated functioning of the fission product removal system.

FISSION PRODUCT REMOVAL SYSTEMS

There are basically two concepts of fission product removal used in containment system design; they are removal within the containment structure boundary and removal outside of that boundary, but within a secondary containment boundary.

The radioactive source terms assumed to exist in the containment following a LOCA for purposes of fission product removal system design, calculation of exposures to the public, and calculation of radiation level for purposes of component and equipment design, are not in any way related to the LOCA. They are arbitrarily established for purposes of site safety evaluation. They are published in TID-14844 ([3]), which is referenced by Title 10, Code of the Federal Regulations, Part 100. These source terms assume the following releases from the reactor core:

a) 100% of the noble gases
b) 50% of the halogens
c) 1% of the particulates

In all containment concepts, wet or dry, credit is given for plate out of half of the halogen source term on containment internal surfaces. Other detailed assumptions to be used in evaluating offsite doses are contained in Regulatory Guides 1.3 ([4]) and 1.4 ([5]) for a BWR and PWR, respectively.

For those containment systems using internal fission product removal mechanisms, two such mechanisms have been employed:

a) removal by air recirculation through charcoal filter beds, and
b) removal through the containment spray process.

The charcoal filter removal system was used fairly extensively in the 1966 to 1970 vintage PWR plants. This system made use of the existing containment fan coolers which are discussed later, and simply added a HEPA/charcoal adsorber filter train in series. This method relies on fan forced recirculation of the virtually 100% humidity post-accident containment atmosphere through the adsorber beds to remove radioactive halogens, particularly iodine. The removal rate is dependent on the rate of containment air recirculation through the filters, as well as the filter removal efficiency. This results

VALVE ARRANGEMENT NO.	INSIDE CONTAINMENT	OUTSIDE CONTAINMENT	PENE-TRATION NO.
1	(check valve)	⎯⎯⎯⎯ (M) valve	17,18,19, 20,21,22
2	(check valve)	⎯⎯⎯⎯ (M) valve	30 80,81
3	valve ←⎯⎯⎯	⎯⎯⎯⎯ valve	50
4	(check valve)	⎯⎯⎯⎯ (A) valve	32,38,54 82
5	(M) valve ⎯⎯⎯	⎯→ (M) valve	31,34,35
6	(A) valve ⎯⎯⎯	⎯→ (A) valve	44,51

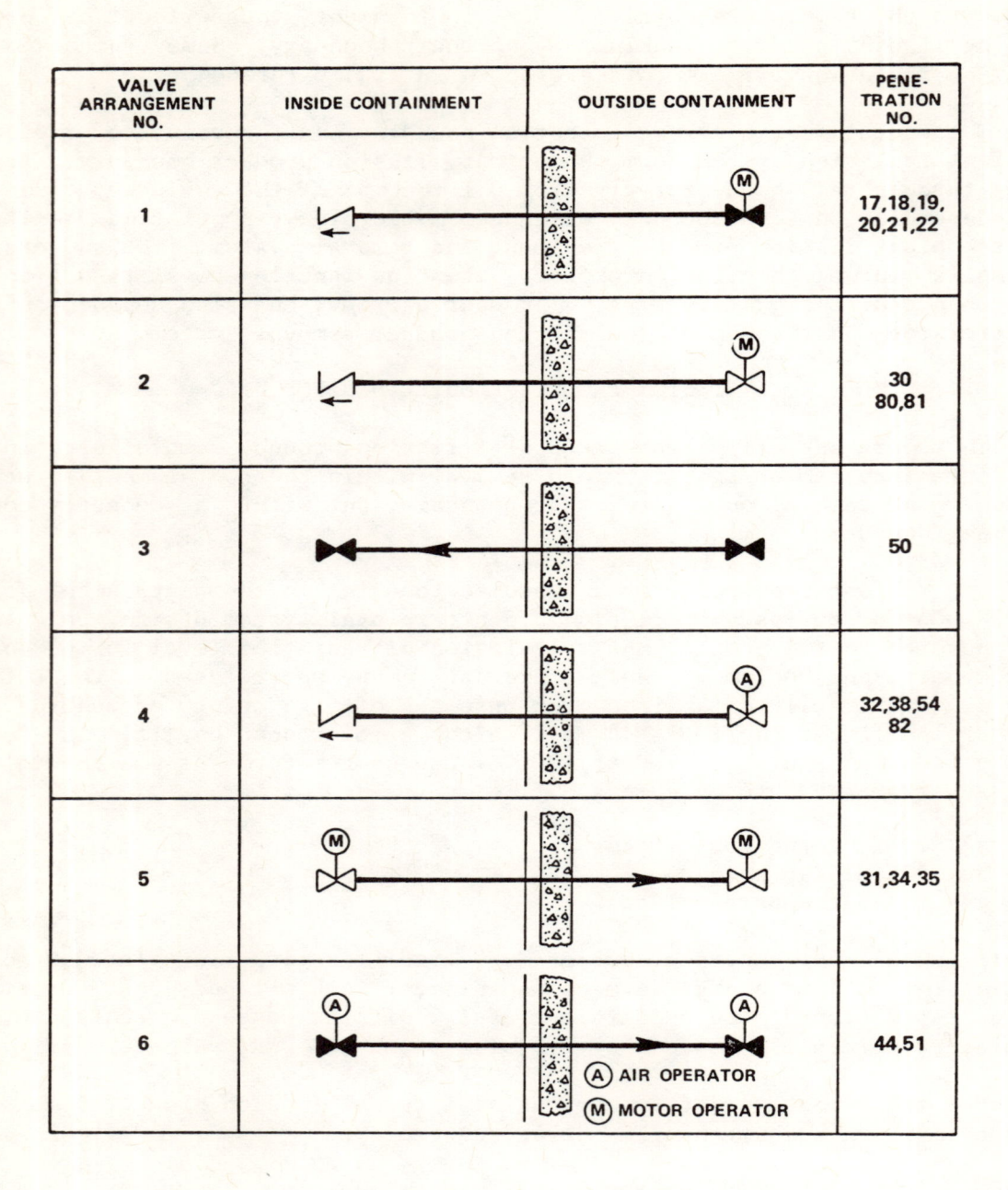

(A) AIR OPERATOR
(M) MOTOR OPERATOR

Fig. 6. Typical Containment Penetration Arrangements

VALVE ARRANGEMENT NO.	INSIDE CONTAINMENT	OUTSIDE CONTAINMENT	PENE-TRATION NO.
7			13,14,15,16
8			41

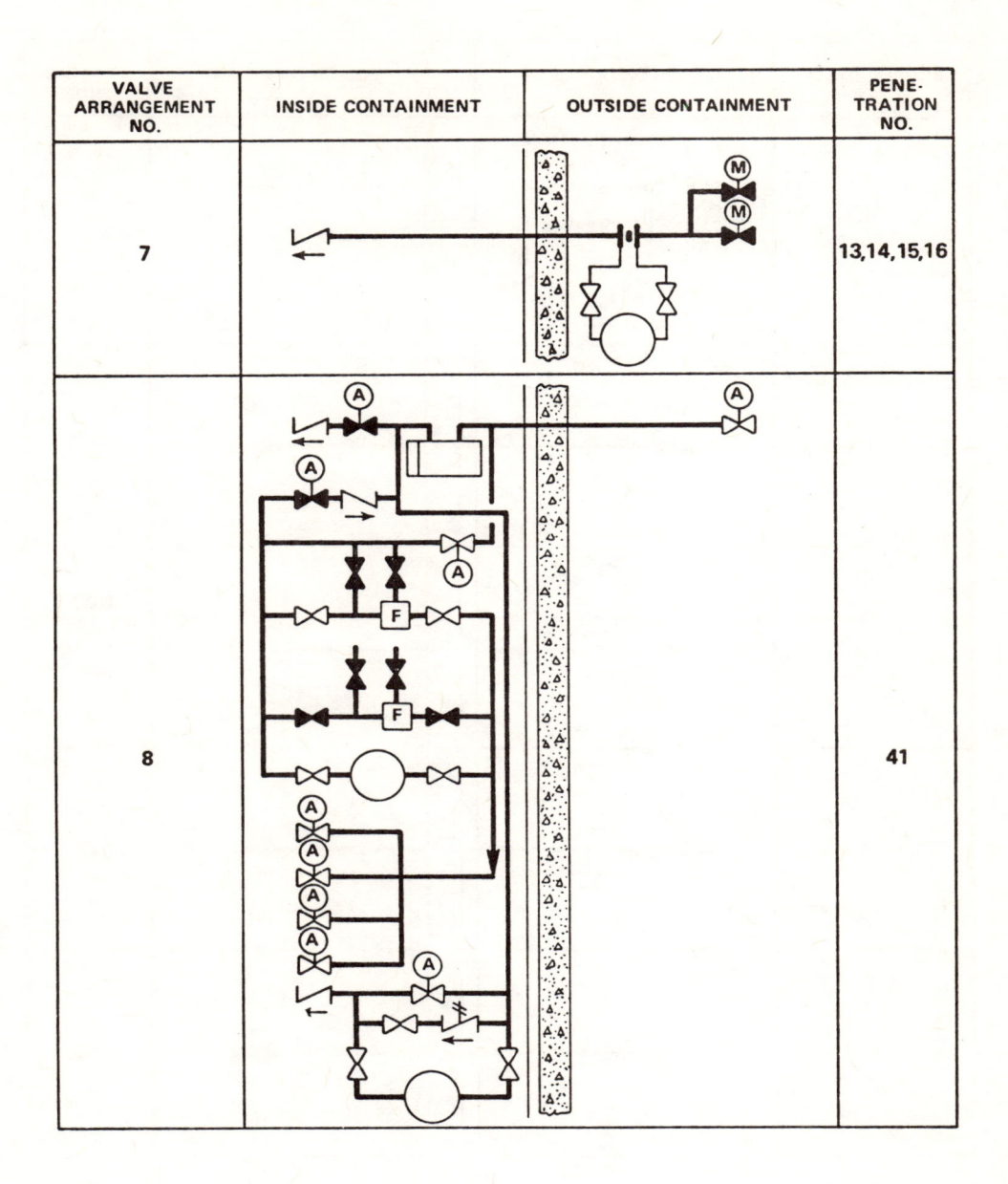

Fig. 6A. Typical Containment Penetration Arrangements

VALVE ARRANGEMENT NO.	INSIDE CONTAINMENT	OUTSIDE CONTAINMENT	PENETRATION NO.
9	TEST CONNECTION / "O" RINGS		53
10			27, 28
11			5,6,33, 42,43,48, 49,52,55, 56,83,86, 91
12			29
13			45

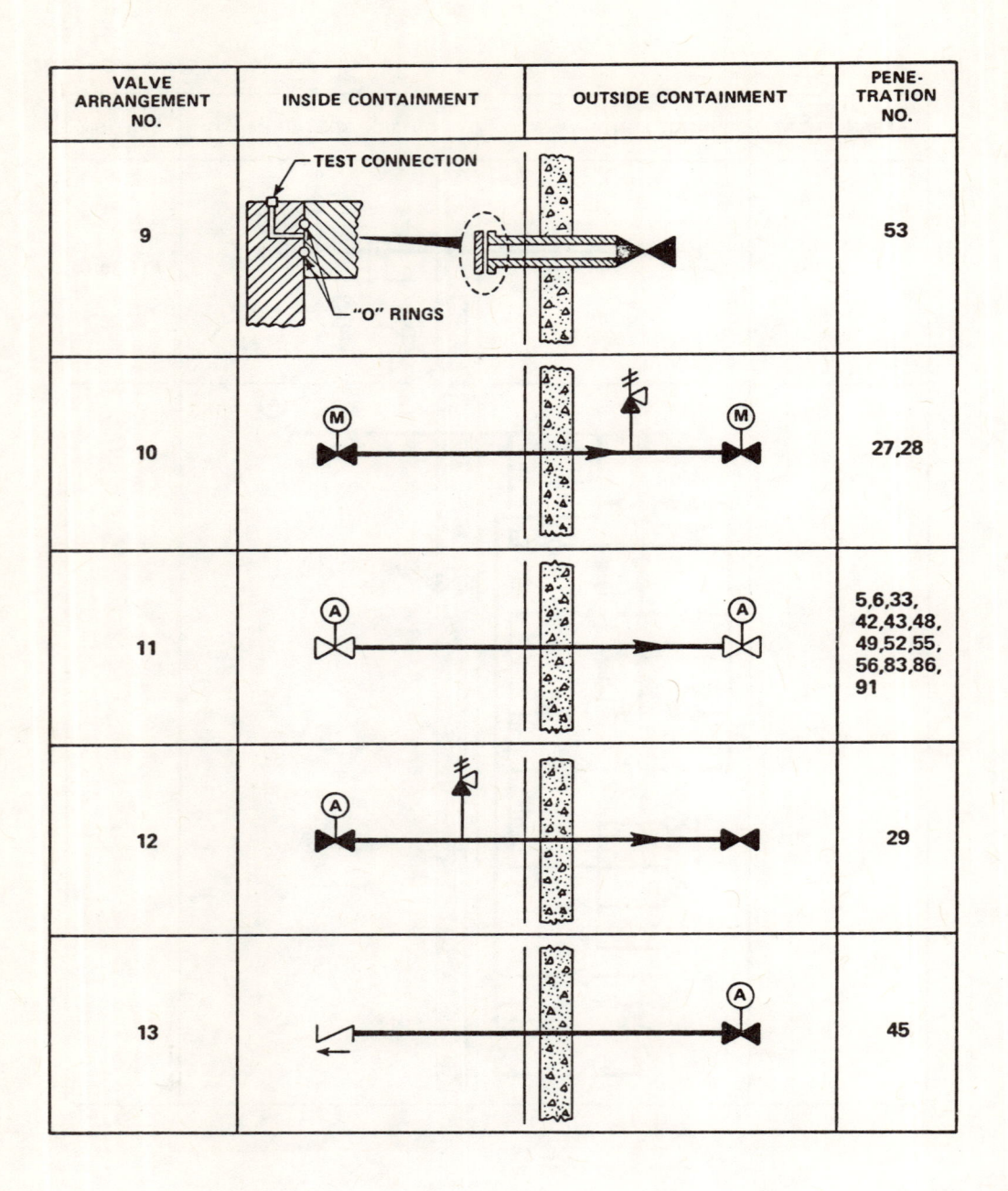

Fig. 6B. Typical Containment Penetration Arrangements

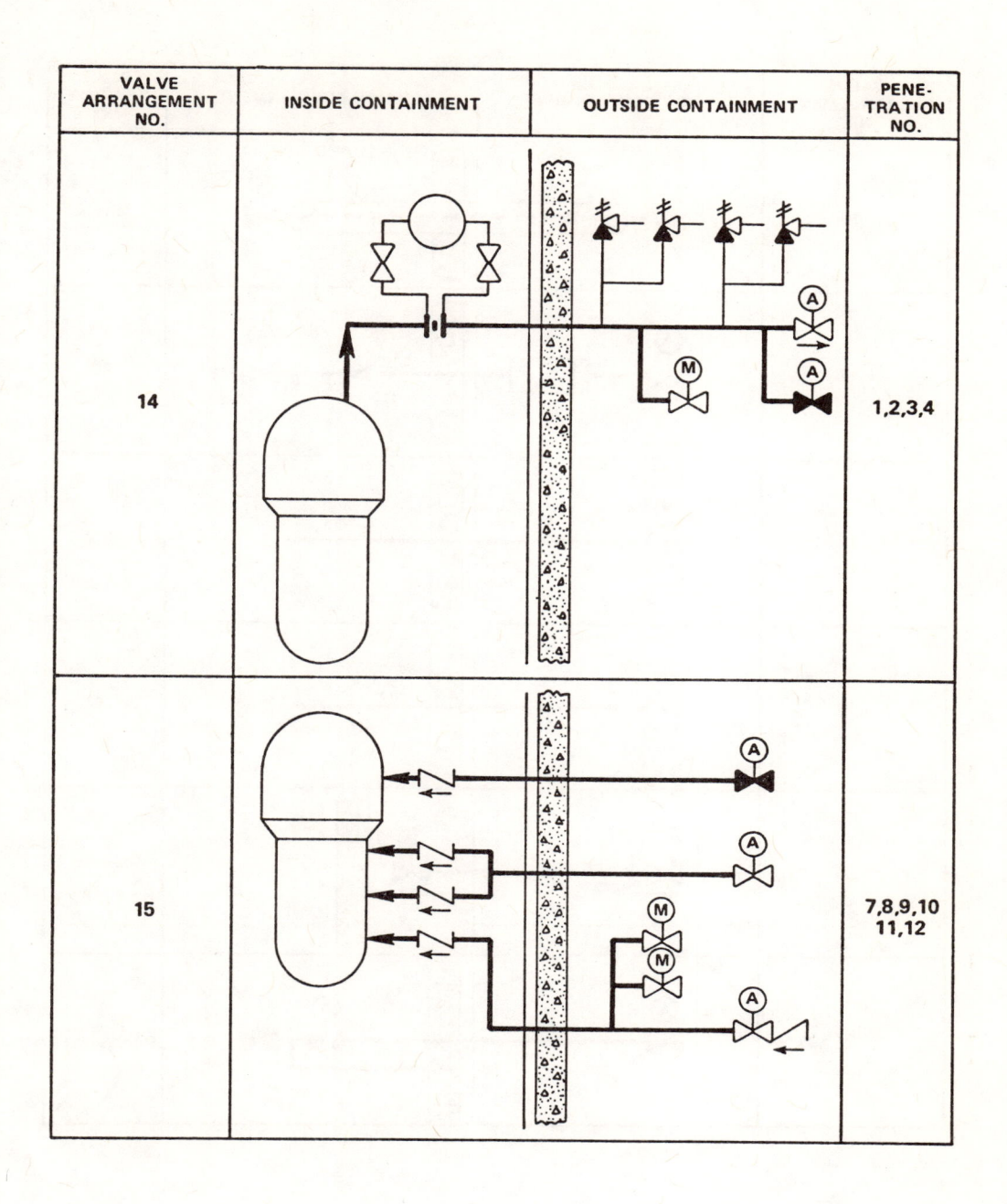

Fig. 6C. Typical Containment Penetration Arrangements

VALVE ARRANGEMENT NO.	INSIDE CONTAINMENT	OUTSIDE CONTAINMENT	PENETRATION NO.
16		GUARD PIPE	23,24
17			40
18			36,37
19			39
20			65,75
21			63,68,73,78

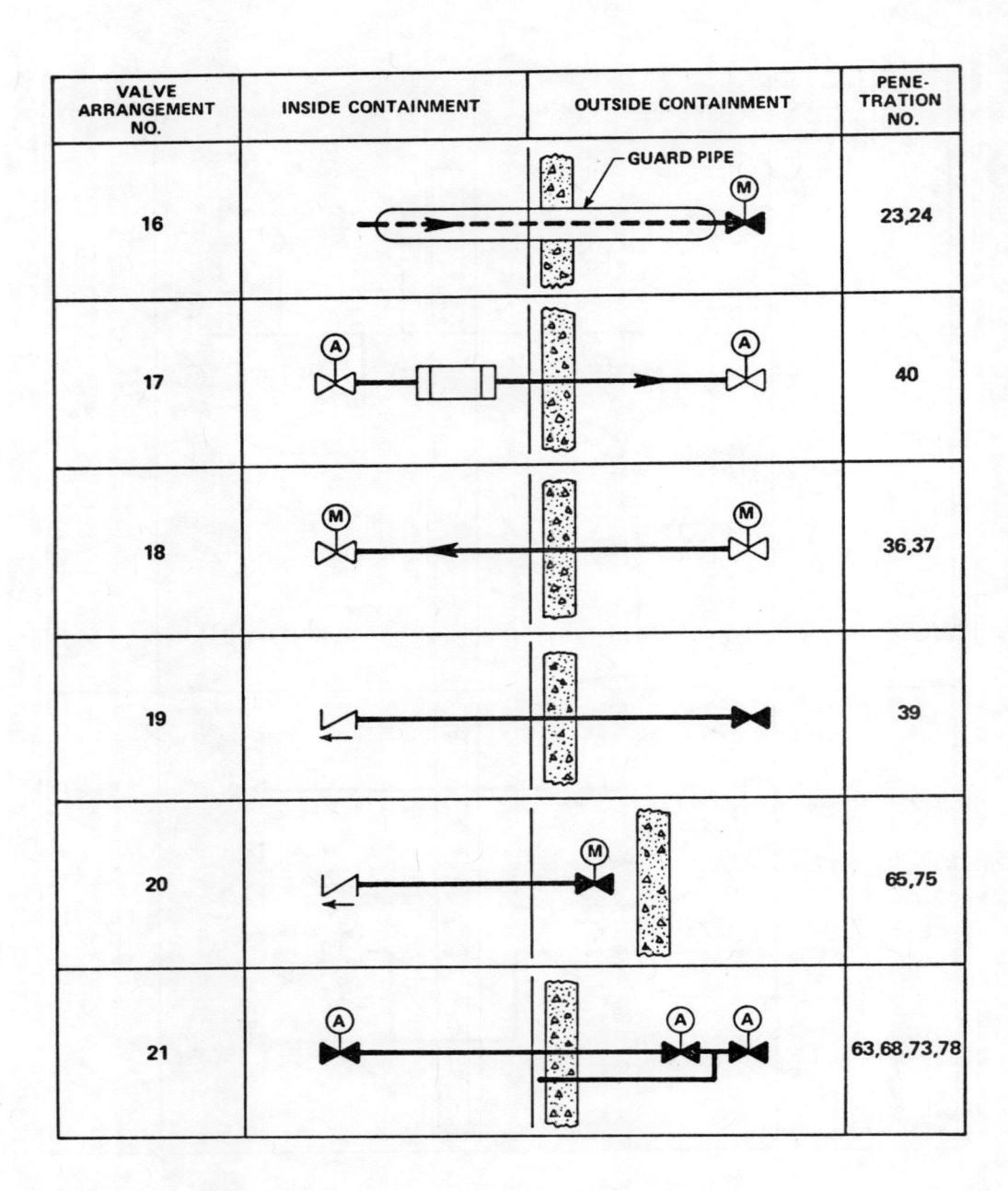

Fig. 6D. Typical Containment Penetration Arrangements

VALVE ARRANGEMENT NO.	INSIDE CONTAINMENT	OUTSIDE CONTAINMENT	PENE-TRATION NO.
22	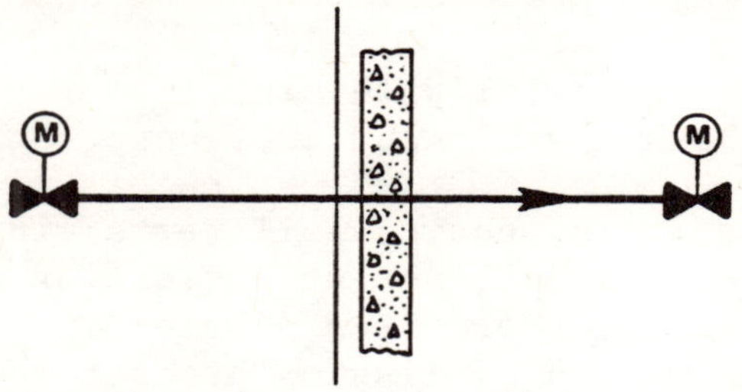 (M valve)	(M valve)	60,70
23	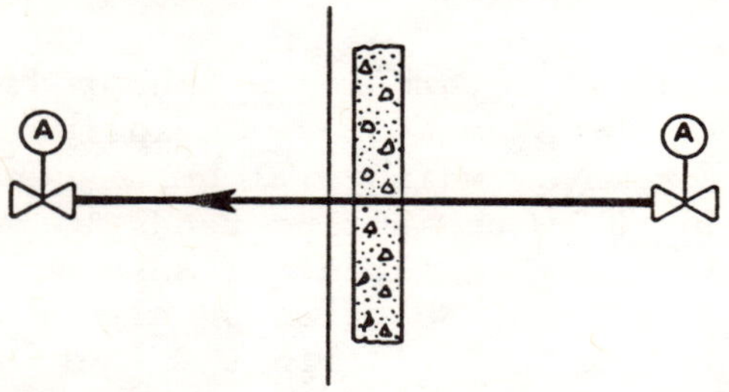 (A valve)	(A valve)	90
24	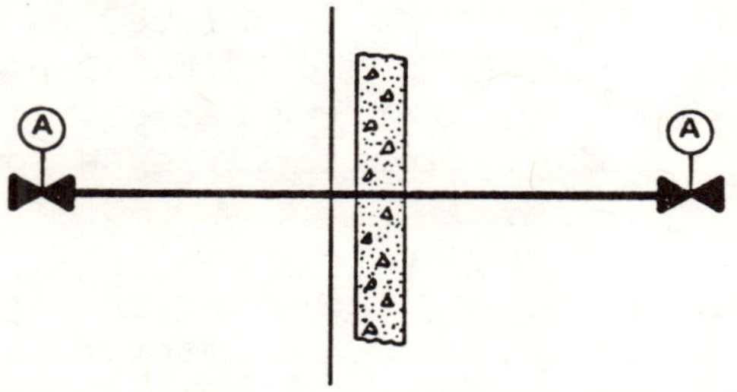 (A valve)	(A valve)	94,95

Fig. 6E. Typical Containment Penetration Arrangements

in the following requirements:

   a) very large fan motors qualified to operate in the post-accident environment conditions of temperature, pressure, and humidity;

   b) large additions to diesel generator capacity requirements;

   c) seismic category I containment duct systems to assure fairly uniform sweeping of the containment atmosphere; and

   d) charcoal filter material with high filter efficiencies at high humidity for both organic and inorganic forms of iodine.

Despite developmental efforts, charcoal filter efficiencies drop off dramatically at 100% humidity conditions, and at humidities even below that for organic iodines, especially if there is a possibility of filter water logging. These factors make this method of fission product removal very expensive and the last choice of most designers.

A more popular method of internal fission product removal from PWR's uses the containment spray system (CSS). The primary function of this system is post-accident energy removal which will be discussed in more detail later. The very fine mist produced by the sprays tends to adsorb both airborne particulate and gaseous iodine. The extent of adsorption depends upon droplet size, free fall height, and pH. Solution pH also affects the capability to retain the adsorbed iodine in solution in the containment sump.

The spray water used in PWR ECCS is borated to maintain reactivity shutdown margin in the reactor core. The addition of either sodium hydroxide or sodium thiosulfate was found to greatly enhance the iodine getter properties of the spray systems, with the hydroxide giving a somewhat higher removal rate.

During normal plant operation, the water in the refueling water storage tank is used during refueling. This same water is used for emergency core cooling system functions and containment spray functions. Therefore, if an hydroxide additive is used as an iodine getter, then it must be stored in a separate container and mixed with the spray solution when it is called upon for LOCA conditions. Figure 7 shows a typical system of this type.

One other caution relating to systems interactions should be noted. The resulting caustic spray solution reacts with aluminum and zinc galvanized surfaces to generate hydrogen. This must be taken into account in the design of post-accident hydrogen control systems which are discussed later.

Fission product removal systems external to the primary pressure containment boundary generally consist of a secondary containment boundary enclosing the primary boundary. Figure 8 shows a typical double barrier PWR containment. Leakage from the primary boundary is permitted at a very low rate of the order of 1/2 of 1% per day. The secondary volume is maintained at a slightly negative pressure with respect to the outside atmosphere through operation of a fan system. This fan system exhausts to the atmosphere via a filter system. In secondary structures where inleakage is kept low and where

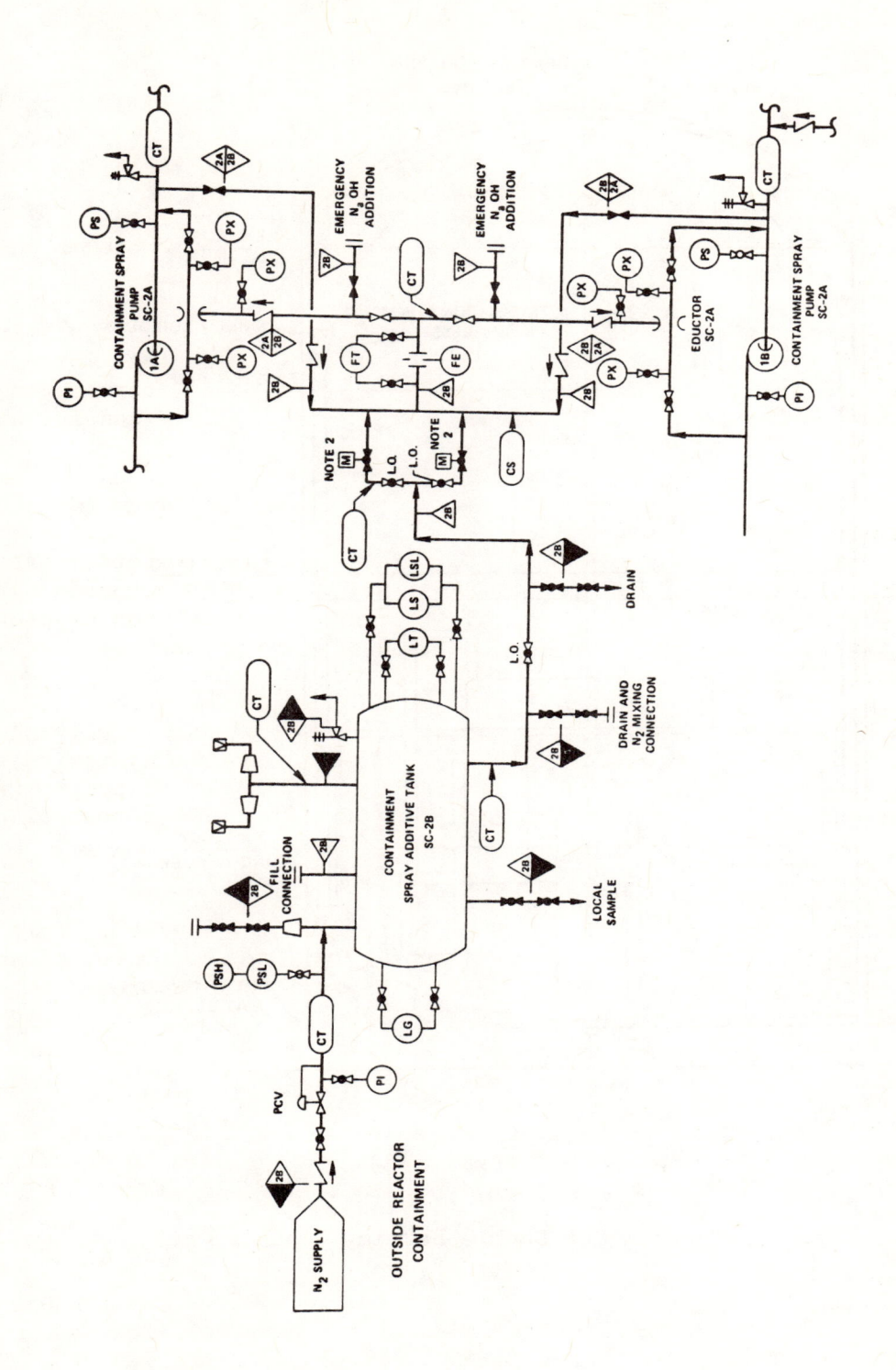

Fig. 7. PWR Spray Additive Arrangement

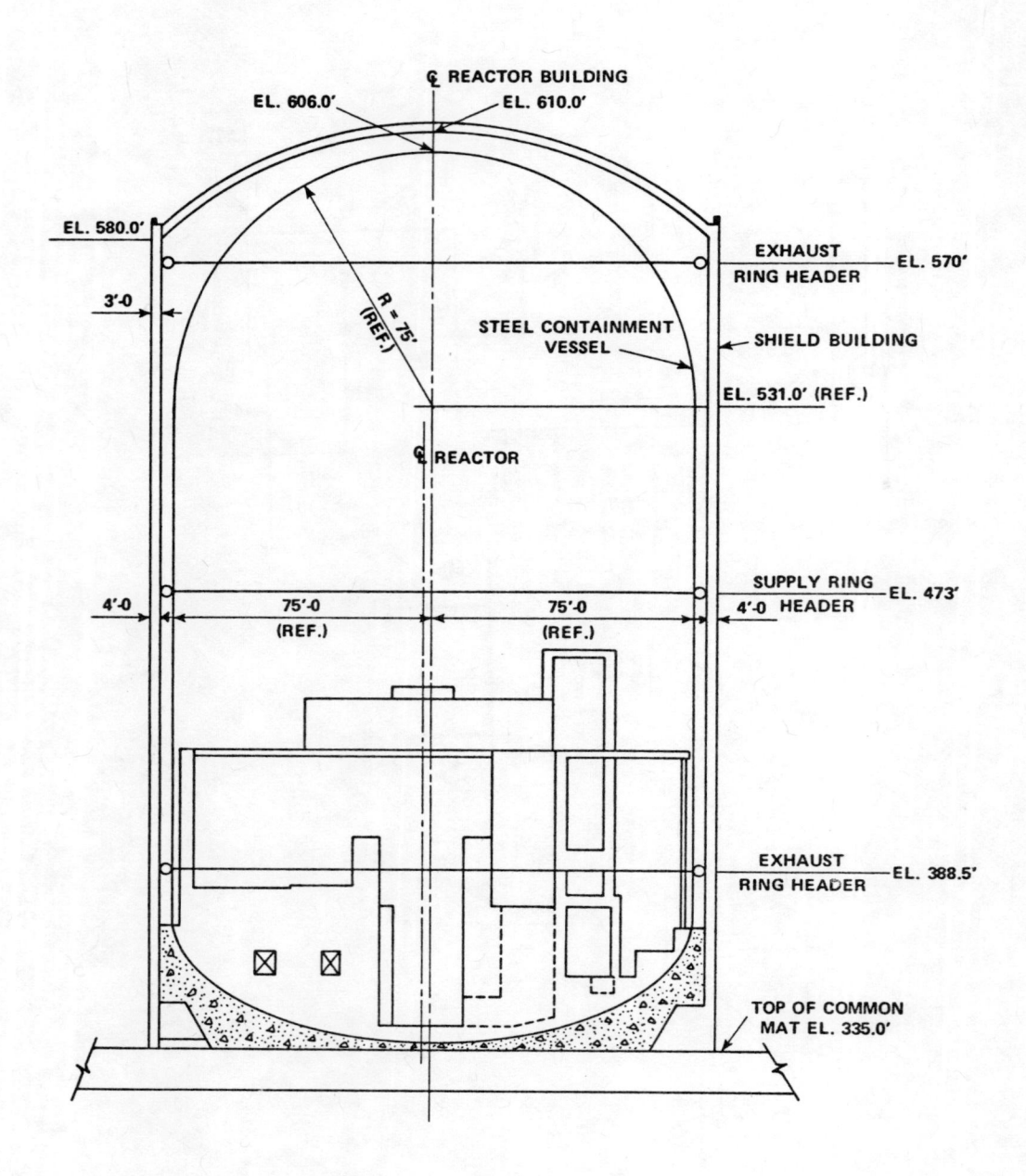

Fig. 8. Typical PWR Double Barrier Containment

site conditions warrant, the exhaust is only maintained at that quantity necessary to maintain a negative pressure. In this latter case, large fans recirculate the secondary containment volume through the filters resulting in very high net removal rates. Figure 9 shows this type of system.

Several cautions must be exercised in the employment of these systems.

1) Where the secondary volume is small, and a primary shell structure thin, provision must be made for removal of energy transfer through the primary boundary. This energy takes the form of both thermal energy and energy transferred via the expansion of the primary boundary under pressure.

2) A duct system must be employed in the secondary containment which provides assurance of mixing of the air in that space.

3) Several containment penetration lines communicate with processes outside of the secondary containment. Leakage through these lines following an accident will bypass the secondary containment and escape to the atmosphere untreated. These bypass leakage quantities must be kept low, since they can easily become dominant in calculation of offsite dose effects.

Some more recent containment concepts where additional dose conservatism was desired have employed both internal containment cleanup systems and the double barrier concept.

CONTAINMENT HEAT REMOVAL SYSTEMS

So far we have discussed how to handle the initial LOCA blowdown energy. We have also gone through several ways in which the public is protected from the postulated radiation released from a LOCA. Two major containment system functions related to LOCA mitigation remain to be discussed:

a) accommodation of reactor decay energy, and
b) reduction of containment pressure and temperature.

The design of the ECCS requires that, within a relatively short period of time after core recovery, decay heat can be removed by the ECCS. In PWR's this is accomplished by continued functioning of either the low pressure or high pressure safety injection system pumps. Water in the containment sump is recirculated through the shutdown cooling heat exchanger and pumped back into the core. Cooling water to the shutdown cooling heat exchanger is provided from another safety grade, plant cooling water system, usually a closed cycle component cooling water or an open cycle emergency service water system. The heat eventually is rejected through what is known as the ultimate heat sink.

If the emergency core cooling system is designed and functioning properly, then the water in the reactor core remains subcooled and little if any energy is transferred to the containment. Despite this, however, the containment cooling systems are still designed as if all the reactor decay heat is released to the containment. PWR containment cooling can be accomplished by two means: (a) a containment fan cooling system, or (b) a containment spray system. Early designs used both of these systems, with each sized to

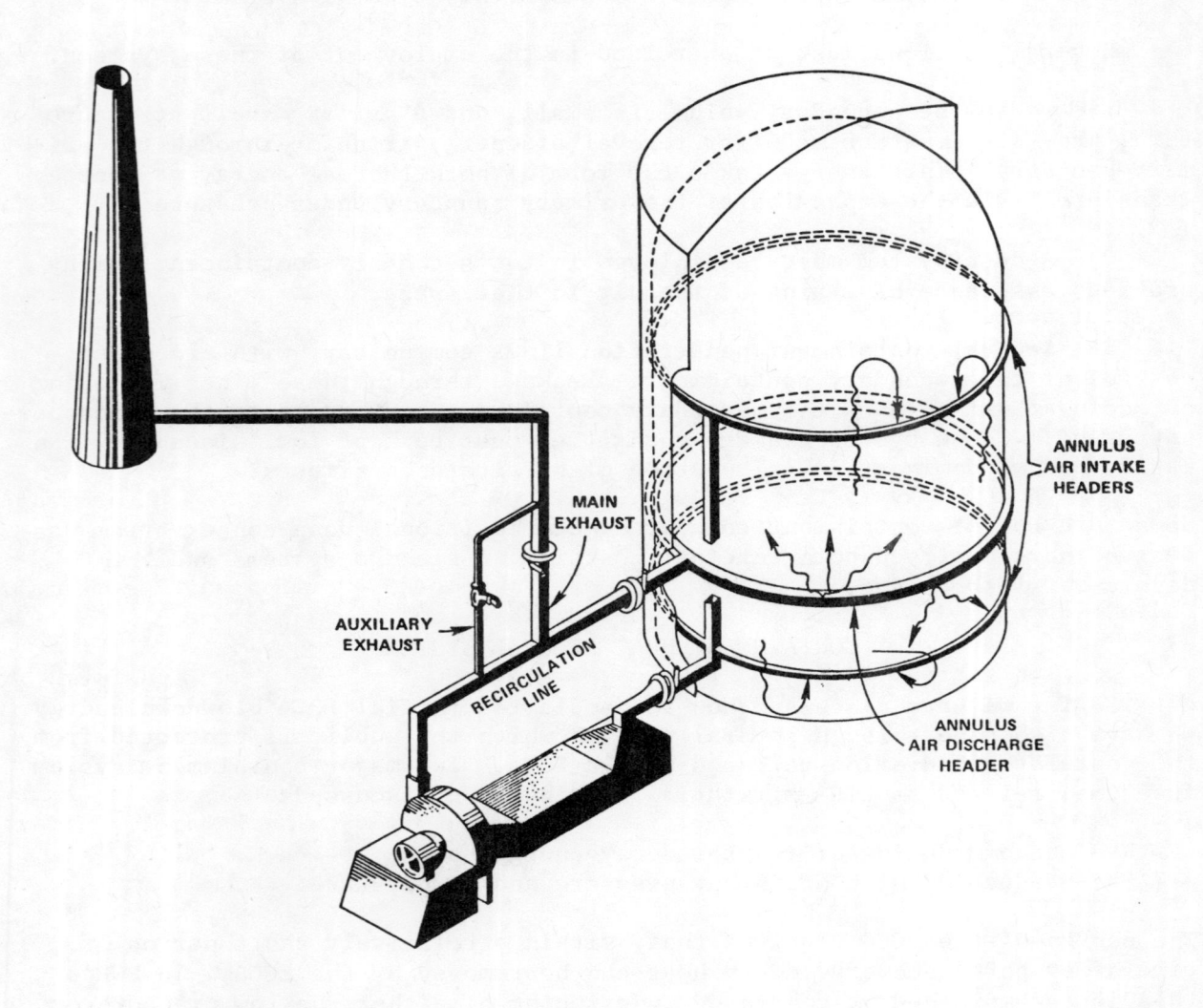

Fig. 9. PWR Secondary Containment Ventilation System

both handle decay heat loads and reduce temperature and pressure at a fairly rapid rate. For pressurized water reactors, if the pressure can be decreased by a factor of 4 in 24 hours, then the NRC permits design leakage rates to be cut in half for the remainder of the dose calculation time interval, i.e., from 24 hours to 30 days.

Because of size, cost, and power requirements of the fan cooler, and the problems associated with qualifying them for operation in the post-accident environment in the containment, the trend today is towards building the added redundancy into the containment spray system, and relying solely on that mechanism for reducing post LOCA temperature and pressure. Figure 10 shows a typical containment spray system with header arrangement in the containment, connection to the refueling water storage tank, the chemical additive tank, and the shutdown cooling heat exchanger. Here again, as with ECCS, the successful accomplishment of the design function depends on the supporting cooling water schemes and eventually the ultimate heat sink.

Post-accident containment heat removal for a boiling water reactor is somewhat more complex for several reasons.

1) The BWR containment suppression pool serves as a heat sink to accommodate heat rejection following a reactor/turbine trip.

2) It is necessary to prevent containment pressure from getting much higher than drywell pressure (drywell vacuum breakers are provided).

3) A single system, the residual heat removal system (RHR), performs several functions.

Figure 11 shows a schematic of the residual heat removal system. Figures 12, 13, 14, and 15 show the line-up of the system operating in its four modes, low pressure coolant injection (LPCI), suppression pool cooling, steam condensing (during hot standby), and nuclear reactor shutdown cooling. A brief description of each mode of operation is included on each of the figures.

One notes that examination of the BWR Mark III pressure suppression system reveals that, if the containment back pressure were to become somewhat higher than the pressure in the drywell, the water in the drywell on the inside would get higher than the weir wall, and flow over into the reactor cavity.

Because of the large containment volume of a Mark III containment as compared to the earlier Mark I or Mark II, an automatically operated spray system has been added. This provides a steam condensing medium for the containment which not only cools that volume, but also condenses any steam which might bypass the drywell.

POST-LOCA HYDROGEN

One other LOCA related phenomenon has to be considered and that is the problem of hydrogen generation. Three sources of hydrogen generation exist: a) Zirconium-water reaction; b) radiolysis of water; c) chemical corrosion of materials in the containment.

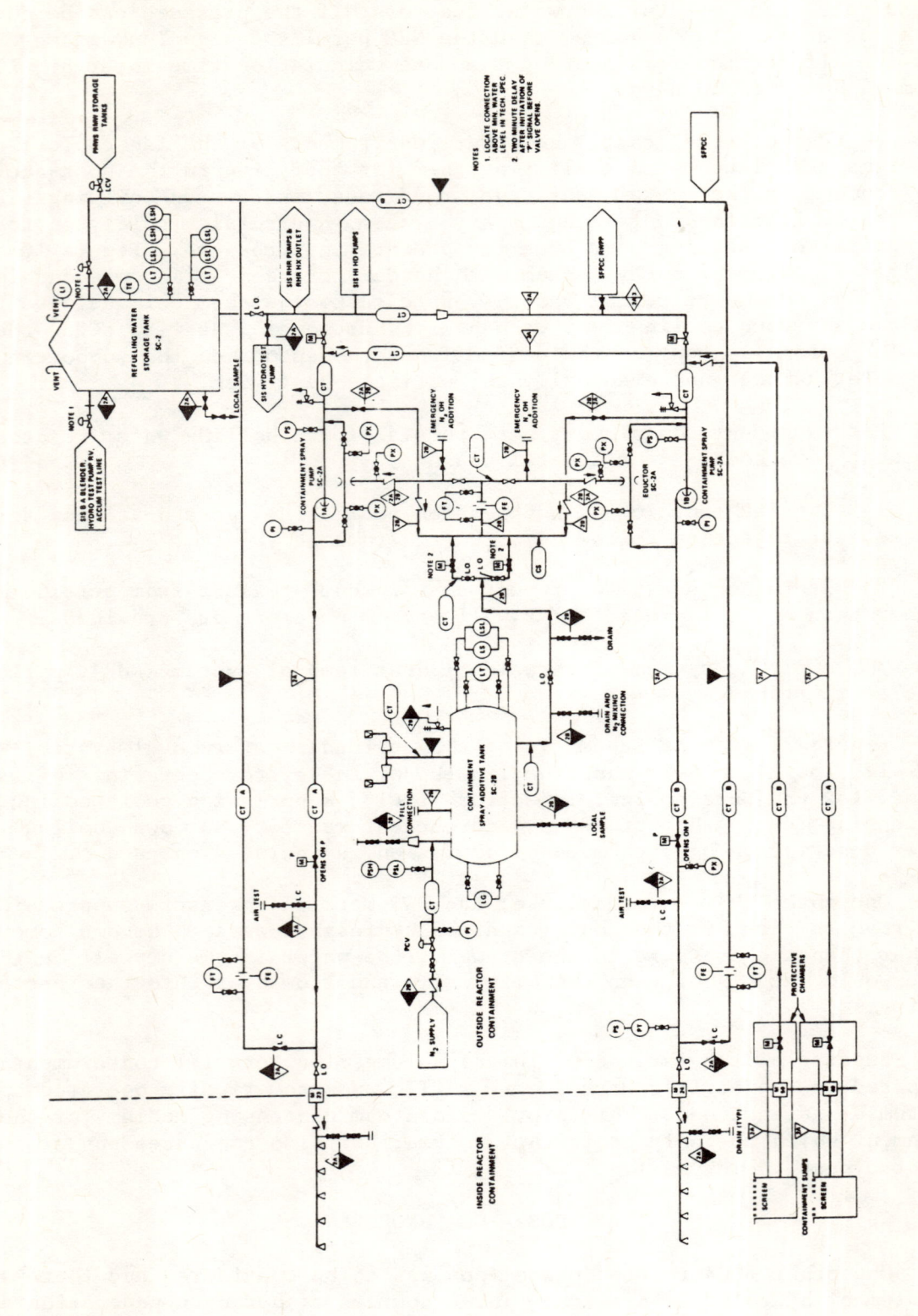

Fig. 10. PWR Containment Spray System

RESIDUAL HEAT REMOVAL (RHR) SYSTEM

The residual heat removal (RHR) system (Figure 4-11) removes post-power energy from the nuclear boiler system under normal (including hot standby) and abnormal conditions. The low pressure coolant injection function of the residual heat removal system is an integral part of the ECCS.

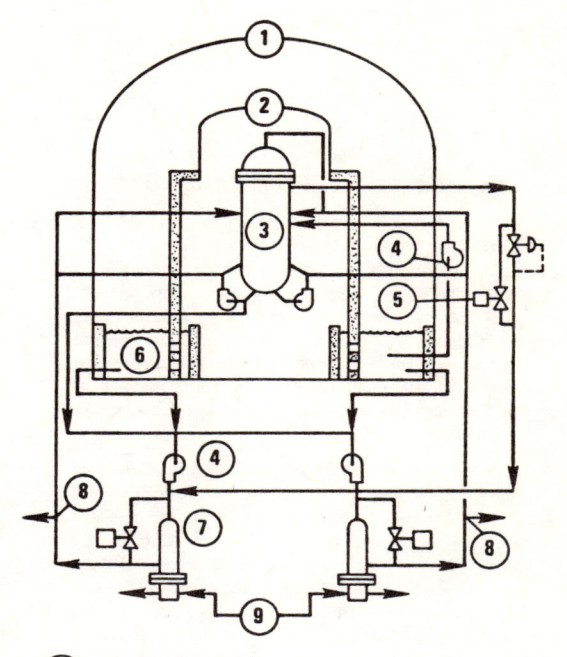

1. CONTAINMENT
2. DRYWELL
3. RPV
4. SYSTEM PUMPS
5. FULL FLOW BYPASS
6. SUPPRESSION POOL
7. HEAT EXCHANGERS
8. RCIC PUMP SUCTION
9. SERVICE WATER

Fig. 11. BWR Residual Heat Removal System (RHR)

LOW PRESSURE COOLANT INJECTION (LPCI)*

The low pressure coolant injection function (Figure 4-12) in conjunction with the low pressure core spray system, the high pressure core spray system, and/or automatic depressurization of the nuclear boiler system (depending upon operability of the high pressure core spray system or level of depletion of reactor vessel water) will restore and maintain the desired water level in the reactor vessel required for cooling after a loss-of-coolant accident.

In conjunction with the low pressure core spray system, redundancy of capability for core cooling is achieved by sizing the RHR pumps so that the required flow is maintained with one pump not operating. Using a split bus arrangement for pump power supply (standby power system), two RHR pumps are connected to one bus and the third RHR pump and a low pressure core spray pump are connected to the second bus to obtain the desired cooling capability. The pumps deliver full flow inside the core shroud when the differential pressure between the reactor pressure vessel and the primary containment approaches 20 psi.

The availability of the LPCI function is not required during normal nuclear system startup or cooldown when the reactor vessel pressure is less than 135 psig.

*Part of the core emergency cooling network.

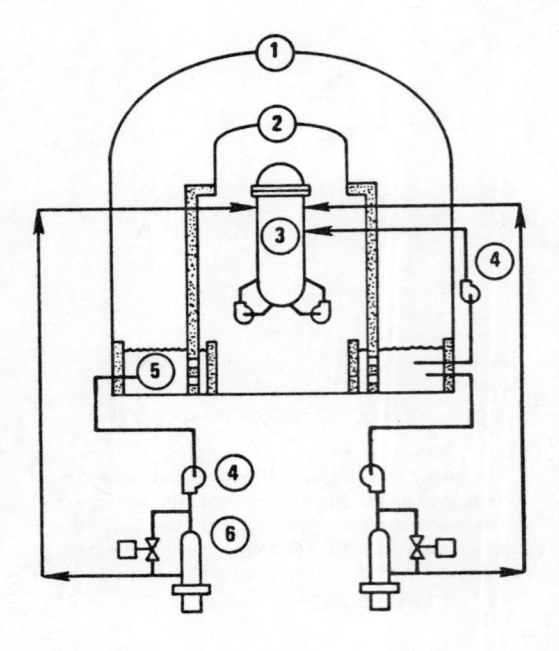

(1) CONTAINMENT
(2) DRYWELL
(3) RPV
(4) SYSTEM PUMP
(5) SUPPRESSION POOL
(6) HEAT EXCHANGERS

Fig. 12. RHR System - LPCI Mode

SUPPRESSION POOL COOLING

The suppression pool cooling function of the residual heat removal system (Figure 4-13) ensures that the temperature in the suppression pool immediately after blowdown, and when the reactor vessel pressure is greater than 135 psig, does not exceed a predetermined limit (generally 170 F*). Suppression pool water is pumped from the pool through either or both of two completely independent loops, including pump and heat exchanger, and returned to the pool. The heat removed by the heat exchanger is transferred to the residual heat removal system service water. Suppression pool cooling is manually initiated.

*This may be exceeded following blowdown in the event of a design basis accident.

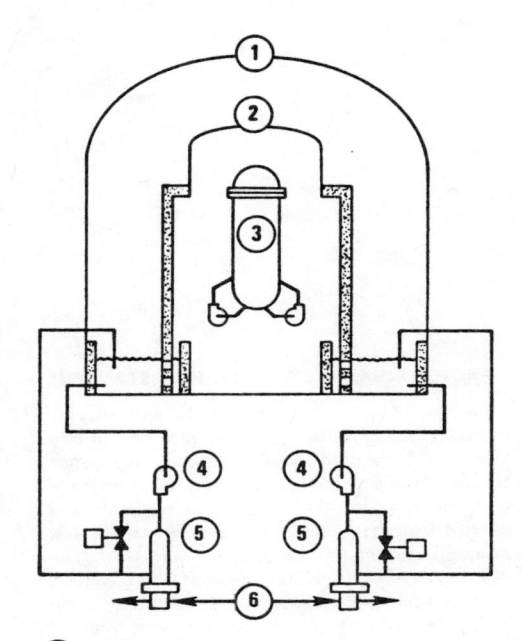

(1) CONTAINMENT
(2) DRYWELL
(3) RPV
(4) SYSTEM PUMP
(5) HEAT EXCHANGERS
(6) SERVICE WATER

Fig. 13. RHR System - Supression Pool Cooling Mode

REACTOR STEAM CONDENSING (HOT STANDBY)

During nuclear boiler system isolation and in conjunction with the operation of the reactor core isolation cooling system and steam blowdown to the suppression pool, steam at reduced pressure and temperature is directed from the main steam lines to the residual heat removal system heat exchanger (see Figure 4-14). Condensate at a temperature not exceeding 140 F is directed to the reactor core isolation system for return to the nuclear boiler system. Noncondensible gases from the heat exchangers are vented to the suppression pool. Steam condensing is manually initiated.

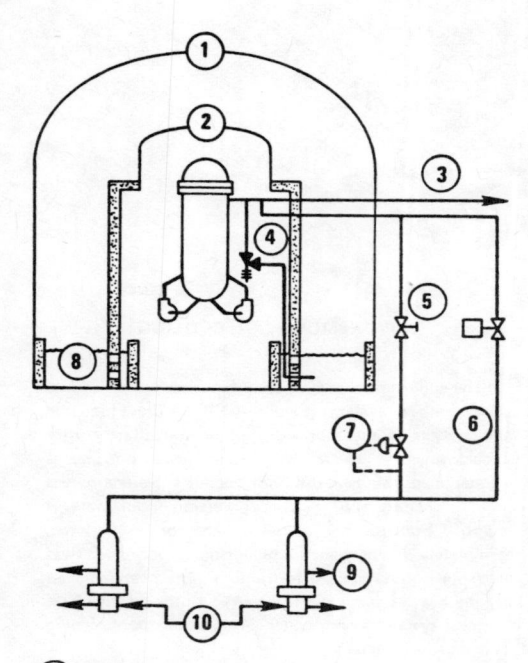

1. CONTAINMENT
2. DRYWELL
3. MAIN STEAM LINE TO TURBINE
4. SAFETY/RELIEF
5. MANUAL REMOTE CONTROL VALVE
6. FULL-FLOW BYPASS
7. PRESSURE REDUCING SYSTEM
8. SUPPRESSION POOL
9. TO RCIC PUMP SUCTION
10. SERVICE WATER

Fig. 14. RHR System - Hot Standby Mode

NUCLEAR BOILER SHUTDOWN COOLING

The shutdown cooling function of the residual heat removal system (Figure 4-15) removes residual heat (decay heat and sensible heat) from the nuclear boiler system after reactor shutdown in preparation for refueling or nuclear system servicing. When the reactor vessel pressure is reduced to 135 psig after shutdown, the shutdown cooling function is manually initiated by first draining the loops of inhibited water and flushing with condensate. (At reactor pressure above 135 psig, the low pressure coolant injection function of the residual heat removal system, which shares equipment and piping with the shutdown cooling function remains functional. The shutdown cooling function has the capability of reducing the reactor vessel to a temperature of 125 F, including draining and flushing, within 20 hours after the control rods are inserted for shutdown and then to maintain this maximum temperature. Reactor water is taken from one of the reactor water recirculation loops, pumped through the heat exchanger, and returned to the reactor vessel by way of the reactor water recirculation loop. Flow from the residual heat removal system, during the shutdown cooling function, can be diverted to the spray nozzle located in the head of the reactor pressure vessel to condense steam while the vessel is being flooded. Shutdown cooling is manually initiated.

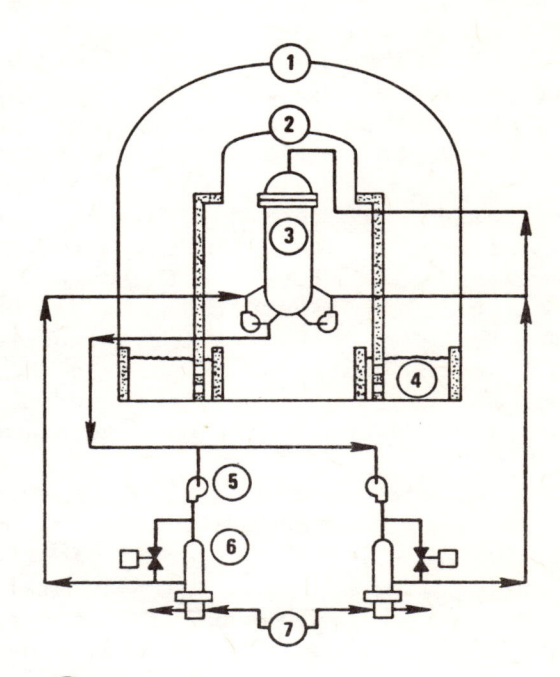

(1) CONTAINMENT
(2) DRYWELL
(3) RPV
(4) SUPPRESSION POOL
(5) SYSTEM PUMP
(6) HEAT EXCHANGERS
(7) SERVICE WATER

Fig. 15. RHR System - Shutdown Cooling Mode

Zirconium-water reaction occurs when the fuel element clad temperature rises; first at a slow rate beginning at about 1800°F, and then increasing exponentially. Since it is an exothermic reaction, at some temperature above about 2300°F, it becomes self-sustaining. The NRC emergency core cooling criteria set a maximum clad temperature limit of 2200°F, conservatively calculated (6). Since power is not distributed uniformly across the core, only a small portion of the cladding is calculated to attain that value, and less than 1% of the core's zirconium reacts to produce hydrogen (a more realistic calculation would show even less hydrogen produced).

Following core recovery, hydrogen generation continues as a result of the radiolysis of water. This reaction is a function of water temperature and gamma radiation level in the water. The principal conservatism in this calculation is the assumption that all halogens released from the core are presumed to be in the water for purposes of radiolysis computations. It should be noted that, for purposes of off-site dose calculations, the same halogens are assumed to be dispersed in a containment atmosphere, and then released via leakage to the outside atmosphere.

For PWR dry containments, where the volume is relatively large, the hydrogen level in the containment builds up at a rate slow enough to be handled. Regulatory Guide 1.7 (7) specifies that redundant recombiners be utilized with purging as a backup. Where several reactors are located on the same site, or sites in close proximity to each other, the same recombiners, if designed for outside containment use, can be used for more than one reactor.

The Mark I and Mark II BWR containments with their smaller volumes were a problem. Hydrogen could be calculated to build up to flammability limits very rapidly, and so atmospheric inerting was required. Since radiolysis also produces oxygen, recombiners were required to handle the long term hydrogen generation problem. The inerting requirement has recently been removed from some BWR's.

When the Mark III containment concept was introduced, it was believed that the larger volume would obviate the need for inerting. The more conservative core design (lower peak linear heat rate) results in a lower calculated peak accident temperature, and a conservatively calculated initial zirconium-water reaction well under 1%. For a long time, however, the regulatory staff stuck to the arbitrary Regulatory Guide 1.7 assumption of 5% zirconium-water reaction. At this level, the Mark III design, because of the compartmentalization between the drywell and the containment ran into trouble. In order to prevent hydrogen pocketing in the drywell, a fan system was installed. In order to prevent hydrogen from reaching flammability limits, a mixing/bypass system had to be installed.

A mixing/bypass system large enough to meet a 5% zirconium-water reaction criterion presented other problems--failure of system valves could open a large bypass area thereby permitting drywell steam to bypass the suppression pool, enter the containment directly, and result in exceeding containment design pressure.

General Electric has been successful in convincing the Advisory Committee on Reactor Safety and the regulatory staff that a 1% zirconium-water reaction assumption is conservative enough. A small bypass/mixing system is provided to accommodate that. Hydrogen recombiners are still required to handle the long term radiolysis problem.

It is my own personal opinion that, for most sites where a double barrier containment is employed, hydrogen recombiners are unnecessary. During long term post-accident phases, hydrogen levels in the containment can be controlled utilizing a bleed system to the secondary containment. Recirculation of that bleed in the filter system can reduce radioactivity concentrations to a miniscule value, so that upon release to the atmosphere, the resulting off-site doses are inconsequential.

HTGR CONTAINMENT DESIGN DIFFERENCES

Before moving into a discussion of containment interfaces with the environment, this would seem to be a good time to briefly touch on some of the key differences between design considerations for an HTGR containment as compared to a light-water-reactor containment.

1) The reactor coolant medium is helium, a fluid following the perfect gas laws.

2) The loss-of-coolant accident is a relatively small break resulting from a postulated blow out of a penetration through the prestressed concrete reactor vessel.

3) The large mass of the core graphite moderator provides a large heat sink for coast down and decay heat; therefore, large quantities of fission products are postulated to be released later in the accident sequence.

4) Emergency core cooling is by means of an auxiliary helium recirculation system and cooling channels built into the walls of the prestressed concrete reactor vessel.

5) Containment cool down and pressure reduction are accomplished only by direct heat removal from the helium and by compressing the helium and storing it.

6) The helium in the containment is used in a recirculation mode by the auxiliary circulator system. Therefore, containment pressure cannot be reduced to atmospheric but must be maintained at a level sufficient to assure proper circulator function.

CONTAINMENT FUNCTIONS AS ENVIRONMENTAL PROTECTION FOR THE REACTOR COOLANT SYSTEM

One of the key functions identified for the containment was protection of the reactor coolant system from environmental effects during normal operations. The containment also provides shielding and other protective functions for the public and plant staff during normal operations. Because of the manner in which the loads are combined to arrive at a containment design, we will discuss briefly the key environmental factors used in design, then discuss how they are used.

Allowable stresses for various loading combinations are established by the applicable codes--ACI 359, or ASME Section III, Part 2 for concrete containment vessels, and ASME Section III, Class MC, for steel containment vessels. Concrete shield vessels surrounding steel containments in the double barrier concepts are designed to ACI 349, with some conservative modifications.

In addition to the accident loads of pressure and temperature previously discussed, the following natural and man-made phenomena have to be considered in the design of a containment system. Not all need be considered at all sites. Appropriate combinations should also be considered as specified in the General Design Criteria, 10 CFR 50, Appendix A.

1) Safe Shutdown Earthquake (SSE)--The SSE is an earthquake which is the largest that could reasonably be expected to occur at the site based on historic and geologic considerations. In no case can it be less than 0.1 g. In many cases it ends up at 0.2 g or greater. It is expected that something of the order of 0.25 to 0.3 g's would be adequate for about 80% of the sites in the U.S. The seismic loading is applied via a sophisticated soil structure interaction analysis, applying the response motion of the earth to the building at the base mat. All building analyses are dynamic, as are most safety class equipment analyses.

2) Operating Basis Earthquake (OBE)--The OBE is that earthquake that one could expect to have a fairly high probability of occurrence during the life of the plant, possibly one in 100 years recurrence. It is never taken at less than 1/2 the SSE value, which usually results in a longer predicted recurrence interval. This value of 1/2 SSE is being seriously questioned by the industry because code allowable stress and loading combination requirements for this OBE condition generally control the design.

For the Safe Shutdown Earthquake some structural deformations in equipment might be permitted, but no loss of safety function is allowed. For the containment structure, stresses must be within code allowances.

3) Design Bases Winds--Three levels of wind are considered in design:

   a) the lowest is a 100 year recurrance wind for which all structures are designed;
   b) the next level is the probable maximum hurricane (PMH) or other appropriate storm conditions for the site for which all safety features are designed;
   c) the most severe is the tornado wind for which safety related equipment must be protected (the level of tornado wind and pressure drop are specified in NRC publications).

4) Design Bases Missiles--The containment structure must be designed for both external and internal missiles. For internal missiles it is presumed that they may be accident generated, i.e., valve stems or shrapnel. Therefore, the design bases for internal missiles is that the fission product barrier integrity must be maintained--this usually results in "design by impact prevention." Two sources of external missiles are considered:

   a) turbine generator missiles--usually avoided by turbine generator

building orientation
b) tornado missiles--the design must protect the reactor coolant system and its essential safety features from a broad spectrum of tornado generated missiles ranging from an automobile to multiple reinforcing rods or steel pipes. The design requirement is to prevent a loss-of-coolant accident and to protect features necessary to effect a safety shutdown.

5) Other abnormal loads which have to be considered in the design include:
   a) pressure buildup and temperature within a compartment in the containment due to a pipe rupture in the compartment;
   b) reactor forces on containment penetrations due to both operating and pipe rupture effects;
   c) impingement effects due to the jet force from a broken pipe;
   d) external hydraulic pressures due to extremely high water tables, flood conditions, or wave action; and
   e) the effects of nearby transportation or industrial facilities, such as:
      i) liquified natural gas storage or shipment,
      ii) nearby natural gas pipe lines,
      iii) railroads, highways, and shipping, and
      iv) airports.

The loads on the structure for each of these design conditions must be determined. Part of the design bases for the structure and its supporting systems must be a determination of what combination of loadings are appropriate to consider as occurring simultaneously. No firm criteria currently exist for some of these loads, but they are being developed by the NRC and by an ANS Standards Group. As an example, it has become common practice to combine a LOCA with an SSE (some European practice does not require this). However, it is not required to assume that a tornado occurs at the same time as a LOCA. A maximum probable flood does not occur simultaneously with an SSE, but it has been the practice to combine a lesser flood, say once in a thousand years, with an OBE.

Once appropriate load combinations are established, they are expressed for concrete structures in the form of factored load combinations. For example, accident pressure may be multiplied by 1.5, PMH winds by 1.25, and tornado winds by 1.0. Many of the appropriate factors are now defined by the applicable codes. They are usually established based on judgment as to the state of the art, and desired margins, and we will not delve into them any further here.

SUMMARY

This chapter has presented an overall view of the safety considerations that go into the design of containment systems for nuclear power plants. The approach has been not to view the containment as a structure, but to emphasize all of the principal functions and system interrelations required to be considered in containment design. It is necessary to view the containment as a system in order to emphasize the necessity of considering system interfaces in the design of containments for nuclear power plants.

REFERENCES

1. "Reactor Site Criteria," Code of Federal Regulations, Title 10, Part 100, as amended.
2. P. DRAGOUMIS, S. J. WEEMS, and W. G. LYMAN, "Ice Condenser Reactor Containment System," Proceedings of the American Power Conference, Vol 30, 347-355 (1969).
3. J. J. DiNUNNO, F. D. ANDERSON, R. E. BAKER, and R. L. WATERFIELD, "Calculation of Distance Factors for Power and Test Reactor Sites," USAEC Rept. TID-14844 (1962).
4. "Assumptions Used for Evaluating the Potential Radiological Consequences of a Loss of Coolant Accident for Boiling Water Reactors," NRC Regulatory Guide 1.3 (June 1974).
5. "Assumptions Used for Evaluating the Potential Radiological Consequences of a Loss of Coolant Accident for Pressurized Water Reactors," NRC Regulatory Guide 1.4 (June 1974).
6. "New Acceptance Criteria for Emergency Core-Cooling Systems of Light-Water-Cooled Nuclear Power Plants," Nuclear Safety, 15, No. 2, 173-183 (March-April 1974).
7. "Control of Combustible Gas Concentrations in Containment Following a Loss of Coolant Accident," NRC Regulatory Guide 1.7 (Safety Guide 7) (March 1971).

POWER PLANT SITING

Geoffrey G. Eichholz
School of Nuclear Engineering
Georgia Institute of Technology
Atlanta, Georgia 30332

INTRODUCTION

The location of a nuclear plant can both contribute to the safety of the operation and determine to what extent such a plant may be considered safe by outsiders. Safety can be defined in several ways:
   (a) the risk of damage to the integrity of the system from any cause;
   (b) the risk of harm to operating personnel from normal or abnormal functioning of the plant; and
   (c) the risk or hazard from any cause, mechanical, biological, or radioactive, to the surrounding population.

The first two safety criteria are usually met by suitable plant design and layout, and are affected by the selected site only in as far as unfavorable geological or meteorological phenomena, such as earthquakes, subsiding ground, flooding or high winds might interfere with the proper functioning of the facility. It is the third criterion that is uppermost in the mind of most intervenors and the general public when they raise objections to the location of a nuclear plant in a given locality. To a large degree a further distinction must be made in distinguishing between routine operation, with low but continuous radioactivity levels in effluents, and accident or emergency conditions leading to a potentially higher-level, but transient emission of contaminating materials. All of these, nowadays, are covered by the requirement that they must lead to activity levels that are "as low as practicable," which in practice may be interpreted as levels below 1% of "maximum permissible" levels.

In selecting a plant site one must satisfy or optimize three categories of selection factors:
   (a) technical factors
   (b) environmental factors
   (c) radiation factors.

Technical Factors

Let us consider the technical factors first. These include many factors that are common to all power plants, whether fossil-fueled or nuclear, such as:
   1. plant size and load characteristics,
   2. distance from load centers,

3. transmission line requirements,
4. cooling water requirements,
5. water storage needs,
6. transportation access, by road, rail or water,
7. plant safety, both for reactor and "conventional" components,
8. manpower availability,
9. land availability and cost,
10. waste disposal,
11. economic factors.

The plant size and load characteristics determine the type of system, whether mixed nuclear and fossil, whether hydro supplementation is possible, and how much waste heat must be dissipated (1). The distance factors for transmission lines to existing load centers or the place of the plant in a regional grid network may rule out certain remote locations on economic grounds. Cooling water requirements dictate the need for ready access to an adequate water source, either fresh water or ocean water, possibly supplemented by a storage lake, for which the site also has to be suitable. In many earlier analyses, the availability of cooling water tended to be a dominant selection factor, especially with once-through cooling. With greater emphasis on closed-loop cooling, and for gas-cooled reactors, the need for large amounts of water has diminished somewhat but it has not disappeared by any means; in fact, it is one of the arguments used in favor of off-shore sites.

Transportation is of importance both during the construction phase and for fuel shipment during the operational phase. Some of the larger plant components, notably the reactor vessel and the turbine rotor may be too large for shipment by road or rail, and barge movement may be the only feasible mode of transportation. This obviously excludes many potential sites and, in the long run, may spur the assembly of such large components on site or the adoption of site-assembled concrete structures, as are in use already for the HTGR. Surprisingly, a large number of reactors under construction have no rail link and this may complicate the logistics of spent fuel shipments over the next decade; it also affects the estimated population dose, though only marginally. The site topology may affect the lay-out of the plant, the pipe runs and the location of the switch yard, all of which may influence operational safety, ease of maintenance, and easy access in case of emergencies. The plant location may also determine the height of stacks and cooling towers and other, largely esthetic, factors.

The economic impact on local communities is important and may take many forms. The plant may provide employment, both during construction and during operation, in a wide variety of jobs and skills, from welders to gatekeepers, from mechanics to kitchen help. It will increase the need for housing, food and usual services, and infuse money into the local economy. It will demand utility services, water, gas, sewerage, on a fairly large scale and increase the demand for schooling and improved local highways. By attracting additional industry it may affect the local pattern of employment and the population structure. All of these need to be taken into consideration, since such economic, social and political factors may determine the local pattern of acceptance or resistance.

Finally, the availability of suitable land of adequate acreage at reasonable cost is an important factor. At this time Congress is considering several bills on land use policy, which are designed to encourage early selection and designation of land for potential power plant sites. The political aspects of this are formidable and the potentials for speculation are obvious. The past pattern of land acquisition, often by right of eminent domain, is coming increasingly under criticism, particularly in view of the very large land areas currently demanded for nuclear power plants. An interesting summary of applicable laws was published as an SINB report in 1972 (2). In many cases the land acquired by the utility has served as an example in agricultural practices to surrounding farm communities. The question of waste disposal is attracting increased attention. The waste tips at coal-fired stations have long been considered a particularly obnoxious feature. For nuclear stations, most high-level waste consists of spent-fuel elements which are stored under water for 3-6 months, and then shipped out at regular intervals. The intermediate-level waste, such as spent demineralizer resin, also is shipped out at intervals, and the site probably need not provide for more than temporary waste storage for solid material.

Engineered Safeguards. If we are concerned with the impact of the plant on the surroundings under all conditions, and vice versa, it is appropriate to look at the plant itself. All the nuclear generating plants under construction, whether LWR, HWR or GCR types, have certain features in common. They have all been designed for the ultimate in reliability, achieved by a previously unheard-of level of quality assurance and by duplication and triplication of all important control instrumentation. The reactors have all been designed to have a negative temperature coefficient so that shutdown is assured during any excess power transient or any other malfunction that could raise the core heat level. Since the fission product inventory in the core and the gamma decay heat produced by it are viewed as the major sources of contamination, several barriers have been designed to contain the fission products, both under normal conditions or in case of accidents. From the inside out, these barriers are provided by the fuel cladding, the reactor vessel and the containment building. To allow for minimum expected failure, it is usually postulated that up to 0.25% of the cladding may be leaking after several months of burnup, but this does not necessarily entail any significant loss of fission products to the coolant. Reactor vessels and containment structures are described in the preceding chapter. Figures 1-5 of this chapter illustrate current containment systems. To this must be added various interlocks designed to contain any radioactive releases within the containment building during all conceivable and inconceivable accident conditions. It must be borne in mind that there is no such thing as a "foolproof" system; all one can do is to design the system to reduce the consequences of human error and instrument failure to the greatest extent possible and site selection is but one of the aspects of this fact.

Environmental Factors

To select a site capable of supporting the large structures required and to minimize the "environmental impact" as defined by the National Environmental Policy Act of 1969 (NEPA) and other state and federal regulations, the site or sites must be evaluated with respect to several classes of geographic, geological, meteorological and ecological factors. These have been summarized in the General Environmental Siting Guides for Nuclear Power Plants, issued by the Atomic Energy Commission in December 1973, and subsequent revisions (3). Table I shows the content page of that report which summarizes in convenient form the various aspects of any proposed site that have to be considered. One can divide these factors into three categories, those related to plant safety, those related to public safety, and those of social and esthetic significance.

Questions of bearing strength of the ground, the hydrology, and the seismicity of the site clearly have a bearing on the ultimate safety and integrity of the plant itself. Atmospheric dispersion, water flow and impoundment, plant life and fauna, and radiation exposure levels are all related to public safety by direct movement of radioactive effluents or by contamination via the food chain. Finally, esthetic impact of the plant and transmission lines, land use and recreational opportunities, water resource planning and disturbance of historical and archaelogical sites all fall into the last category. In time and with experience, the relative weighting of these factors may change. An example of this is the question of the significance of geologically old tectonic fault lines and other geological discontinuities and their likely effect on plant design and integrity; yet a thorough study of soil stability and rock structure will probably always be essential. Several recent instances where such a study had been insufficient have resulted in very costly delays, and expensive requirements for remedial grouting of foundation rocks.

Although seismic qualification of equipment, structures and instrumentation may be considered primarily as plant design features (4,5) and would be covered in the Safety Analysis Report, the choice of a plant site should obviously exclude any seismically active fault locations and the preliminary site review must include a detailed geological examination to assess the probability and maximum expected intensity of any ground movement. Experience in earthquake-prone regions such as Mexico or Japan has shown that it is possible to design foundations and structures to withstand groundshocks of appreciable severity. Since moderate earth tremors can conceivably occur at any point on land or sea, additional design emphasis has been placed on the response characteristics for pipes, cables and closures to eliminate any possible failure due to resonances and vibrations from single modes or combined effects over a range of shockwave spectra (6). Table II shows recommended damping values for use in specifying types of support for various plant components (5).

Similar guides have been issued for other site-related design features to protect the plant against flooding, icing or tornadoes. Most of these aspects would lie well within the competence of any experienced architect-engineer team.

Radiation Factors

This aspect has received the most attention by the public, since it tends to attract the largest number of emotional reactions. Given the fact that a reactor is not a bomb, and I hope the other chapters in this book will have at least convinced the readers of that one fact, there is still the possibility of escape of radioactive materials, both during normal operation and in case of accidents. The relevant siting philosophy, as contained in

Table I. Table of Contents from "General Environmental Siting Guides for Nuclear Power Plants" (3)

	Page
INTRODUCTION	1
SUMMARY OF GENERAL ENVIRONMENTAL SITING GUIDES	5
GEOLOGY AND SOILS	14
General Seismic and Geological Characteristics	14
Soil Stability and Topography	15
Subsidence	17
ATMOSPHERIC FACTORS	20
Air Pollution Standards	20
Dispersion Climatology	21
Atmospheric Dispersion - Valley and Canyon Sites	23
Atmospheric Dispersion - Shoreline Sites	25
Vortex Phenomena and Extreme Winds	28
Fogging and Icing	31
HYDROLOGY	33
Water Use Policies	33
Adequate Water Supply (Quantity)	39
Groundwater	43
Inadvertent Loss of Water	46
Water Quality	48
Icing and Sedimentation	54
Mixing Zones	57
Stratified Waterbodies	59
Impoundments	61
ECOLOGY	63
Temperature Sensitive Aquatic Species	63
Breeding Habitats	70
Species Migration	73
Terrestrial Vegetation	75
Rare or Endangered Species	77
PUBLIC EXPOSURE TO RADIATION	82
LAND USE	91
Land-Use Compatibility	91
Land-Use Planning	92
Coastal Zone Planning	95

Table I. Concluded

	Page
Watershed Planning	96
Transmission Line Corridors	97
HUMAN INTEREST FACTORS	98
Unique Natural Resource Areas	98
Historical Areas	100
Archaeological Sites	101
Fossil and Rock Deposits	102
ESTHETICS	103
View of Transmission Facilities	103
View of Power Plant Site	105
OTHER CONSIDERATIONS	107
Transportation Provisions	107
Construction Impact	109
REFERENCES	113

Table II. Damping Values[1] (Percent of Critical Damping) ([5])

Structure or Component	Operating Basis Earthquake or 1/2 Safe Shutdown Earthquake[2]	Safe Shutdown Earthquake
Equipment and large-diameter piping systems[3], pipe diameter greater than 12 in.	2	3
Small-diameter piping systems, diameter equal to or less than 12 in.	2	3
Welded steel structures	1	2
Bolted steel structures	2	4
Prestressed concrete structures	4	7
Reinforced concrete structures	2	5
	4	7

[1] Table II is derived from the recommendations given in Reference 1

[2] In the dynamic analysis of active components as defined in Regulatory Guide 1.48, these values should also be used for SSE.

[3] Includes both material and structural damping. If the piping system consists of only one or two spans with little structural damping, use values for small-diameter piping.

REFERENCE: 1. Newmark, N. M., John A. Blume, and Kanwar K. Kapur, "Design Response Spectra for Nuclear Power Plants," ASCE Structural Engineering Meeting, San Francisco, April 1973.

10-CFR-100, implies two different criteria: one related to the radiation hazard in case of a "design-base accident" (DBA), the other related to the chronic radiation dose to the general public during routine operation of one or more reactors at a given plant site (7).

The release of radioactive material in case of a DBA is estimated on the basis of several assumptions. These include:
  (a) the fission product and activation product inventory in the reactor core at the time of the accident;
  (b) assumptions regarding the fraction of that inventory released to the building air;
  (c) assumptions regarding the fraction of gaseous and particulate material that may escape from the containment building; and
  (d) worst-possible meteorological conditions that would minimize uniform dispersion or lofting in the atmosphere.

The fission product inventory in the core depends on the fuel burnup and fuel management up to the time of the accident with a relative buildup of the longer-lived fission products with burnup time. Site calculations assume maximum burnup and hence maximum buildup of longer-lived fission products. It is then assumed that up to 100% of noble gases, 10% of halogen and 1% of other fission products escape into the containment air, the rest being chemically reacted or contained within the hot fuel. This fission product mixture then constitutes the source term for calculations on their subsequent fate and dispersion. Failure probabilities using a fault tree approach have been calculated in detail in the course of the "Rasmussen study" (8).

Escape from the containment building up the stack or through a breach in the wall is then assumed to permit escape of 10-50% of the airborne activity, both in gaseous volatile form and as fine particulates. Site meteorological data are then utilized to calculate ground doses at varying distances downwind under typical and least favorable conditions, using fractionation curves of the type shown in Fig. 1. On the basis of this information, one defines an "exclusion area" which is fully under the control of the plant operator, a "low-population zone" surrounding it, with a population density low enough to make warning measures or evacuation feasible if necessary, and "population centers," as towns containing more than about 25,000 residents. These terms are described more fully in 10 CFR 100, part of which is shown as Table III (7).

Any site selected must make provision for an exclusion area large enough so that an individual located at any point on its boundary for 2 hours immediately following a postulated fission product release would not receive a total radiation dose to the whole body in excess of 25 rem or a total dose to the thyroid from iodine exposure in excess of 300 rem.

The low population zone should extend far enough that an individual at any point of its outer boundary should not receive, during the total period of passage of the fission product cloud, a whole-body dose of 25 rem or a thyroid dose from iodine exposure over 300 rem. The nearest population center should be at least 1.33 times the distance to the boundary of the low population zone, as defined above, in that direction.

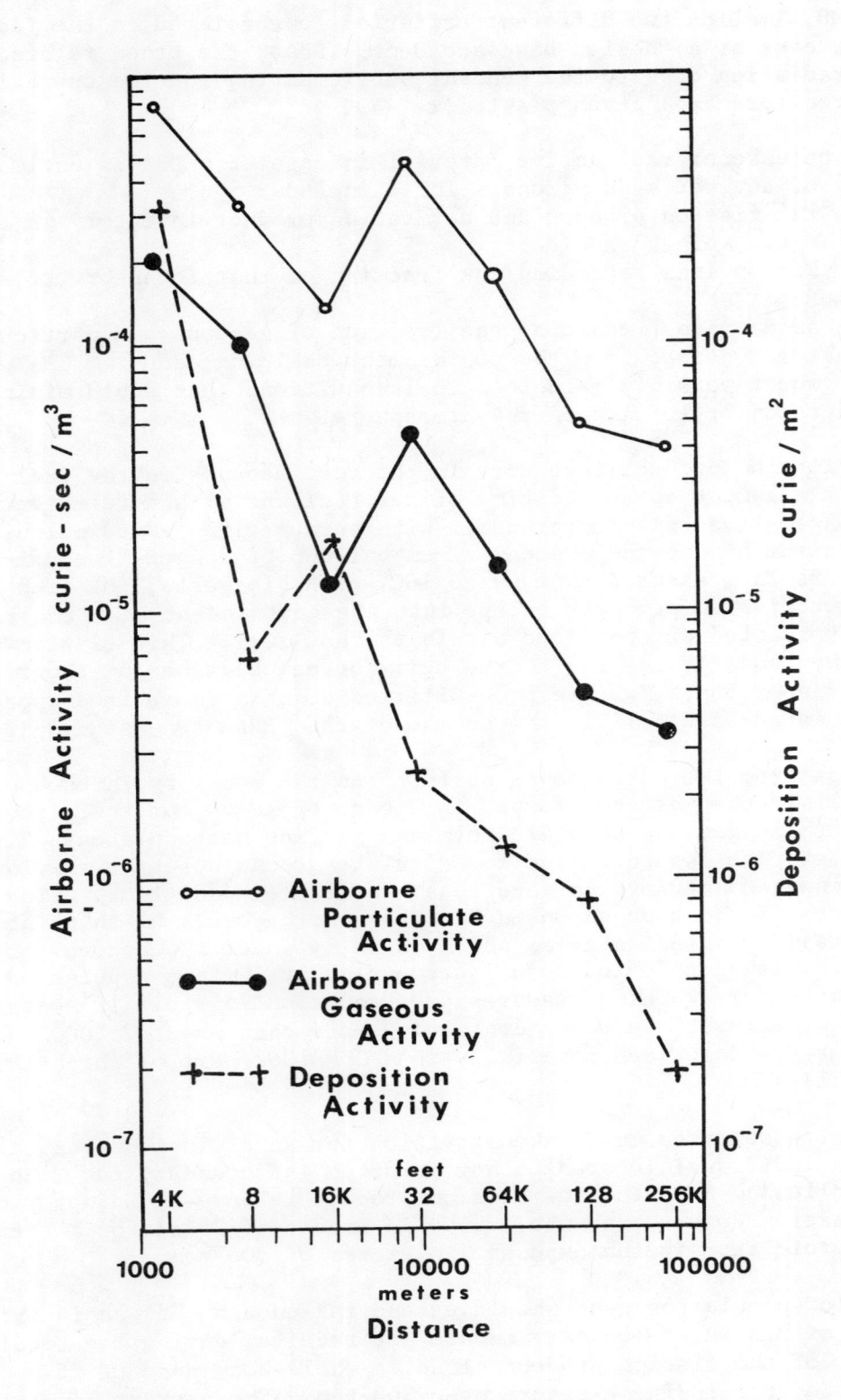

Fig. 1. Atmospheric Fractionation with Distance for Gaseous and Particulate Airborne Activities

Table III. U. S. NRC Reactor Site Criteria (extracted from 10 CFR 100 (7))

100.3 Definitions.

As used in this part:

(a) "Exclusion area" means that area surrounding the reactor, in which the reactor licensee has the authority to determine all activities including exclusion or removal of personnel and property from the area. This area may be traversed by a highway, railroad, or waterway, provided these are not so close to the facility as to interfere with normal operations of the facility and provided appropriate and effective arrangements are made to control traffic on the highway, railroad, or waterway, in case of emergency, to protect the public health and safety. Residence within the exclusion area shall normally be prohibited. In any event, residents shall be subject to ready removal in case of necessity. Activities unrelated to operation of the reactor may be permitted in an exclusion area under appropriate limitations, provided that no significant hazards to the public health and safety will result.

(b) "Low population zone" means the area immediately surrounding the exclusion area which contains residents, the total number and density of which are such that there is a reasonable probability that appropriate protective measures could be taken in their behalf in the event of a serious accident. These guides do not specify a permissible population density or total population within this zone because the situation may vary from case to case. Whether a specific number of people can, for example, be evacuated from a specific area, or instructed to take shelter, on a timely basis will depend on many factors such as location, number and size of highways, scope and extent of advance planning, and actual distribution of residents within the area.

(c) "Population center distance" means the distance from the reactor to the nearest boundary of a densely populated center containing more than about 25,000 residents.

100.11 Determination of exclusion area, low population zone, and population center distance.

(a) As an aid in evaluating a proposed site, an applicant should assume a fission produce release from the core, the expected demonstrable leak rate from the containment and the meteorological conditions pertinent to his site to derive an exclusion area, a low population zone and population center distance. For the purpose of this analysis, which shall set forth the basis for the numerical values used, the applicant should determine the following:

(1) An exclusion area of such size that an individual located at any point on its boundary for two hours immediately following onset of the postulated fission product release would not receive a total radiation dose to the whole body in excess of 25 rem or a total radiation dose in excess of 300 rem to the thyroid from iodine exposure.

(2) A low population zone of such size that an individual located at any point on its outer boundary who is exposed to the radioactive cloud resulting from the postulated fission product release (during the entire period of its passage) would not receive a total radiation dose to the whole body in excess of 25 rem or a total radiation dose in excess of 300 rem to the thy-

Table III. Concluded

roid from iodine exposure.

(3) A population center distance of at least one and one-third times the distance from the reactor to the outer boundary of the low population zone. In applying this guide, due consideration should be given to the population distribution within the population center.

Where very large cities are involved, a greater distance may be necessary because of total integrated population dose consideration.

(b) For sites for multiple reactor facilities consideration should be given to the following:

(1) If the reactors are independent to the extent that an accident in one reactor would not initiate an accident in another, the size of the exclusion area, low population zone and population center distance shall be fulfilled with respect to each reactor individually. The envelopes of the plan overlay of the areas so calculated shall then be taken as their respective boundaries.

(2) If the reactors are interconnected to the extent that an accident in one reactor could affect the safety of operation of any other, the size of the exclusion area, low population zone and population center distance shall be based upon the assumption that all interconnected reactors emit their postulated fission product releases simultaneously. This requirement may be reduced in relation to the degree of coupling between reactors, the probability of concomitant accidents and the probability that an individual would not be exposed to the radiation effects from simultaneous releases.

It is clear that the extent of the low population zone is somewhat arbitrary and it is important to project population trends for that area over the period of plant operation, since only rarely would it be subject to stringent zoning or development controls. It is also evident that any release calculated will depend on the nature of the DBA assumed and the extent to which breach of containment is postulated. At this time, most of these calculations are probably over-conservative and hence exclusion areas may be much larger than required.

In the Rasmussen Study, the "safety" of the reactor was ultimately defined in terms of the risk of injury or death to the surrounding population due to the release of radioactive materials following a core melt accident. This risk is made up of several probabilistic components: the probability that a certain sequence of events will lead to a core melt, the probability of escape of a given fraction of the coolant and fuel inventory of radionuclides, the dispersion and precipitation distributions, the various uptake pathways by inhalation or ingestion by man, and the population distribution around the plant site. The site-related factors of such an analysis include meteorological parameters, such as wind speed and wind direction as a function of time of day or season, the local ecology, local food habits, the absence or existence of a dairy industry or fisheries, and a knowledge of the present and projected population distribution.

Assuming "representative" conditions for current U. S. light-water reactors and rather pessimistic assumptions regarding the actual escape of radionuclides from a plant following a core-melt, risk assessments for 100 reactors were expressed in comparison to other natural and man-caused risks in the form shown in Figs. 2 and 3, which in rough terms equate such risks with those of getting killed by a meteorite. While this is doubtlessly reassuring to the general public, such an analysis does not cast any light on the site selection process for any single power station except to re-affirm that the current, highly conservative approach does lead to risk values well within the usually accepted risks for other activities as long as a plant site does not diverge unduly from the "standard" conditions as to meteorology or population density.

In this connection it is interesting to note that the U. S. Supreme Court has re-affirmed recently that a "population center," as defined in 10 CFR 100, is to be taken to be located at any actual clustering of population regardless of legal or municipal boundaries that might distort the apparent "distance" to the plant site.

RISK CRITERION

By using the limiting dose due to the most severe postulated accident, however improbable, to define exclusion area dimensions, the USAEC criteria tend to require very large exclusion areas. An alternative method to evaluate the suitability of a potential reactor site with respect to radiological exposure has been advocated by Farmer (9). In this approach the site criterion is not based on limiting dose to individuals and the general public in the region surrounding the reactor site, but on the overall risk of exposure, principally to radioiodine accumulated in the thyroid, taking into account the probability of any given accident or the design accident actually happening. Once the relationships of radiation dose vs. probability have been determined for the distances of interest, the relationships proposed to correlate mortality risk and dose can be used to obtain mortality risk as a function of accident probability for several distances. The method is not restricted to use of the linear risk-dose assumption used here,

$$M_{t/wb}(P,s) = D_{t/wb}(P,s) m_{t/wb}, \qquad (1)$$

where

$M_{t/wb}(P,s)$ = mortality probability (thyroid or whole-body), a function of accident probability (P) and distance (s) from the reactor,

$D_{t/wb}(P,s)$ = dose (thyroid/whole-body) in rads, a function of accident probability (P) and a distance (s) from the reactor,

$m_{t/wb}$ = mortality probability (thyroid or whole-body) per rad of radiation

The total risk to an individual is found by integrating the mortality risk of Eq. (1) over all accident probabilities.

$$R(s)_{t/wb} = \int_{P_1}^{P_2} M_{t/wb}(P,s) \, dP, \qquad (2)$$

where

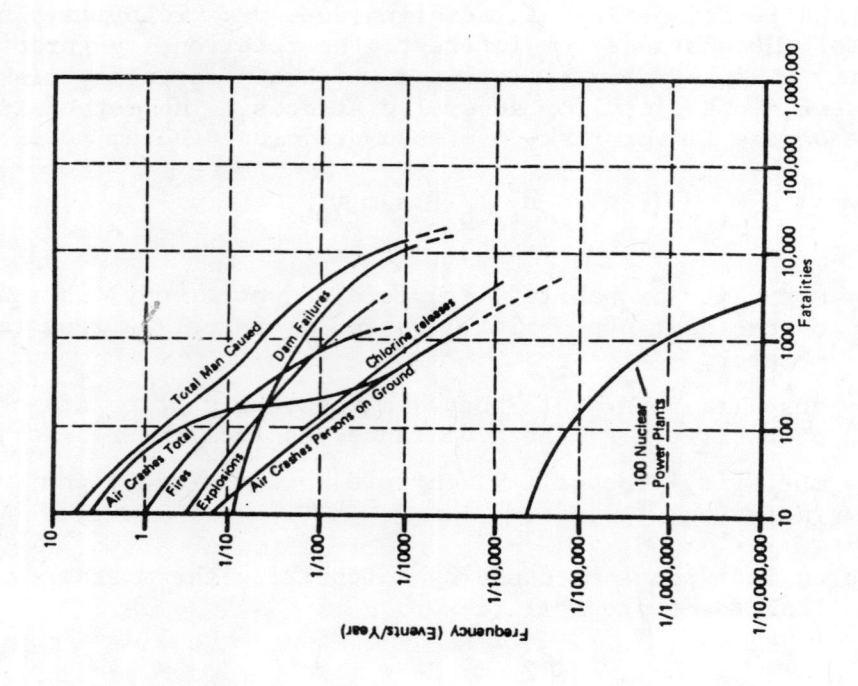

Fig. 2. Comparison of the Frequency of Fatalities from Man-caused Events (8)

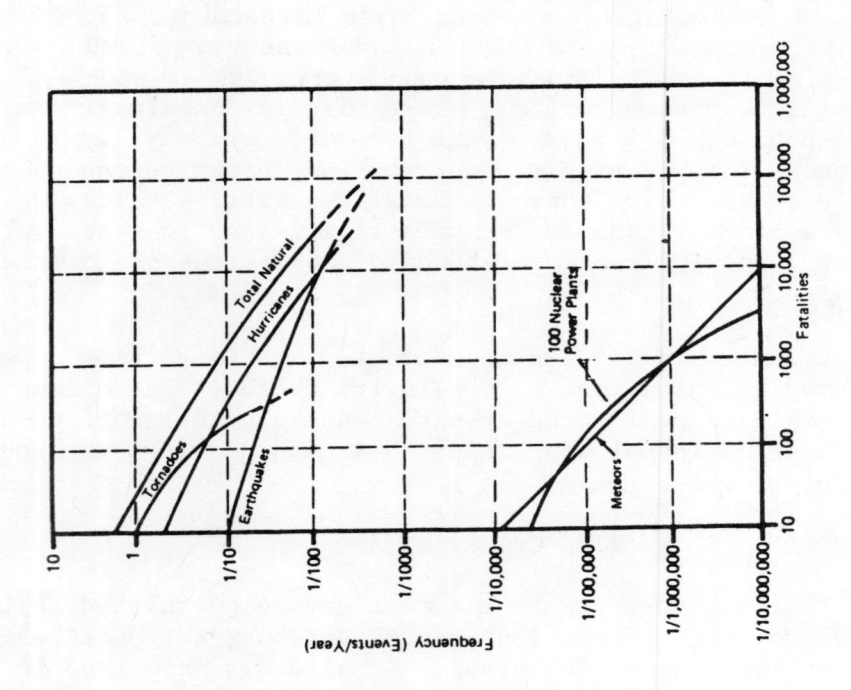

Fig. 3. Comparison of the Frequency of Fatalities from Natural Events (8)

$R(s)_{t/wb}$ = individual mortality risk per year (thyroid or whole-body), as a function of distance (s).

This yields the total risk to an individual as a function of a specific distance from the reactor; then this risk is used as the criterion for determining exclusion-radius requirements so that risk can be set at any desired value.

The total risk to the population around the reactor site in terms of deaths per year or total deaths during the projected reactor lifetime can be found by integrating the individual risk (weighted by the number of individuals at risk) over distance. The only additional datum needed is the population distribution as a function of area covered by the fission-product cloud. We may assume that the cloud occupies a 30° arc at 100 m and decreases linearly with distance to a 15° arc at 10,000 m. The population density is conservatively assumed constant over this distance. The total risk is given by

$$R_t = \int_{s_1}^{s_2} R(s) \, P(s) \, ds, \qquad (3)$$

where
R_t = total yearly death risk to the population in the affected area,
$R(s) = R_i(s) + R_{wb}(s)$ as defined by Eq. (2), and
$P(s)$ = population distribution within the affected area as a function of radial distance from the reactor.

The limits of integration of Eq. (3) are normally from the exclusion radius to infinity, or to a point where doses are below the threshold. Figure 4 indicates the form the criterion takes. By plotting the equivalent ground release of radioiodine against the number of reactor-years elapsing between iodine releases of given magnitude one obtains a number of contours like the line shown, which represent conditions of comparable risk. For Great Britain for a "standard site" there was assumed to be a uniform population density of 13,000 per square mile in all directions from 0.5 to 10 miles from the site. The line drawn in Fig. 5 then represents the condition that the aggregate risk of casualties from the operation of a GCR reactor under those conditions on an urban site is of the order of 0.01 per year, or an individual risk of 1×10^{-7} per year, several orders of magnitude lower than the risks of incurring leukemia or accidental death from other causes. Similar risk criteria can be derived for other sites using fault-tree methodology of safety analysis. Otway has developed the risk approach and analyzed the contribution of design factors, site-related factors, and of assumptions regarding the effects of single exposures to high radiation fields on the risk calculation (10). He concluded that for a 1000 MWe PWR an exclusion radius of 350 m (1000 ft) would keep the individual mortality risk at the site boundary below a value of 10^{-7} per person per year and the total number of deaths expected from a 30 yr operating life would be 0.003. This was based on an assumed mortality risk from I-131 irradiation of the thyroid of 1×10^{-6} cancer cases per person per rad, with a 1-rad threshold. Any reduction in the release consequences of a design-base accident by improvement in engineered safeguards would progressively shrink the required site area or further reduce the mortality risk at the site boundary.

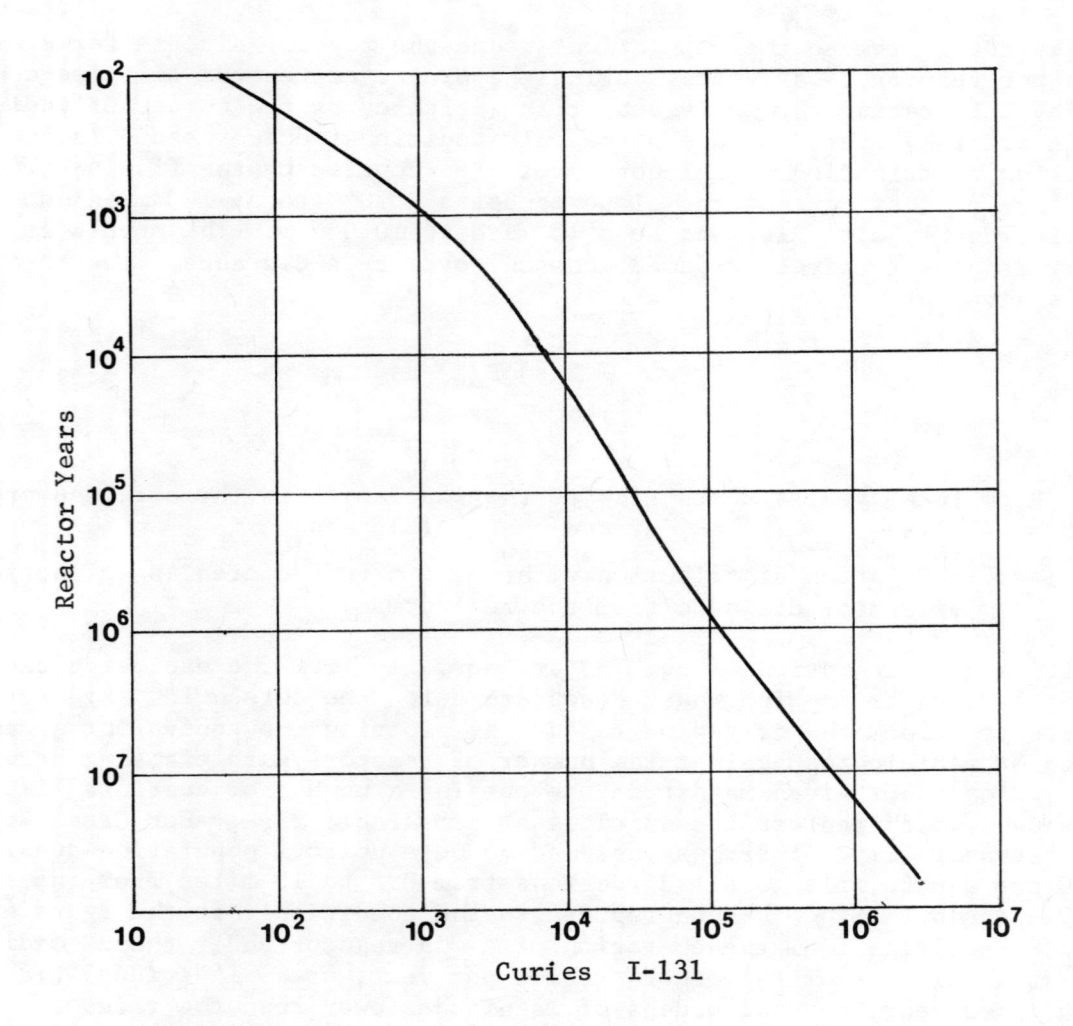

Fig. 4. Proposed Risk Criterion from I-131 Activity

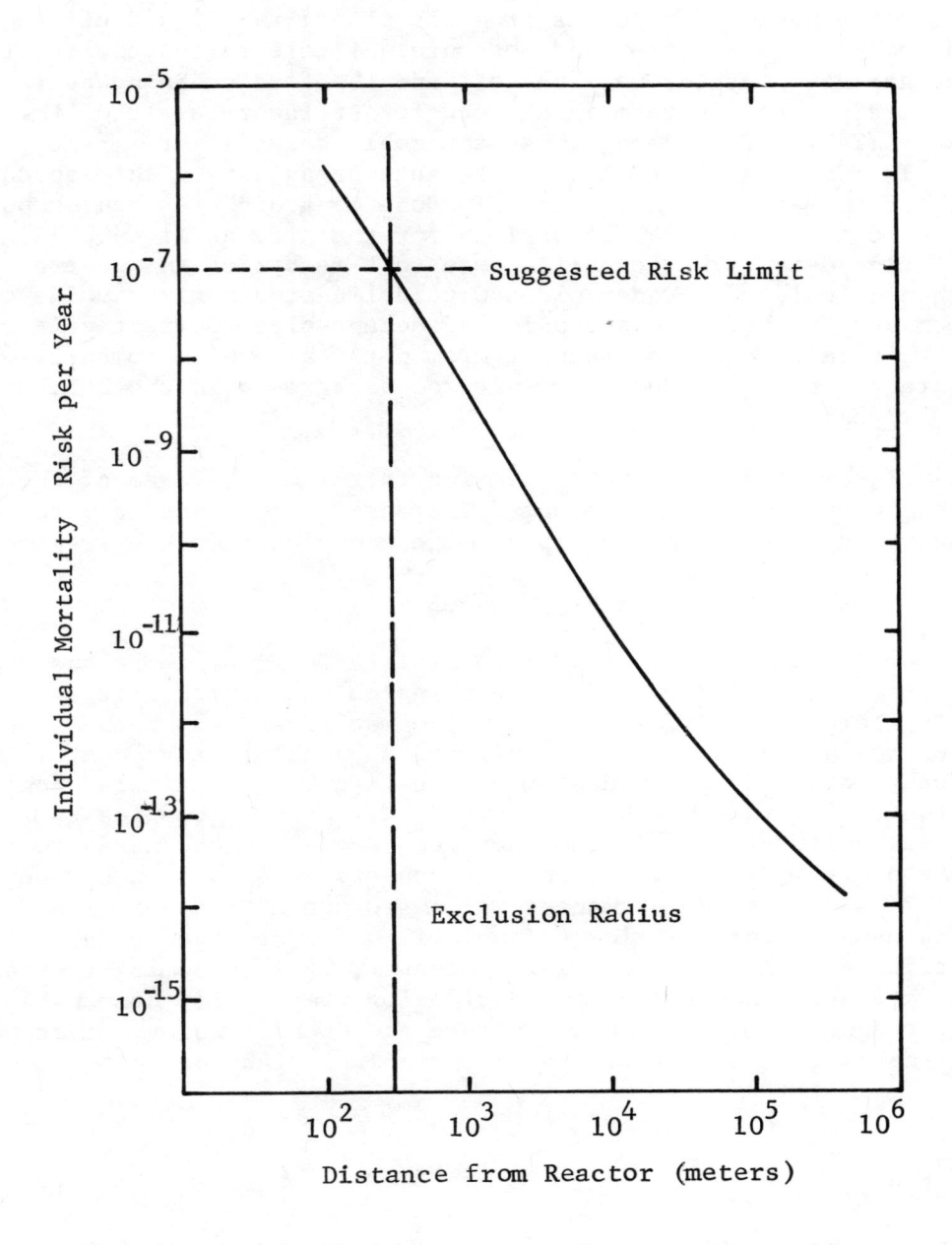

Fig. 5. Use of Risk Assessment to Determine Acceptable Site Dimensions

Routine release of radioactive effluents predicates another set of site criteria. It is these which are intended to be "as low as practicable," and which are referred to in the AEC's guidelines, WASH-1258 (11). Basically, they relate to the gaseous effluents from the plant; the liquid effluents are assumed to be fully purified and to contain little radioactivity, though their uptake and movement through the soil and the food chain needs to be considered. Again a source term is assumed for effluent levels of the various components (Table IV). Among these the noble gases (^{85}Kr, ^{133}Xe), the halogens (^{129}I, ^{131}I) and tritium are most noteworthy, since they account for the bulk of the escaping activity. Isodose contours are then computed for the expected meteorological conditions for the site as given by windrose and precipitation data. The fenceline dose to a member of the general population is then calculated. Under old AEC guidelines, a maximum value of 170 mrem per year per individual was considered acceptable. Current criteria have pushed this value down to around 2 mrem per year, which compares well with the average population dose from natural background in the U.S. of about 125 mrem/year.

In general this would not be a limiting factor on the area of the exclusion zone; however, if it turns out differently, the plant operator will need to acquire additional land or spend more money on radwaste treatment.

WASTE STORAGE

Storage or disposal of low- and intermediate-level waste at the reactor site may constitute a minor factor in deciding on site suitability. This storage would include temporary containment of solid wastes, such as spent demineralizer resins and contaminated piping, charcoal filters containing iodine and other airborne effluents, and liquified gas containers, containing mainly krypton-85. Although waste treatment at reactors occupies ever larger structures, waste storage need not occupy much space and one needs to select merely a reasonably out-of-the-way spot to construct suitable concrete bins. One would not anticipate any permanent waste storage at reactor sites, though the hold-up characteristics of the soil under storage or delay tanks for liquid effluents may need scrutiny. However, it is becoming increasingly evident that below-surface bin accommodation for temporarily held wastes is important to reduce plant personnel exposure and such structures must meet the same stringent design criteria as the principal plant buildings.

SPECIAL SITES

Off-Shore Sites

Most of the current generation of U.S. reactors are planned for inland sites on major streams or lakes or at ocean beach locations. Because of the supposed difficulties of finding additional suitable locations that meet all environmental guidelines, a considerable effort is being devoted to the development of standardized reactor designs to be located on artificial islands or floats in 60 ft of water off-shore. The basic assumption underlying these plants is that ocean water is free and the sea can dissipate heat more harmlessly than bodies of fresh water. In practice the availability of such sites opposite their needed beach front area, which would accommodate the switchyard and loading docks, is by no means unlimited either. The design of sea-

Table IV. Representative Data on Effluent Release

Parameters Used in Radioactive Effluent Analysis for PWR

	WASH-1258 Assumptions	Westinghouse Recommended Assumptions
Reactor power, MWt	3500	3500
Plant capacity factor, %	80	80
Fraction of fuel releasing radioactivity to primary coolant, %	0.25	0.05
Primary coolant to containment, lb/day	240	40
Turbine building steam leak rate, lb/hr	1700	1700
Auxiliary building leak of reactor coolant, lb/day	160	160
Steam generator leak rate, lb/day	110	110
Filtration of containment/reactor building effluent, % efficiency	90	99
Filtration of auxiliary building effluent, % efficiency	90	99
Filtration of air ejector effluent, % efficiency	90	99

Expected Gaseous and Liquid Effluents from a Typical Westinghouse PWR During Normal Operation (WASH-1258 Assumptions)

Gaseous Effluent (curies/yr/unit)

	Containment Purge	Waste Gas Proc. Sys.	Aux. Bldg.	Turbine Bldg.	Steam Gen. Blowdown Vent	Condenser Air Ejector	Total
Noble Gases	21	980	160	---	---	160	1300
Iodine (I-131 only)	0.00045	---	0.0073	0.025	---	0.011	0.044

Liquid Effluent (curies/yr/unit)

	Clean Waste	Dirty Waste	Steam Gen. Blowdown	Turbine Drains	Total
Corrosion and Activation Products	---	0.003	0.0015	0.00015	0.0046
Fission Products	---	0.067	0.059	0.00067	0.13
Tritium	---	96	76	86	258

based reactors poses a new set of safety considerations, not all of which have been fully developed at this time.

Figure 6 shows an artist's conception of two such reactors located within a breakwater over above 20 m (60 ft) of water and Fig. 7 shows a proposed layout of such a plant. The environmental impact of such plants has been studied in detail (12-14) and seems to lead to acceptable conditions. However, it still leaves some ethical questions whether small but finite contamination of the oceans is more acceptable than that of smaller, better-defined bodies of fresh water.

Underground Sites

Periodically, suggestions are made to locate nuclear power plants in natural or artificial caves underground, because such locations are supposedly "safer" than surface sites (15-19). In fact, small reactors have been built and operated in mountain caves in Sweden and Norway. If one analyzes this concept, the advantages of underground sites are more apparent than real. Underlying it is still the notion that the reactor is a potential bomb that needs to be confined in a massive enclosure. If one discounts this assumption, there are few real advantages left. A cave of appropriate size would be very expensive to excavate. It would need to be lined with a "containment vessel" to protect against rockfalls. Structural and watertable problems may be quite difficult and construction in a deep hole is manifestly more complicated than surface construction. Since biological shielding of the reactor, close in, presents few problems and the reactor operating personnel needs to be protected anyhow, little is gained by the shielding effect of the surrounding rock or soil.

Figures 8 and 9 show two conceptual layouts for such plants. The excavation needed would be substantial; a shell liner would certainly be required to eliminate rockfalls. If release of radioactivity in case of a core melt is considered the principal safety hazard, enough vents and access shafts exist to permit such releases on a scale not too different from that of a surface plant. Table V shows some cost estimates for the additional cost of underground plants in terms of 1971 dollars (18). It is seen that this additional cost is substantial, though not excessive in terms of overall plant costs. However, this extra cost must still be justified by a corresponding reduction in population dose.

In summary, until additional factors are brought forward, there seems to be no real advantage in placing nuclear power plants underground.

CONCLUSION

In summary, site selection and proper safety design of nuclear power plants play an important part in the public acceptance of such systems and in minimizing population exposure to radiation in case of major plant failures.

Fig. 6. Artist's Conception of Two FNP's within Breakwater

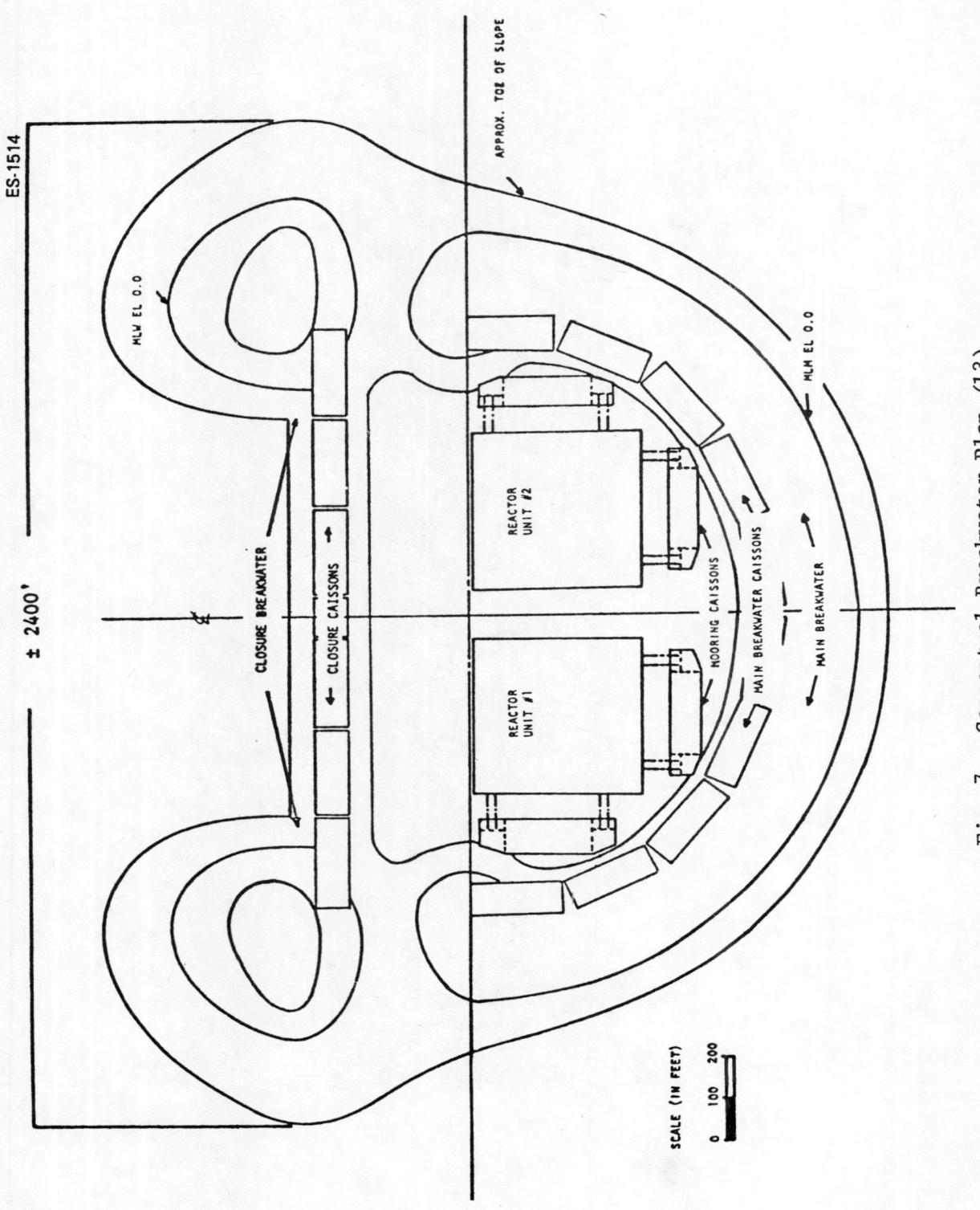

Fig. 7. Conceptual Breakwater Plan (13)

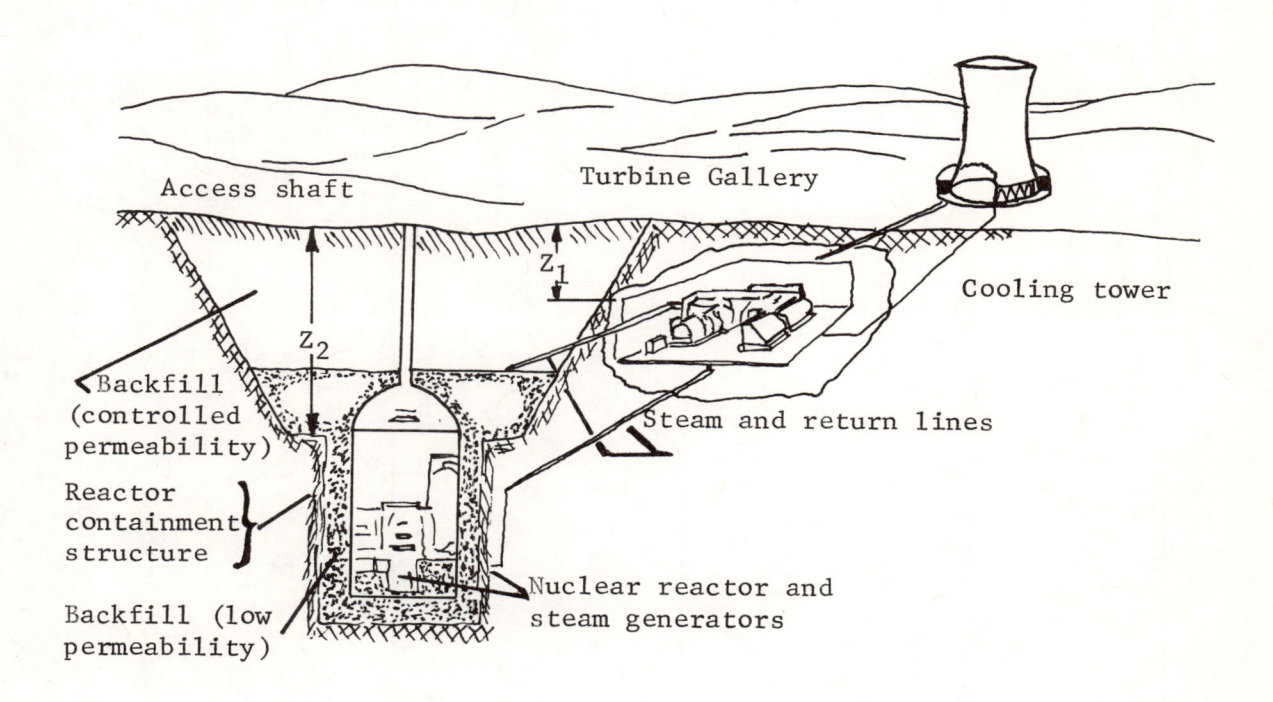

Fig. 8. Possible Layout of an Underground Nuclear Power Plant Constructed in Rock (19)

54

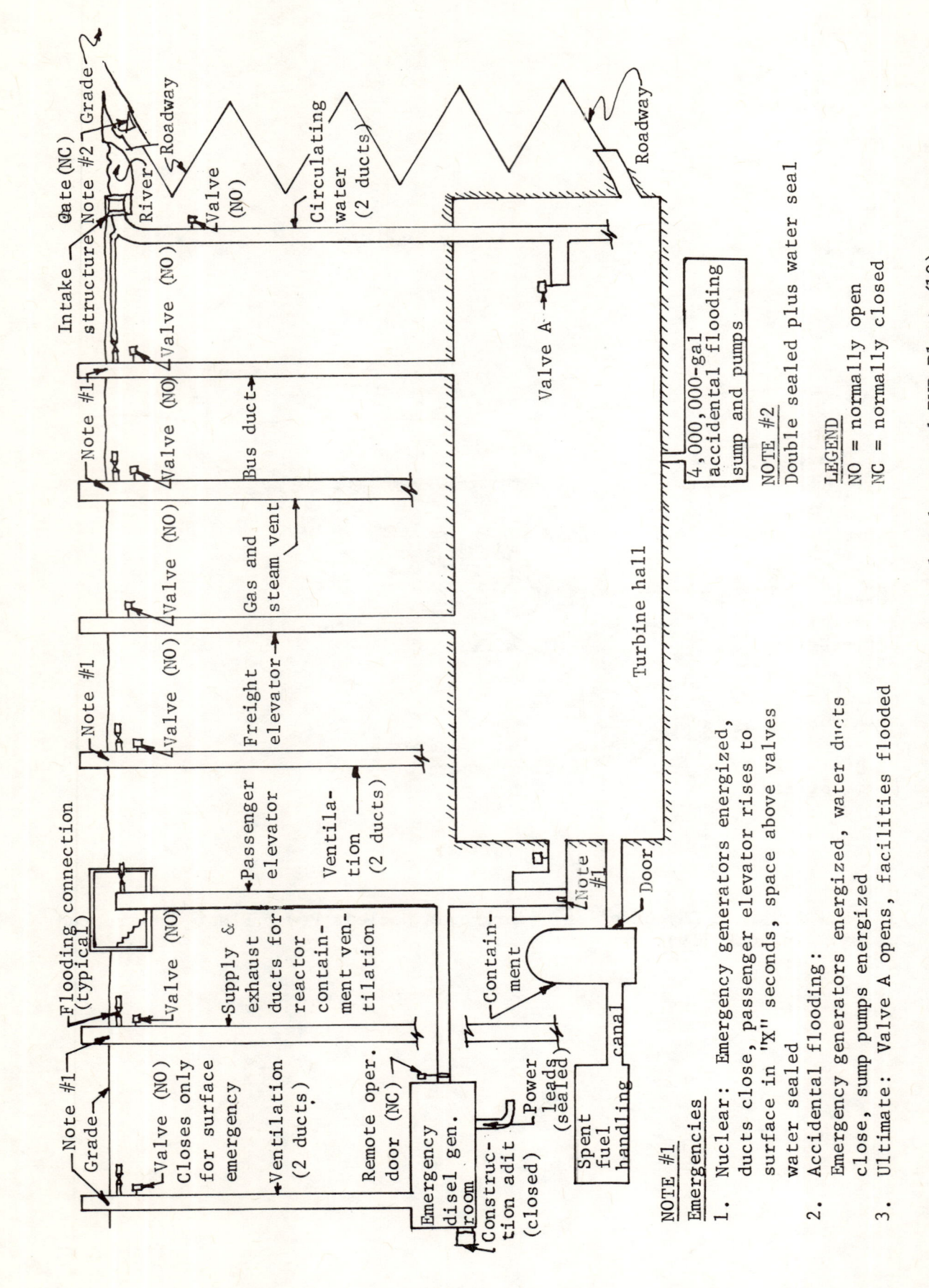

NOTE #1
Emergencies
1. Nuclear: Emergency generators energized, ducts close, passenger elevator rises to surface in "X" seconds, space above valves water sealed
2. Accidental flooding: Emergency generators energized, water ducts close, sump pumps energized
3. Ultimate: Valve A opens, facilities flooded

NOTE #2
Double sealed plus water seal

LEGEND
NO = normally open
NC = normally closed

Fig. 9. Vertical Section of Conceptual Underground PWR Plant (19)

Table V. Excavation Cost Estimate

Item	Volume,* yd³	Unit cost,† $	Excavation cost, $
Access tunnel	155,000	27.50	4,263,000
Vertical conduits and shafts	10,000	75.00 (50.00)	750,000
Horizontal conduits	15,000	(35.00) (22.00)	525,000
Reactor containment	130,000	35.00	4,550,000
Balance of plant			
Heading	92,000	30.00 (15.00)	2,760,000
Benching	248,000	20.00 (10.80)	4,960,000
Other	120,000	25.00	3,000,000
Total	770,000		20,808,000

*Excavation volume is based on requirements for placing a 1000-MW(e) PWR underground.

†Unit costs in parentheses are from the Churchill Falls experience (1971).

SELECTED REFERENCES

1. "The Safety of Nuclear Power Reactors (Light-water-cooled) and Related Facilities," USAEC Rept. WASH-1250, Washington, D. C. (1973).

2. D. G. JOPLING, "Power Plant Siting in the United States: 1972--A State Summary," Southern Interstate Nuclear Board, Atlanta (1972).

3. "General Site Suitability Criteria for Nuclear Power Stations," USNRC Regulatory Guide 4.7 (Rev. 1) (1975).

4. "Seismic Design Classification," USAEC Regulatory Guide 1.29 (Rev. 1) (1973).

5. "Damping Values for Seismic Design of Nuclear Power Plants," USAEC Regulatory Guide 1.61 (1973).

6. "Combination of Modes and Spatial Components in Seismic Response Analysis," USAEC Regulatory Guide 1.92 (1974).

7. "Reactor Site Criteria," Code of Federal Regulations, Title 10 Part 100, as amended.

8. "Reactor Safety Study: An Assessment of Accident Risks in U.S. Commercial Nuclear Power Plants," USAEC Rept. WASH-1400, Washington, D. C. (Draft 1974, Final 1975).

9. F. R. FARMER, "Reactor Safety and Siting: A Proposed Risk Criterion," Nuclear Safety, 8, 539-548 (1947).

10. H. J. OTWAY, "The Application of Risk Allocation to Reactor Siting and Design," USAEC Rept. LA-4316, Los Alamos Scientific Lab (1969).

11. "Numerical Guides for Design Objectives and Limiting Conditions for Operation to Meet the Criterion 'As Low As Practicable' for Radioactive Material in Light-water-cooled Nuclear Power Reactor Effluents," USAEC Rept. WASH-1258 and attachments, Washington, D. C. (1973,1974).

12. "Draft Environmental Statement: Manufacture of Floating Nuclear Power Plants, Part II," USNRC Rept. NUREG-75/113, Washington, D. C. (1975).

13. J. A. FISCHER and F. L. FOX, "Siting Constraints for an Offshore Nuclear Power Plant," Engrg. Bull. 42, Dames & Moore, Inc., Los Angeles, Calif. (1973).

14. O. H. KLEPPER and T. D. ANDERSON, "Siting Considerations for Future Offshore Nuclear Energy Stations," Nucl. Technol., 22, 160-169 (1974).

15. F. M. SCOTT, "Locating Nuclear Power Plants Underground," Envir. Letters, 9, 333-353 (1975).

16. E. JAMNE, "Underground Siting of Nuclear Power Plants in Norway," Proc. 4th Geneva Conf. Peaceful Uses of Atomic Energy, 3, 359-373 (1971).

17. F. C. ROGERS, "Underground Nuclear Power Plants," Bull. Atomic Scientists, 27(8), 38-51 (1971).

18. J. H. CROWLEY, P. L. DOAN and D. R. McCREATH, "Underground Nuclear Plant Siting," Nucl. Safety, 15, 519-534 (1974).

19. J. M. CARDITO, E. V. SOMERS and J. H. McWHIRTER, "Hydrodynamic Containment for Underground Nuclear Power Plants," Nucl. Technol., 28, 119-126 (1976).

NUCLEAR REGULATORY COMMISSION REGULATIONS AND LICENSING

Robert B. Minogue and Abraham L. Eiss*
U. S. Nuclear Regulatory Commission
Washington, D. C. 20555

INTRODUCTION

While the country was in the midst of commemorating its 200th birthday, the Nuclear Regulatory Commission celebrated only its first--the Commission was formed on January 19, 1975.

This chapter discusses the background of the NRC as successor to the regulatory portion of the Atomic Energy Commission, the mission assigned to the NRC by the Congress, and the way in which it is organized to carry out this mission. Next is a more detailed discussion of the regulatory program of the agency: licensing and enforcement, confirmatory research, and standards development.

BACKGROUND OF NRC

In 1974 Congress passed the Energy Reorganization Act, which separated the regulatory functions of the former Atomic Energy Commission from its developmental and promotional activities. The central Federal responsibility for energy research and development now lies with the Energy Research and Development Administration, while other Federal bodies--such as the Federal Energy Administration and the Energy Resources Council--play key roles in advising the President on national energy policies, the adequacy of energy resources, and programs for dealing with energy shortages, including conservation measures.

Placing nuclear regulatory responsibilities in an independent Commission and separating those responsibilities from developmental considerations were watershed events in the Nation's nuclear experience. They recognized the fact that nuclear energy had reached a stage in the United States where its governmental regulation demanded the full attention of a separate and autonomous agency.

NRC's Mission

NRC's role, as a regulator of the use of nuclear energy in the civilian sector, is to ensure that the construction and operation of nuclear facilities and the use of nuclear materials are carried out in a manner which

---
*Robert B. Minogue, Director, Office of Standards Development.
Abraham L. Eiss, Technical Assistant to the Director, Division of Engineering Standards, Office of Standards Development.

provides reasonable assurance of public health and safety and is consistent with national security considerations and with a proper regard for environmental values. In keeping with the President's call for regulatory reform, the Commission has established a policy of preparing a formal value/impact statement for all major programmatic activities. These statements will ensure that the Commission's decisions will continue to be consistent with its responsibility to ensure the public health and safety and to protect the environment. Moreover, the value/impact statements contain a careful analysis of economic costs of regulatory actions to ensure that the actions are fully cost effective.

In addition, in our regulatory program we are placing greater emphasis on recognizing that there are many alternative ways of achieving safety goals and to providing a full opportunity to applicants and licensees to select the approach that is most cost effective for their individual plant in achieving the basic goal of safe performance.

Organization

Organizationally, the NRC consists of five Commissioners, each appointed by the President with the advice and consent of the Senate. One Commissioner is designated by the President as Chairman and acts as the Commission's executive agent and official spokesman. Each Commission member, including the Chairman, has equal responsibility and authority and exercises one vote in Commission decisions. The organization of the NRC is shown in Fig. 1.

Under the terms of the Energy Reorganization Act of 1974, three regulatory offices report directly to the Commission. The Office of Nuclear Reactor Regulation performs a wide range of functions relating to the licensing and regulation of nuclear reactors. The Office of Nuclear Material Safety and Safeguards is responsible for carrying out licensing and safeguards functions related to the supporting nuclear fuel cycle--from milling operations to waste management.

The third major component, the Office of Nuclear Regulatory Research, is responsible for planning and carrying out research necessary for the performance of the Commission's licensing and related regulatory responsibilities. Finally, we have an Executive Director for Operations, who is charged with coordinating and directing the day-to-day operational and administrative activities of the agency.

Two other major components complete the line organization structure. The Office of Standards Development develops regulations, criteria, guides, standards, and codes pertaining to the protection of the public health and safety and the environment in all stages of a nuclear facility's life. And the Office of Inspection and Enforcement ensures that licensees comply with the provisions of their license and with the Commission's rules.

These five line offices carry out the major regulatory functions of the NRC which are as follows:

(a) Standards-Setting and Rule-Making
(b) Technical Reviews and Studies

NUCLEAR REGULATORY COMMISSION

```
                    ┌──────────────┐
                    │     THE      │
                    │  COMMISSION  │
                    └──────┬───────┘
                           │
                    ┌──────┴───────┐
                    │  EXECUTIVE   │
                    │  DIRECTOR    │
                    │     FOR      │
                    │  OPERATIONS  │
                    └──────┬───────┘
                           │
    ┌──────────┬───────────┼───────────┬──────────┐
    │          │           │           │          │
┌───┴────┐ ┌───┴────┐ ┌────┴────┐ ┌────┴────┐ ┌───┴────┐
│OFFICE  │ │OFFICE OF│ │OFFICE OF│ │OFFICE OF│ │OFFICE  │
│  OF    │ │NUCLEAR  │ │NUCLEAR  │ │NUCLEAR  │ │  OF    │
│STANDARDS│ │MATERIAL │ │REACTOR  │ │REGULATORY│ │INSPECTION│
│DEVELOP-│ │SAFETY   │ │REGULATION│ │RESEARCH │ │  AND   │
│ MENT   │ │AND      │ │         │ │         │ │ENFORCE-│
│        │ │SAFEGUARDS│ │         │ │         │ │MENT    │
└────────┘ └─────────┘ └─────────┘ └─────────┘ └────────┘
```

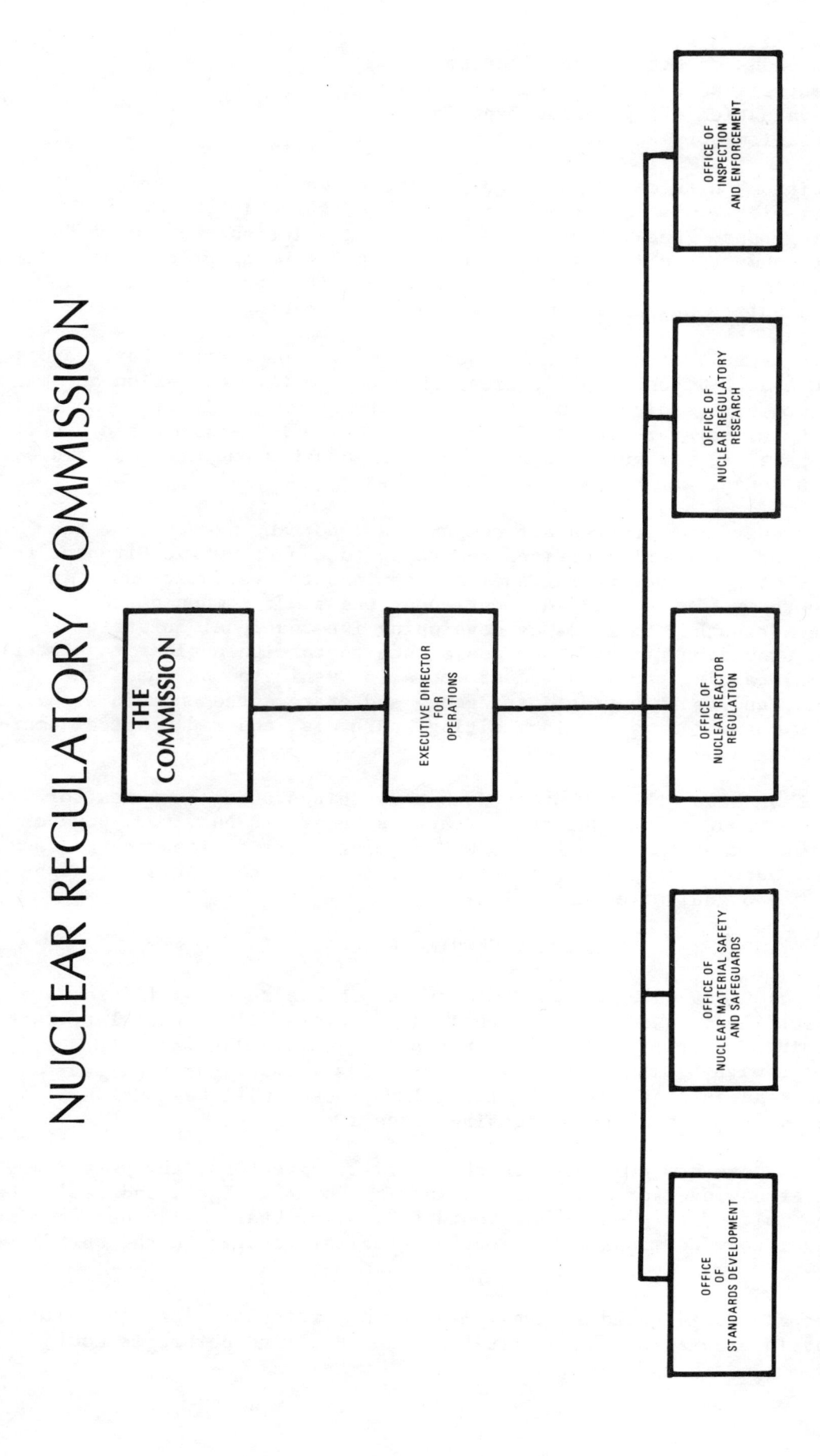

Figure 1

(c) Actions on License Applications
(d) Surveillance and Enforcement
(e) Evaluation of Operating Experience
(f) Confirmatory Research

In addition, nine major staff components shown in Fig. 2 report directly to the Commission. One of the most important of these is the Atomic Safety and Licensing Board Panel. Out of this group, appointments are made to Boards that conduct hearings in connection with licensing proceedings. A related component is the Atomic Safety and Licensing Appeal Panel, from which Atomic Safety and Licensing Appeal Boards are chosen.

A third organization is the Advisory Committee on Reactor Safeguards, which reviews each license application submitted to the Commission and makes a recommendation regarding issuance of a construction permit or an operating license. Its members are technical experts who are full-time employees of major universities, national laboratories, or industry and are employed by the NRC on a consultant basis.

There are several other staff components reporting directly to the Commission, and a number of offices reporting to the Executive Director for Operations that carry out operational and administrative functions. In addition to the activites already mentioned, the staff components are responsible for such activities as developing international information agreements, providing training and assistance to foreign nuclear regulatory agencies, maintaining public document rooms in Washington and near the reactor site, and coordinating interagency and state agreements in such areas as environmental monitoring, siting approvals, and radiological emergency planning.

As we have seen, the NRC staff does many things other than reactor licensing and ensuring the safety of reactors under its purview. However, the main emphasis of this chapter is NRC regulations and licensing procedures related to reactor safety. The components related to that area are highlighted on the organization chart in Fig. 2.

REACTOR LICENSING PROCESS

The regulations under which NRC carries out its responsibilities are found in Title 10 of the Code of Federal Regulations. In particular, Part 50 (1) provides the principal requirements for reactor licensing, and Part 100 (2) deals with siting. Some of the specific regulations that bear directly on reactor safety and the licensing process will be mentioned as part of the discussion of the licensing process.

Let's suppose now that you are the board of directors, the president, and the chief engineer of a company called XYZ Power & Light, and let's step through the multi-year process you would follow to obtain a license from the Nuclear Regulatory Commission to operate a nuclear reactor in the United States.

Before XYZ could build a power plant at a particular site, it would have to obtain a construction permit from the NRC. And before it could

NUCLEAR REGULATORY COMMISSION

THE COMMISSION

ADVISORY COMMITTEE ON REACTOR SAFEGUARDS

ATOMIC SAFETY AND LICENSING BOARD PANEL

ATOMIC SAFETY AND LICENSING APPEAL PANEL

OFFICE OF INSPECTOR AND AUDITOR

OFFICE OF POLICY EVALUATION

OFFICE OF THE GENERAL COUNSEL

OFFICE OF THE SECRETARY

OFFICE OF PUBLIC AFFAIRS

OFFICE OF CONGRESSIONAL AFFAIRS

EXECUTIVE DIRECTOR FOR OPERATIONS

OFFICE OF THE EXECUTIVE LEGAL DIRECTOR

OFFICE OF THE CONTROLLER

OFFICE OF EQUAL EMPLOYMENT OPPORTUNITY

OFFICE OF PLANNING AND ANALYSIS

OFFICE OF INTERNATIONAL AND STATE PROGRAMS

OFFICE OF MANAGEMENT INFORMATION AND PROGRAM CONTROL

OFFICE OF SPECIAL STUDIES

OFFICE OF ADMINISTRATION

OFFICE OF STANDARDS DEVELOPMENT

OFFICE OF NUCLEAR MATERIAL SAFETY AND SAFEGUARDS

OFFICE OF NUCLEAR REACTOR REGULATION

OFFICE OF NUCLEAR REGULATORY RESEARCH

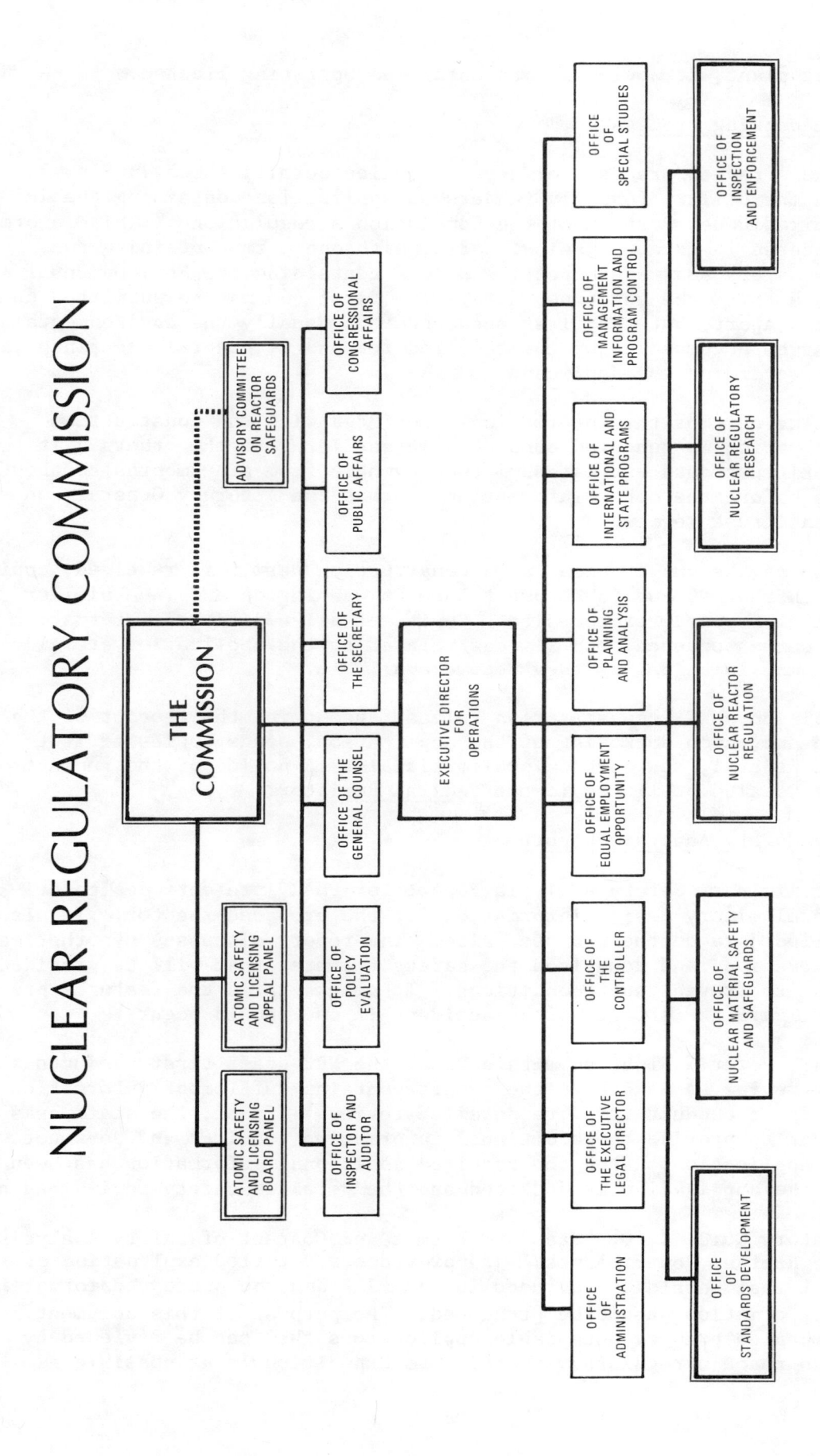

OFFICE OF INSPECTION AND ENFORCEMENT

Figure 2

operate that plant, it would have to obtain an operating license.

Review of Licensing Process

The NRC's review process for a construction permit, shown in Fig. 3, begins when the utility formally tenders an application containing the information required under Part 50 of the Commission's regulations. This information is tendered in three principal parts, with one part containing the Preliminary Safety Analysis Report, a second containing the Environmental Report, and a third the antitrust data. The first two may be submitted up to six months apart, but the first one tendered, usually the Environmental Report, must be accompanied by the required fee and by general and financial information regarding the applicant.

The third part of the information associated with the construction permit application is tendered considerably earlier than the others. It contains antitrust data and is submitted anywhere from nine months to about three years before the other information so that the Attorney General can begin the antitrust review.

As soon as the application for a construction permit is received, copies are placed in the NRC Public Document Room in Washington and in a similar room near the proposed reactor site. Copies of all additional information submitted, correspondence, and filings related to the application are also placed in these locations as they become available.

In addition, NRC issues a press release announcing the receipt of the application and, upon docketing of the application, sends copies of it to appropriate federal, state, and local officials. A notice of the receipt of application is also published in the Federal Register.

Preliminary Safety Analysis Report

The Preliminary Safety Analysis Report, or PSAR, presents design criteria and preliminary design information for the proposed reactor. It also gives detailed data on the proposed site. The report discusses hypothetical accident situations and describes the safety features that will be provided in the plant to prevent such conditions. It also details the features provided to mitigate the effects of an accident if one should occur.

When XYZ Power & Light submits a PSAR, the NRC staff first conducts a preliminary review to make sure the report contains sufficient information for the staff to conduct its more detailed review. If not, the staff asks the applicant to provide the additional information required and does not docket the application. When the required additional information has been submitted, the application is docketed and the detailed safety review begins.

Regulatory Guide 1.70, "Standard Format and Content of Safety Analysis Reports for Nuclear Power Plants," (3) provides a detailed explanation of the material that should be included in the PSAR and the preferred format in which the information should be presented. The purpose of this document is to help companies prepare acceptable applications that can be reviewed by the staff to reach a regulatory decision as expeditiously as possible.

CONSTRUCTION PERMIT REVIEW PROCESS

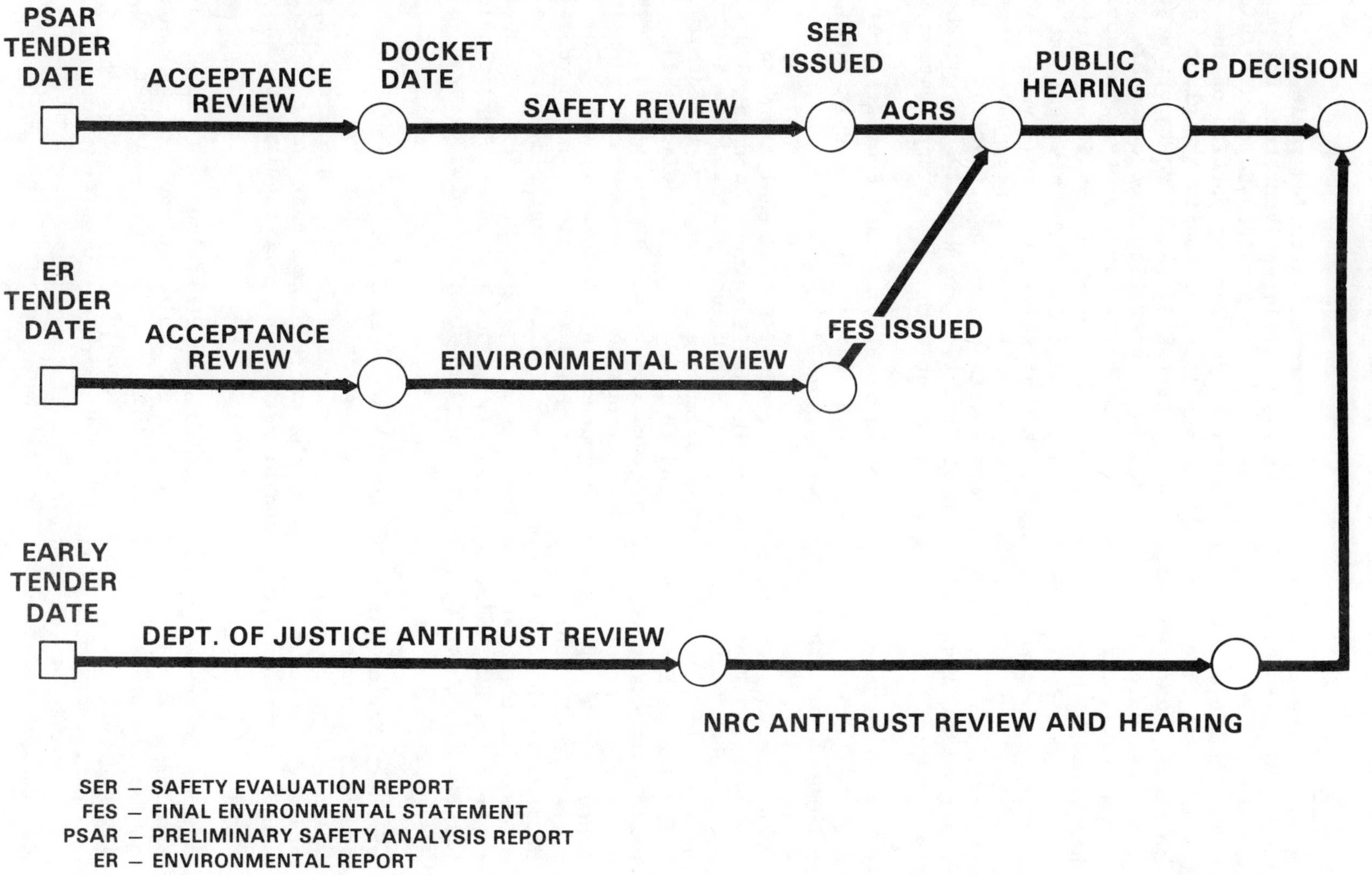

SER — SAFETY EVALUATION REPORT
FES — FINAL ENVIRONMENTAL STATEMENT
PSAR — PRELIMINARY SAFETY ANALYSIS REPORT
ER — ENVIRONMENTAL REPORT

Figure 3

The review of the Environmental Report proceeds in much the same fashion, but I'll first go along the path of the PSAR and then discuss the environmental information.

Going on, ...the staff conducts a complete and detailed review of the PSAR to determine if the proposed plant will sufficiently protect the public health and safety. They may ask the applicant to provide additional information on certain subjects. If the staff finds that any portion of the application is inadequate, they ask the applicant to modify the plant to correct the items that are considered unacceptable. At this stage the staff may determine that special design features or operating procedures are necessary to sufficiently protect the public health and safety. The applicant is then faced with three choices. He may accept the staff's position and make the recommended changes. He may choose to pursue his application without the changes and run a strong risk that it will eventually be rejected. Or, he may decide that the proposed changes will impose unacceptable difficulties or limitations and will therefore withdraw the application. As a practical matter the first and third alternatives are usually chosen; that is, the applicant will agree to the changes or withdraw the application.

The staff conducts this review from a regulatory base consisting of the Commission's regulations in Title 10 of the Code of Federal Regulations, NRC regulatory guides, and other regulatory requirements.

The "Standard Review Plan for the Review of Safety Analysis Reports for Nuclear Power Plants" (4) describes how the staff reviews incoming SAR's. This document, known as the "SRP," was published last year and has two primary purposes: First, it is expected to improve the quality, uniformity, and predictability of staff reviews and to present a well-defined base from which proposed changes to regulatory requirements can be evaluated. Second, it carries out the Commission's policy of openness in regulation by making information on regulatory matters and procedures widely available. The SRP has been placed in the Public Document Room and is available for purchase from the National Technical Information Service in Springfield, Virginia. The SRP is basic to the stabilization of the regulatory process. Further discussion about the SRP and the regulatory guides that discuss particular methods that have been found acceptable by the staff to solve certain safety problems will appear later.

The standard review performed by the staff includes an examination of the proposed design methods, procedures, and calculations to establish that they are valid and appropriate for the particular facility and site. Actual calculations are checked to determine whether the applicant has conducted his analysis in sufficient depth and breadth to support required findings with respect to safety.

In addition to reviewing plant design, the staff looks at the applicant's plans for conducting plant operations. This includes an examination of the organization structure, the technical qualifications of operating and technical support personnel, and the proposed industrial security measures. A particularly important aspect of this review is a look at the applicant's planning for emergency actions in case of an accident that might affect the general public. In addition, the staff checks the applican't proposed program for quality assurance in both design and procurement.

The staff also evaluates the proposed site, including a consideration of population density and use characteristics of the surrounding area. The site's physical characteristics, including the seismology, meteorology, geology, and hydrology, are examined to ensure that they have been adequately evaluated and given appropriate consideration in the plant design. Additionally, the staff checks to see that the site characteristics are in accordance with the siting criteria in Part 100 of the Commission's regulations.

In looking at potential effects of plant operation, the staff reviews the design of systems to be provided for control of radiological effluents from the plant. They consider whether these systems will be able to control the release of radioactive wastes from the plant within the limits specified by Appendix I to Part 50 (5) of the regulations and ensure that the level of radioactive releases from the plant will be as low as is reasonably achievable.

The licensing review also includes a consideration of the applicant's programs for testing and research and development work on features or methods whose safety the applicant cannot completely demonstrate when the PSAR is presented. In accordance with Part 50 of the regulations, the results of these testing programs must be presented to the NRC and satisfactorily evaluated before an operating license can be issued.

The NRC's review of your XYZ Power & Light's PSAR might take about a year. As much as possible, this process is expedited by using previous evaluations of other reactors approved for a construction permit. Thus, to the extent that the plant's design and procedures are standardized to follow the design and procedures of other similar plants previously approved, the review may move along more quickly.

When the evaluation of the PSAR has progressed to the point where the staff is satisfied that acceptable criteria, preliminary design information, and other data are documented in the application, the licensing staff prepares a Safety Evaluation Report. This report summarizes the staff's review and evaluation with respect to the proposed facility's anticipated effects on the public health and safety. It also points out any unresolved issues that will have to be settled before the plant can be built or operated.

Advisory Committee on Reactor Safeguards

The next step in the licensing process for XYZ Power & Light is a review of the application by the Advisory Committee on Reactor Safeguards. As mentioned earlier, the ACRS is an independent statutory committee established to provide advice to the NRC on reactor safety. This committee reviews every application for a construction permit or an operating license for a commercial nuclear power plant. It has a maximum of 15 members appointed by NRC on a consultant basis for four-year terms. The members are selected from various technical disciplines and have applicable experience in industry, research, and universities. The ACRS also calls in additional consultants in particular specialized areas when needed.

Copies of the PSAR are given to the ACRS when the application for a construction permit is docketed. The application then goes to a project subcommittee made up of about five ACRS members. However, if the plant proposes a "standard design" and if the site appears to be generally acceptable, the ACRS subcommittee does not begin its review until the staff has nearly completed its review. On the other hand, if the plant involves new or modified concepts or special site considerations, the ACRS subcommittee begins its review earlier in the licensing process. In either case, as the staff is conducting its review, it keeps the ACRS informed of requests for additional information from the applicant and of meetings held, so that the subcommittee's information will remain current and so that the committee will be continually aware of any new developments that may indicate the need for a change in the plant. As appropriate, the subcommittee sets up meetings with the applicant and the staff to discuss pertinent safety issues concerning the application. It then prepares a report for the full committee.

The ACRS committee as a whole does not normally review the application until it receives the staff's Safety Evaluation Report. This report, together with that of the subcommittee, then forms the basis for the committee's full review of the proposed plant. The review places special emphasis on items that are of particular safety significance for the reactor involved and on any new or advanced features proposed. The committee meets with the NRC staff and the applicant at least once during the review period. Such meetings are open to the public.

When the ACRS has completed its review, it reports to the NRC via a letter to the Chairman of the Commission. This report is also made public.

The staff then prepares a supplemental Safety Evaluation Report that addresses the safety issues raised by the ACRS report. The supplement also includes any new information that has become available since the original SER was issued.

This brings us up to the public hearing on the safety review side, so let's drop back now and consider the environmental review and the steps in that procedure leading to the public hearing.

Environmental Report

As noted earlier, the environmental part of the information submitted in an application for a construction permit is often tendered before the safety information. The main reason for this is that you officers at XYZ Power & Light probably want to begin construction activities just as soon as you can, and it is sometimes possible to obtain a limited work authorization to begin certain early site activities before an actual construction permit is granted--provided the environmental review is far enough along.

When the Environmental Report is tendered, the staff first conducts a preliminary acceptance review, just as it does for the PSAR, to make sure that the report contains sufficient information for the staff to conduct its detailed review. Then the formal environmental review process begins.

Regulatory Guide 4.2, "Preparation of Environmental Reports for Nuclear

Power Stations," (6) describes in detail the information that should be included in the Environmental Report and the format in which this information should be presented. Although conformance with the guide is not required, conformance with the suggested format and content will expedite staff review.

The environmental review is performed by the staff and its consultants to evaluate the potential environmental impact of the proposed plant. The staff also conducts a value/impact analysis to provide a comparison between the benefits to be derived from the plant and its potential effects on the environment.

Once the environmental review of XYZ Power & Light has been completed, the staff prepares a Draft Environmental Statement containing its initial conclusions on the proposed site. The draft statement is circulated among appropriate federal, state, and local agencies, and is made public so that individuals and organizations representing the public can also review it. When comments on the draft statement have been received and outstanding issues have been resolved, the staff issues a Final Environmental Statement, which is also made public.

At this point, the environmental issues go before a public hearing and converge with the safety review.

Public Hearing

Let us now examine the public hearing. The law requires that a public hearing be held before a construction permit is issued for a nuclear power plant. Soon after an application is docketed the NRC issues a notice of the hearings which will be held after completion of the safety and environmental reviews. In addition the hearing is advertised in local newspapers near where the plant is to be built. This hearing allows interested members of the public to participate in the licensing process. An intervenor can participate in one of three ways: he may submit written statements to be entered into the hearing record; he may appear at the hearing to give a direct statement as a limited participant; or he may petition for leave to intervene as a full participant in the hearing and would then gain the right of cross-examining all direct testimony in the proceeding.

Intervenors in past hearings have included individuals and groups of local citizens who were concerned about the effects of a large plant in their localities, national and local environmental organizations, and local and state governments. Their participation has injected a healthy element into the hearings.

Issues raised by intervenors have been taken into account in the development of NRC's regulations and have thereby directly influenced the levels of safety at almost all nuclear plants. I am thinking in particular of the ECCS Evaluation Criteria (7) in Appendix K to 10 CFR Part 50 and the new Appendix I (5) to Part 50 dealing with effluents from nuclear power plants.

In many cases, intervenors have raised substantive issues that have had a major effect on plant configuration or operation. One utility, for example, was forced to eliminate two of its four proposed reactors on a site because of a successful environmental intervention that required a reduction in the size of the cooling lake. At another installation, a protest by the Izaak Walton League against open cycle cooling discharge to the river and a suit to hold up plant operation resulted in a conversion to closed cycle cooling. As a footnote, the last chapter in this particular story may not have been written, since the spray canals installed to achieve the closed cycle cooling may cause the unit to lose up to 50% of its power rating during the peak loads of hot weather, with resulting high costs of replacement power to be borne by the consumer.

The clash of views at contested hearings, is, as I said, a healthy process. It may bring out facts that the hearing board may not be aware of and often places known facts in a new perspective. It also, I hope, encourages applicants and the NRC staff to avoid complacency and to do their jobs thoroughly.

The public hearing itself is conducted by a three-member Atomic Safety and Licensing Board Panel. The Board is composed of one lawyer, who acts as chairman, and two other technically qualified persons. The NRC staff presents the Safety Evaluation Report and its supplements and the Final Environmental Statement as evidence at the public hearing. The applicant and intervenors (if there are any) may add additional testimony. The hearing can consider both safety and environmental matters at one time or may be split into two separate hearings. The Board considers all the evidence presented and issues its decision. If the decision regarding environmental and safety matters is favorable, the Director of Nuclear Reactor Regulation issues a construction permit to the applicant. That is not necessarily the end of the matter, however. If any of the parties to the hearing takes exception to the Board's decision, they may request a review by the Atomic Safety and Licensing Appeal Board, or the Appeal Board may, on its own, decide to review the application. The Board decisions can be appealed to the courts. You may have read about one recent case in the newspaper. A federal court in the Midwest overturned the construction permit given to a utility for a plant near Gary, Indiana, on the grounds that the NRC had not followed its own regulations regarding siting requirements. This ruling was itself overturned when the Supreme Court agreed that the Commission was qualified to interpret its own rules.

As mentioned earlier, even before a decision on a construction permit is made, the Director of Nuclear Reactor Regulation may allow the XYZ Power & Light to carry out limited amounts of work. The authority under which this is done is known as a "Limited Work Authorization" or "LWA". Two types of LWA can be given. Under one type, site preparation work, installation of temporary support facilities, excavation, construction of service facilities, and certain other construction not subject to quality assurance requirements, can be allowed. Under a second LWA, NRC may authorize the installation of structural foundations.

An LWA can be granted only after the hearing board has found that the plant meets all the environmental requirements for issuance of a construction permit and has determined that there is reasonable assurance that the proposed site is suitable from a radiological health and safety standpoint. Before the second type of LWA can be granted, the Hearing Board must also find that there are no unresolved safety issues relating to the specific work to be authorized. So, if the NRC staff, for reasons of public health and safety, have recommended changes in the design or construction of any of the structural foundations to be covered by the LWA, XYZ Power & Light cannot get this authorization without first agreeing to the changes or convincing us that they will achieve an equivalent level of safety.

As stated earlier, there is a third part of the license application. This part is not taken up at the Atomic Safety and Licensing Board hearing, nor does it bear on plant safety. The law requires that antitrust aspects of the nuclear power plant license application must be considered in the licensing process. When XYZ Power & Light submits information on antitrust matters, it is sent to the Attorney General for his advice on whether activities under the proposed license would create or maintain a situation inconsistent with the antitrust laws. This generally relates to whether smaller power companies, rural cooperatives, and municipal utilities that may want to participate in the project in order to get a share of its power, will be able to do so under reasonable terms. The Attorney General's advice is published promptly when received and opportunity is provided for interested parties to raise antitrust issues. When necessary an antitrust hearing is held.

Operating License Review

XYZ Power & Light has reached the point where they have their construction permit and have gone off to build their plant. The next step in the licensing process comes sometime later, although, as will be discussed later, the NRC does not drop out of the picture during this period.

When construction of the facility has progressed to the point where most of the final design information and plans for operation are ready, the applicant submits his Final Safety Analysis Report in support of an application for an operating license. Where the PSAR in many places presented design criteria, the FSAR provides important details on the final design of the facility including final containment design, design of the nuclear core, and details of the waste handling system. The FSAR also supplies plans for operation and procedures for coping with emergencies. As they did for the PSAR, the staff makes a detailed review of the information. Amendments to the FSAR may be made as more information becomes available or in response to questions from the NRC staff. The staff again prepares a Safety Evaluation Report, and the ACRS again makes an independent evaluation and presents its advice to the Commission. This second Safety Evaluation Report and its supplement, the ACRS meetings and the ACRS letter to the Commissioners are all available or open to the public.

The law does not require that a public hearing be held prior to the issuance of an operating license. However, soon after acceptance of the operating license application the Commission publishes notice that it is considering issuance of the license. At that point any person or group whose interest may be affected may petition the NRC to hold a hearing. If the hearing is held, the procedures and decision process described for the construction permit hearing are followed.

The operating license, when finally granted, is usually for a period of forty years. Initially, it may have restrictions limiting the plant to loading fuel or performing low power tests. The license contains a set of technical specifications that impose particular safety and environmental protection conditions on the facility and set conditions of operation that must be met to assure protection of the health and safety of the public and protection of the environment.

The NRC staff has developed Standard Technical Specifications for each reactor vendor and these are now incorporated into the SAR's of most applicants. These Standard Tech Specs serve a function similar to the Standard Format and the Standard Review Plan, that is they help assure quality, uniformity, and completeness in the technical specifications.

Even after the license is granted, the NRC continues to take steps to assure that the plant is operated in accordance with the terms of its license and to assure that when the useful lifetime of the plant comes to an end, it is safely shut down and decommissioned.

INSPECTION AND ENFORCEMENT

The discussion of Fig. 2 pointed out several components of NRC that play an important role in assuring reactor safety through the licensing process. In the review of the licensing process you have seen the central role played by the Office of Nuclear Reactor Regulation and the important contributions of the ACRS and the Atomic Safety and Licensing Board and Appeal Panels. The role of the other major groups--the Office of Inspection and Enforcement, the Office of Nuclear Regulatory Research, and the Office of Standards Development--has not been so obvious but is also significant.

The Office of Inspection and Enforcement is the eyes and ears of the NRC. I&E is responsible for assuring that the applicant builds his plant in accordance with the terms of his construction permit and in compliance with all NRC regulations. Some of the areas inspected during the construction phase are as follows:

(a) Quality Control and Testing at Equipment Fabricators' Shops
(b) Component Handling and Inspection Onsite
(c) Compliance with Construction Permit
(d) Compliance with NRC Regulations
(e) Implementation of QA Procedures for Construction Activities

Where violations are found, I&E can order work to be stopped until the violation is corrected and can levy fines against offending permit holders and licensees. This is no empty threat. A construction delay of one day can cost an applicant as much as $100,000 or even more in lost revenue. Utilities have paid fines in the tens of thousands of dollars for such violations as failure to maintain plant security at the required level or violations of environmental technical specifications.

Inspection and Enforcement does not end its participation when construction is complete. It continues making periodic scheduled and unscheduled inspections during startup and operation, particularly covering the following areas:

- (a) Operating Organizational Structure
- (b) Training of Personnel
- (c) Results of Preoperational and Startup Tests
- (d) Performance of Equipment and Personnel during Operation
- (e) Radiation and Effluent Control - Results of Monitoring and Sampling Programs
- (f) Results of Environmental Monitoring
- (g) Plans and Training for Emergencies
- (h) Security Provisions
- (i) Administrative Controls for Safety)

I&E coverage will continue through final decommissioning of the facility.

Inspection and Enforcement is also working to standardize its procedures by defining its inspection procedures and requirements. A computerized management program has been developed for monitoring the status of plants and the schedules for inspection. Written procedures help ensure that all licensees and applicants receive uniform coverage no matter which inspector visits their plant.

Another very significant contribution of I&E is information feedback. Records are kept of defects and violations found at plants, and regular summaries of these data are published and circulated within the NRC. If, for example, frequent reports of valve failures start coming in, not only might I&E start increasing inspections of valves, but NRR might look more closely at the design and analysis of valves selected for plants under review, and Standards Development might step up its work with the national standards program aimed at developing standardized good practices for the design and fabrication of valves.

RESEARCH

While the Office of Nuclear Regulatory Research doesn't participate directly in the licensing process, the results of their confirmatory research may affect the margins that might be applied to certain systems. Those margins as determined may be greater or less than those claimed by the applicant and may result in increased or decreased requirements. Other research programs are carried out to enable the NRC to verify the claims of applicants.

STANDARDS

A final topic to be discussed in the licensing process is the role played by standards in general and the Office of Standards Development in particular in achieving safety through the reactor licensing process. Let me begin by repeating a definition of standards that I have used on other occasions: "Standards are codifications of sound systematic disciplined engineering and the lessons of experience." When invoked by the designer and used by those who build and operate the plant, they can substantially increase the confidence of a plant owner in the safety and reliability of his facility. Availability of a systematic body of well-thought-out standards can also, as a practical matter, significantly decrease design and fabrication costs, and facilitate orderly scheduling of fabrication and construction. This same body of codified good practice gives the regulatory, NRC, a high level of confidence that the performance requirements and design bases on which he has based his safety review will be implemented.

The objectives of NRC's Office of Standards Development are summarized as follows:

(a) Identify and Resolve Generic Issues of Safety, Safeguards and Siting Associated with Regulation of Nuclear Facilities and Materials
(b) Develop a Comprehensive Body of Standards Codifying the Developed Solutions

First we attempt to identify and resolve the generic safety, safeguards and siting issues associated with regulation of nuclear facilities and taking these solutions, codify them into a comprehensive body of standards. Development of a comprehensive body of standards is a key part of the Commission's strategy for getting safe nuclear power plants on line as quickly as possible and for minimizing delays caused by the licensing process.

Several reasons for having an NRC standards program are:

(a) Foundation of Safety, Safeguards, and Environmental Protection (and Reliability)
(b) Define Performance Requirements which Establish Acceptable Levels of Risk
(c) Codify Good Practice
(d) Stabilize Licensing Process
(e) Limit Contested Issues
(f) Basis for Enforcement
(g) Foundation for Standardization and Site Designation
(h) Public Acceptability

Most important are our programs to set broad policy to protect the public health and safety and to establish performance requirements for the nuclear industry. To the extent that we can, we rely on the national nuclear standards program coordinated by the American National Standards Institute (ANSI) to define what must be done to achieve the required level of performance.

It may be apparent by now that achieving all the broad objectives just outlined is a monumental task. We, therefore, have set priority goals in each of our areas of activity. These goals, shown in Table I, are based primarily on NRC needs, but also consider input from ANSI, the nuclear industry and the public.

The Office of Standards Development is now spending about half its time developing the basic safety standards that define acceptable levels of risk. The rest of its effort is spent in defining acceptable methods of implementing these basic standards--much of this effort involves working with the national standards program.

Table I. How Objectives and Priorities are Established

NRC NEEDS
- Policy Direction by NRR, NMSS, and NRC
- Current Licensing Issues
- Inspection and Enforcement Needs
- ACRS and Boards Issues
- Need to Stabilize and Accelerate Regulatory Process

OTHER INFORMATION
- Operating Experience
- Research Results
- Input from ANSI and Industry
- Public Input

Generally speaking, nuclear safety is promoted by three types of standards which are summarized in Table II. NRC regulations define general or specific requirements that must be met. In the area of nuclear reactors many pertinent regulations are found in Part 50 of Title 10 of the Code of Federal Regulations (1). For example, Appendix A to Part 50 defines the General Design Criteria for Nuclear Power Plants; Appendix B sets the Quality Assurance Criteria; Appendix G sets Fracture Toughness Requirements for the reactor vessel and certain other key components; Appendix J establishes requirements for containment leakage testing and so on. A key regulation not in Part 50 is Appendix A to Part 100 (8), which provides the seismic and geologic siting criteria for nuclear power plants.

The NRC issues regulatory guides that provide guidance on meeting the requirements of a regulation and may describe an acceptable solution. An applicant for a construction permit or operating license may simplify his SAR by indicating his decision to follow the solutions described in certain regulatory guides. If he does, generally no further information on the subject is required. Since guides do not have the force of law as do regulations, the applicant can propose alternate solutions; but in such cases he must demonstrate that his approach will result in an acceptable level of safety.

Regulatory guides are issued in the ten divisions tabulated in Table III. Of greatest concern in nuclear power plant licensing are Division 1, Power Reactors, and Division 4, Environmental and Siting.

Certain regulatory guides indicate to applicants the information that should be submitted in support of licensing decisions. Regulatory Guide 1.70, the Standard Format for SAR's mentioned earlier falls into this category. This guide was recently revised so that it specifically identifies the information needed by the staff to complete its review in accordance with the Standard Review Plan.

Other guides provide information on techniques used by the staff in evaluating safety problems. Examples of this type of information are the assumptions used in evaluating the consequences of potential accidents and the assumptions used in calculating potential radiation doses.

Finally there are the important Regulatory Guides that identify good engineering practice. These are the guides in which ideally we use the product of the national standards program to define how the required level of performance is to be achieved. Whenever possible the guides reference ANSI standards as shown at the bottom of Table II.

The national standards program makes a major contribution to reactor safety in licensing. As stated earlier, the availability of a systematic body of well-thought-out standards and a codified body of good practice gives the NRC a high level of confidence that the performance will be implemented. Perhaps the most widely accepted national standard is Section III of the ASME Boiler and Pressure Vessel Code, which deals with the design of nuclear grade pressure containing components. This is a mature standard that has been the basis for industry designs and NRC reviews for several years. As experience with its use increases, minor changes are made and issued as semiannual

Table II. Types of Standards

- **NRC REGULATION**
  - General or Specific Requirement Must be Met
- **NRC GUIDE**
  - Guidance on Meeting Requirements of Regulation
  - Describes Acceptable Solution
  - Not Substitute for Regulation
  - Other Solutions May be Acceptable
- **ANSI STANDARD (ASME, IEEE, ASTM, ANS, INMM, etc.)**
  - Codifies Good Practice in Design, Construction and Operation

Table III. Ten Divisions of Regulatory Guides

1 – Power Reactor Guides
2 – Research and Test Reactor Guides
3 – Fuels and Materials Facilities Guides
4 – Environmental and Siting Guides
5 – Materials and Plant Protection Guides
6 – Product Guides
7 – Transportation Guides
8 – Occupational Health Guides
9 – Antitrust Review Guides
10 – General Guides

addenda. Section XI of the Code treats inservice inspection, and the new Division 2 of Section III covers concrete containments. These sections are relatively new and can be expected to undergo more significant changes as experience in their use accumulates.

Another area in which the national standards program has made a major contribution is Quality Assurance. ANSI Standard N45.2, "Quality Assurance Requirements for Nuclear Power Plants," was issued in 1971 in order to amplify the requirements of Appendix B to CFR Part 50. Since that time, about two dozen daughter standards have been developed or are in preparation to amplify N45.2 and to provide guidance in specific areas such as design, procurement, cleaning of fluid systems, calibration and control of measurement equipment, and auditing.

The national standards program consists of hundreds of individual tasks in a wide range of areas being carried out by thousands of individual contributors. All this activity is coordinated by ANSI (American National Standards Institute) through its Nuclear Standards Management Board. This group periodically issues a report that indicates the status of each nuclear standard that has been issued or is under development. To give you some idea of the breadth of this program, a 1975 edition of the status report required 114 pages of small type to cover all applicable standards.

Just as ANSI is developing standards on the national level, the International Atomic Energy Agency is working to develop nuclear safety standards that would be applied in nations around the world. A Senior Advisory Group and a number of Technical Review Committees have been formed and work has begun on several safety guides. Initially activity is concentrating on siting, design, operation, quality assurance, and government organization. NRC participates in this program at all levels down to and including the working groups.

One of the most important standards in the NRC regulatory base is the Standard Review Plan, which, as mentioned earlier, details the methodology by which the staff reviews safety analysis reports. It and the Standard Format are the benchmarks on which the safety review is conducted.

The mechanism for change in NRC standards is a formal process that includes preparation of a value/impact statement and an upper-level management review. Before making any substantive change in the SRP, issuing a new regulatory guide, or revising an existing guide, the staff prepares a value/impact statement outlining the perceived benefits in terms of improved safety, efficiency of the review process, resolution of open issues, and ACRS generic concerns. On the negative side, the statement sets out the anticipated impacts on plants at various stages in the review process or in more advanced states of completion. It answers such questions as: Do the changes that increase the safety of one system result in reduced safety of other systems? Do changes that improve the plant's response to one type of accident or event adversely affect its response in the case of other accidents or events? What are the additional costs? How will it impact schedules?

The proposed change and associated value/impact statement are evaluated by the Regulatory Requirements Review Committee (RRRC), a group of senior staff members at the Office level, with particular attention given to any backfitting requirements that may be imposed if change is implemented. The RRRC establishes the date on which the staff will begin using the new or revised standard in the review process. This date may be immediately--if the standard is very important to safety or if RRRC finds that the new issuance reflects current licensing practice and is merely being published to permit wide distribution of known methods of licensing evaluation. Or it may range up to several months if the proposed guide or change to the SRP is an actual revision to the regulatory process. In either case, the staff always encourages public comments, asking that the comments on guides be submitted within about 60 days, if possible, so that they can be considered before the next revision to the guide is issued.

STANDARDIZATION

A further refinement of the standards concept is standardization of a complete plant, or large portions thereof, and early site approval. If successfully implemented, these techniques should reduce the time consumed between initial site selection and actual operation, reduce total plant costs, and streamline and expedite the licensing process. In addition, standardization increases reliability because it is based on established and proven designs in which an appreciable degree of confidence exists.

Several types of standardization are envisioned. Option I involves prior approval of standard plant designs, as authorized by Appendix O to Code of Federal Regulations, Part 50. Table IV lists standard plant designs submitted by the reactor vendors and several architect-engineering firms. The standardized nuclear island design submitted by GE as GESSAR-238 falls into this category. Preliminary design approval has been given to this system, which covers all portions of the plant except the turbine island. C. F. Braun has submitted a turbine island design that will combine with GESSAR-238 to give a complete plant.

Several other applications cover standard designs of reduced scope. Included in this group are Combustion Engineering's CESSAR and Westinghouse's RESAR-41, both of which offer a standard design for the nuclear steam supply system (NSSS) and both of which have received preliminary design approval.

SWESSAR 2, from Stone & Webster, is the first balance of plant design matching RESAR-41. A number of other balance of plant designs are also included in Stone & Webster's application, including a balance of plant to match Babcock and Wilcox's BSAR-205. In addition, Fluor Pioneer has recently tendered an application for a BOP matching RESAR-41. The NRC has received and is reviewing several applications based on these reference systems. Table V lists several utility applications.

Option II provides for the awarding of a license to construct and operate multiple reactors of duplicate design at different sites. Applicable authority is given under Appendix N to Part 50. A listing of several utilities using Option II is given in Table VI The SNUPPS group's application for five such units at four sites has passed ACRS and is now in the final stages of review. Duke Power has ordered six units from Combustion Engineering for placement at various sites. Environmental hearings have been completed for this project. Another example of duplicate plant standardization

Table IV. Standardization Applications -- Option I Reference Systems

PROJECT	APPLICANT	DOCKET DATE	ACTUAL OR PROJECTED DATE OF PDA
* GESSAR-238 (NI)	General Electric	7-30-73	Issued 12-22-75
* CESSAR (NSSS)	Combustion Eng.	12-19-73	Issued 12-31-75
* RESAR-41 (NSSS)	Westinghouse	3-11-74	Issued 12-31-75
** SWESSAR (BOPs)	Stone & Webster		
Matched to RESAR-41		6-28-74	February 1976
Matched to CESSAR		10-21-74	April 1976
Matched to RESAR-3S		10- 2-75	August 1976
Matched to B-SAR-205		12-19-75	Not Scheduled as Yet
* C. F. Braun SSAR (TI) Matched to GESSAR-238 (NI)	C. F. Braun	12-21-74	February 1976
GESSAR-251 (NSSS)	General Electric	2-14-75	July 1976
RESAR-3S (NSSS)	Westinghouse	8- 1-75	July 1976
GESSAR-238 (NSSS)	General Electric	9-24-75	May 1976
B-SAR-205 (NSSS)	Babcock & Wilcox	10-24-75 (Tendered)	Not Scheduled as Yet
F-P SSAR (BOP) Matched to RESAR-41	Fluor Pioneer	11-17-75 (Tendered)	Not Scheduled as Yet

*APPROVED   **APPROVAL EXPECTED IN NEAR FUTURE

79

Table V. Standardization Applications -- Utility Applications Using Option I Reference Systems

PROJECT	APPLICANT	DOCKET DATE	ACTUAL OR PROJECTED DATE OF SER SUPPLE.
Cherokee 1-3 (CESSAR)	Duke Power	5-24-74	April 1976
Perkins 1-3 (CESSAR)	Duke Power	5-24-74	April 1976
South Texas 1&2 (RESAR-41)	Houston Light & Power	7- 5-74	10-29-75
WNP-3&5 (CESSAR)	Washington Public Power Supply System	8- 2-74	May 1976
Palo Verde 1-3 (CESSAR)	Arizona Public Service	10- 7-74	12-26-75
Hartsville 1-4 (GESSAR-238-NI)	Tennessee Valley Authority	11-11-74	April 1976
Black Fox 1&2 (GESSAR-238-NSSS)	Public Service of Oklahoma	12-23-75	Not Scheduled as Yet
Phipps Bend 1&2 (GESSAR-238-NI)	Tennessee Valley Authority	11- 7-75	January 1977

Table VI. Standardization Applications -- Duplicate Plants

PROJECT	APPLICANT	DOCKET DATE	ACTUAL OR PROJECTED DATE OF SER SUPPLE.
Byron 1&2/Braidwood 1&2	Commonwealth Edison	9-20-73	8- 1-75
SNUPPS			
Wolf Creek 1	Kansas Gas & Electric Kansas City Power & Light	5-17-74	1-14-76
Callaway 1&2	Union Electric	6-21-74	11-21-75
Tyrone 1	Northern States Power	6-21-74	May 1976
Sterling 1	Rochester Gas & Electric	6-21-74	February 1976
WUPS			
Koshkonong 1&2 (As Many as Six Units Planned on Three Sites)	Wisconsin Electric Power Madison Gas & Electric Wisconsin Power & Light Wisconsin Public Service	8- 9-74	March 1976

is Commonwealth Edison's Byron/Braidwood, with two units each at two sites--all using the same design. Construction permits have been issued for both sites.

Option III is the license to manufacture--authorized by Appendix M (Table VII). Under this umbrella are the plans of Offshore Power Systems to build barge-mounted plants for installation at coastal sites. Their first order is for two units to be located off the New Jersey coast.

Another standardization concept that is in use and covered under a policy and procedures statement is the replicate plant, in which an applicant proposes to construct a plant exactly like one already licensed. Two applications shown in Table VIII were received using this option: Jamesport, by Long Island Lighting, and Marble Hill, by the Public Service Company of Indiana.

A final category of standardization is early site reviews. Under this option, the Commission would review sites independently of the plant and hopefully a bank of pre-approved sites could be established for use upon demand. This approach could shorten the licensing process from the time a utility identifies its need for power to the issuance of an operating license. It also provides for reviews, including public participation, before there is a major commitment of money or natural resources.

The NRC expects soon to issue revised regulations and a policies and procedures statement covering this option and a number of standards providing major details of the review are now being developed. Applications for early site approval can be accepted from all comers and the review carried up to the hearing stage. At that point, however, only bona fide construction permit applicants can be considered under the present regulations. This would exclude the State of Maryland, for example, which has expressed an interest in obtaining early site approvals in its jurisdiction. A bill now pending in Congress would remove this roadblock.

All of these standardization concepts offer an opportunity to avoid snarls and many of the pitfalls of a full-scale review. These concepts are not completely without problems, however. One major difficulty is the interface between the standardized portion and the balance of the plant. NRC and the national standards program are working on solutions to this issue.

OPERATOR LICENSING

Until now I have discussed how the NRC's licensing program for facilities contributes to nuclear safety. Before closing I would like to briefly review another NRC licensing program that has a direct effect on safety--the licensing of reactor operators. The Commission's regulations in this regard are contained in 10 CFR Part 55 (9), which establishes the criteria for issuance of reactor operator licenses and outlines the procedures followed, and in Appendix A to Part 55, which describes the requalification program requirements that an operator must meet to retain his license.

Two classes of operators licenses are issued. An "operator" is an

Table VII. Standardization Applications -- Option III License to Manufacture

PROJECT	APPLICANT	DOCKET DATE	ACTUAL OR PROJECTED DATE OF SER SUPPLE.
Floating Nuclear Plant (FNP) 1-8	Offshore Power Systems	7- 5-73	March 1976
UTILITY APPLICATIONS USING LICENSE TO MANUFACTURE			
Atlantic 1&2	Public Service Electric & Gas	3- 1-74	January 1977

Table VIII. Standardization Applications -- Replication

PROJECT	(BASE PLANT)	APPLICANT	DOCKET DATE	ACTUAL OR PROJECTED DATE OF SER SUPPLE.
Jamesport 1&2	(Millstone 3)	Long Island Lighting	9-6-74	January 1976
Marble Hill 1&2	(Byron/Braidwood)	Public Service of Indiana	9-17-75	November 1976
UEA 1&2	(Vogtle)	Alabama Power	*	
NEP 1&2	(Seabrook)	New England Power	*	

*FUTURE APPLICATION

individual who actually manipulates the controls of a facility (or directs others to do so), and a "senior operator" is a person authorized by the reactor licensee to supervise the activities.

In order to obtain a license, an operator or senior operator must complete an approved training course that includes classroom sessions as well as practical experience, must provide evidence that his physical and mental health are satisfactory, and must pass a written examination and operating test administered by the NRC. The written exam for operators and senior operators covers a range of subjects from nuclear theory through the design and specific safety features and operating procedures of the reactor at which the licensed operator will be working. The senior operator's exam has additional questions dealing with conditions and limitations in the facility license and technical specifications, as well as such things as fuel handling facilities and procedures, and procedures and equipment for treatment and disposal of radioactive wastes. The detailed operating test for both the operator and senior operator covers a full range of situations from pre-operational through startup and full power operation. The operator must show that he understands the reactor and its response to changes in control settings. He must demonstrate an understanding of the meaning of various instrument readings and annunciator signals and a knowledge of the proper responses. He must, in addition, show thorough knowledge of the emergency plans for the facility.

Most operator licensing is carried out by NRC personnel at the reactor facility over a five or six month period beginning about three months before actual operation is scheduled to begin. This permits the operating test to be given in the actual control room under the required broad range of operating conditions.

In order to assure that licensed operators continue to be qualified to safely operate the nuclear facility, licenses are issued for only two years. To obtain a renewal, the operator must show that he has had a reasonable amount of actual operating experience during the term of his license, that he is still in good health, and that he has completed an approved requalification training program. The requalification program, which is described in Appendix A to 10 CFR Part 55, must be conducted in a continuous period up to two years and must be promptly followed by successive requalification programs. The program itself consists of lectures in those areas where, based on annual written exams, need is indicated as well as of on-the-job training, evaluation including written exams and response to simulated emergency conditions, and systematic observation and evaluation by supervisors.

SUMMARY

In summary, the NRC licensing process goes through an orderly chain of events commencing with the receipt of an application for a construction permit and continuing through the staff review of the preliminary Safety Analysis Report and Environmental Report, the ACRS safety review, the public hearing, the awarding of a construction permit, the on-site inspection during construction and pre-operational and startup testing, the staff review of the Final Safety Analysis Report, the second ACRS evaluation, and finally the awarding of an operating license.

Throughout this process, the staff's evaluation of the application is conducted from a written and publicly available regulatory base consisting of the Standard Format, the Standard Review Plan, and NRC guides and regulations. For you out there at XYZ Power & Light, or your own real companies with hard concerns involving stockholders, customers, and your responsibilities to the public, it's a long hard pull that requires some excellent minds and conscientious hearts.

In closing, I'd like to emphasize that each small part of both your job--to design, build, or operate reactors--and ours--to ensure the safe use of nuclear energy--is important. If each of us does his job in an atmosphere of professional integrity, openness, and open-mindedness, we can provide the public the assurance it demands that nuclear power will maintain the outstanding public safety record that it has established since its start with the S1W submarine prototype in 1953.

REFERENCES

1. "Licensing of Production and Utilization Facilities," Title 10, Code of Federal Regulations, Part 50.

2. "Reactor Site Criteria," Title 10, Code of Federal Regulations, Part 100.

3. "Standard Format and Content of Safety Analysis Reports for Nuclear Power Plants," USNRC Regulatory Guide 1.70, Revision 2 (September 1975).

4. "Standard Review Plan for the Review of Safety Analysis Reports for Nuclear Power Plants," Office of Nuclear Reactor Regulations, Nuclear Regulatory Commission.

5. "Numerical Guides for Design Objectives and Limiting Condition Conditions for Operation to Meet the Criterion 'As Low As Practicable' for Radioactive Material in Light-Water-Cooled Nuclear Power Reactor Effluents," Appendix I, Title 10, Code of Federal Regulations, Part 50, Licensing and Production Utilization Facilities.

6. "Preparation of Environmental Reports for Nuclear Power Stations," USNRC Regulatory Guide 4.2, Revision 1 (January 1975)

7. "ECCS Evaluation Models," Appendix K, Title 10, Code of Federal Regulations, Part 50, Licensing and Production Utilization Facilities

8. "Seismic and Geologic Siting Criteria for Nuclear Power Plants," Appendix A, Title 10, Code of Federal Regulations, Part 100, Reactor Site Criteria.

9. "Operators' Licenses," Title 10, Code of Federal Regulations, Part 55.

INSPECTION AND QUALITY ASSURANCE

Hugh C. Dance
Office of Inspection and Enforcement
Nuclear Regulatory Commission
230 Peachtree Street, Suite 818
Atlanta, Georgia   30303

INTRODUCTION

The subject of Inspection and Quality Assurance has been selected to present the inspector's perspective in this Nuclear Power Safety Course. A combination of these two topics is indeed appropriate. The Nuclear Regulatory Commission (NRC) inspection program is interwoven with the quality assurance concept and therefore this title is as descriptive as any which might be chosen. Thus, this brief presentation is intended to provide the reader with an insight into the NRC's reactor inspection program and its relationship to the Quality Assurance Program of an applicant or licensee. Presentations are made elsewhere in this book regarding the organization and function of the Nuclear Regulatory Commission. The author shall therefore minimize discussion of interfaces within NRC and limit discussion to that phase with which he is associated, namely, the Office of Inspection and Enforcement. Of the five program offices within NRC, this is the office that performs field inspections.

OFFICE OF INSPECTION AND ENFORCEMENT

The Office of Inspection and Enforcement is responsible for the development and administration of programs and policies for (a) inspections and investigations necessary to determine whether licensees are complying with license provisions and rules, and to ascertain whether licensed operations are being conducted safely; (b) establishment of bases for the issuance or denial of a construction permit or license; (c) investigation of accidents, incidents, and theft or diversion of special nuclear materials; (d) enforcement actions; and (e) evaluation of licensed operations as a basis for recommending changes to standards and license conditions and for issuance of reports to the nuclear industry and the public, for recommending research for the discharge of these functions by the Commission; for developing policy options for Commission consideration; and for providing management and direction of NRC Regional Offices.

Protection of the public health and safety and safeguarding the environment are paramount in the NRC's activities. The prime safety consideration in the operation of a nuclear facility is control of its radioactive material, both during normal operation and under accident conditions. To perform the task, assigned to the Office of Inspection and Enforcement, there is a Headquarters staff located in Bethesda, Maryland, and five regional offices located in King of Prussia, PA; Atlanta, GA; Glen Ellyn, IL; Arlington, TX; and Walnut Creek, CA. Each of these offices has a technical staff with expertise in the fields of construction, reactor and fuel facility operation, environmental and radiation protection, and nuclear safeguards. As of January 1976, the Office of Inspection and Enforcement was comprised of 478 persons of whom 376 were located in the regional offices. The basic organizational chart is shown on Figure 1.

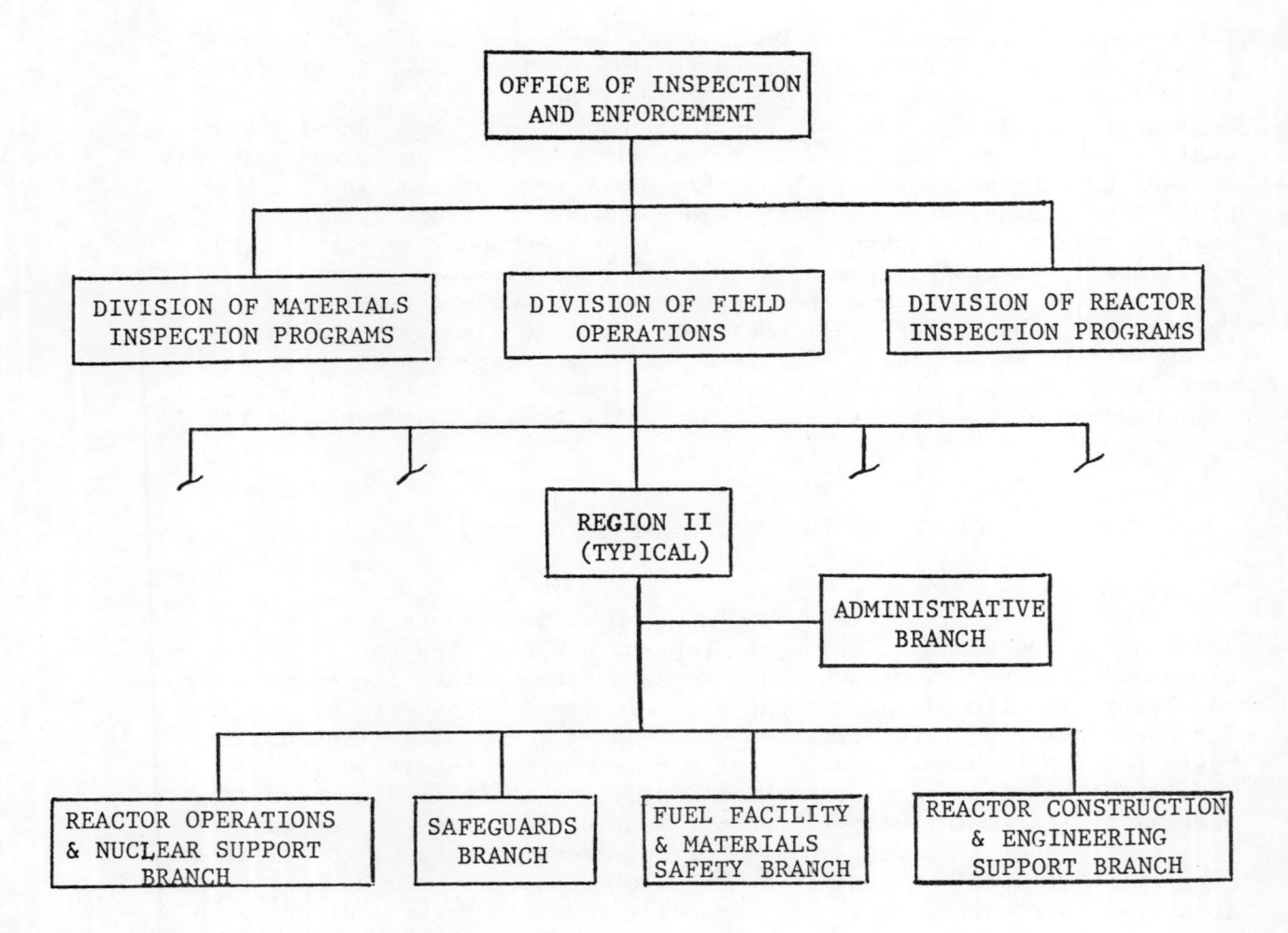

Fig. 1. Office of Inspection and Enforcement Organization

The Region II office located in Atlanta has inspection responsibility for ten states in the Southeast, Puerto Rico, Canal Zone, and the Virgin Islands. This encompasses the inspection of 23 nuclear power plants under review for a construction permit, 35 nuclear power plants under construction, 12 plants in operation, 16 research and test reactors, 10 fuel facilities and numerous other materials licenses authorizing use of radioactive materials (primarily medical, academic, and industrial applications). The Region II staff totals over 90 at the present time.

QUALITY ASSURANCE

Prior to describing the inspection program let us take a look at the subject of quality assurance. Since Appendix B to 10 CFR 50 (1) became a regulation in July 1970, the subject of quality assurance has received almost constant attention both by the NRC and the nuclear industry. Appendix B defines quality assurance as comprising all those planned and systematic actions necessary to provide adequate confidence that a structure, system, or component will perform satisfactorily in service. The 18 quality assurance criteria shown in Table I have been dissected and restated, cussed and discussed.

TABLE I. QUALITY ASSURANCE CRITERIA
10 CFR 50 Appendix B

I.	Organization	IX.	Control of Special Processes
II.	Quality Assurance Program	X.	Control of Special Processes
III.	Design Control	XI.	Test Control
IV.	Procurement Document Control	XII.	Control of Measuring and Test Equipment
V.	Instructions, Procedures, and Drawings	XIII.	Handling, Storage and Shipping
		XIV.	Inspection, Test, and Operating Status
VI.	Document Control		
VII.	Control of Purchased Material, Equipment, and Services	XV.	Nonconforming Materials, Parts, or Components
VIII.	Identification and Control of Materials, Parts and Components	XVI.	Corrective Action
		XVII.	Quality Assurance Records
		XVIII.	Audits

It is not the author's intention to discuss each criterion. However, these are the criteria or elements which need be incorporated into the management system to meet the regulation. No one disputes the need for quality assurance in the nuclear industry. From the very beginning of the nuclear power industry, there has been no argument that nuclear power plants must be designed, constructed, and operated to higher quality standards than those employed at conventional power plants. Quality assurance is the proven vehicle to achieve these standards. Difficulty arises in translating this philosophical harmony into an implemented program that satisfies the standards and the requirements of both the owner operator and the regulator. Although field offices still have disagreement with licensees over the degree of implementation required from time to time, the author has noted in the past two years while examining practicing QA programs that more key elements are being incorporated prior to the first review. This is seen as passing of the word that QA is truly a regulatory requirement and also, hopefully, a recognition that indeed it does have merit. Nevertheless, less missionary type effort is being required of an inspector in the development of a complete program.

Standards

It is appropriate to note that ANSI N18.7-1972, Rev. 1 (Draft No. 5) Administrative Controls and Quality Assurance for the Operational Phase of Nuclear Power Plants (2), is a single standard currently under revision that defines the quality assurance program requirements for operation. This revision was an outgrowth to combine in a single standard the contents of Regulatory Guide 1.33 (Formerly Safety Guide 33) Quality Assurance Program Requirements (operation) (3); ANSI N18.7 - 1972, Administrative Controls for Nuclear Power Plants (4); and ANSI N45.2 - 1971, Quality Assurance Program Requirements for Nuclear Power Plants (maintenance) (5). The latter was one of the original standards related to design, construction, maintenance, and modifications of nuclear power plant structures, systems, and components. ANSI N45.2 also has some twenty daughter standards specifying requirements in different areas of expertise. Most of these have been endorsed by Regulatory Guides as acceptable means for implementing NRC requirements. The NRC provided the "Gray Book," Guidance on Quality Assurance Requirements During Design and Procurement Phase of Nuclear Power Plants (6). The "Gray Book" in essence endorsed the ANSI N45.2 series of standards and Regulatory Guides, even though many were in draft form at the time. The "Green Book," Guidance on Quality Assurance Requirements During the Construction Phase of Nuclear Power Plants (7) and the "Orange Book," Guidance on Quality Assurance Requirements During the Operations Phase of Nuclear Power Plants (8) were similar books endorsing standards and guides related to construction and operations respectively. Thus standards addressing quality assurance are continually being refined as are other industry standards and codes. Development of industry standards has played a large role in defining acceptable means for implementing Appendix B requirements. From an inspectors viewpoint, it is beneficial to have an accepted standard that plants are committed to implement and can focus their attention.

Management and Quality Assurance

Let me diverge here for a moment to discuss the author's personal views of management philosophy as it relates to Appendix B. The basic goal of any well planned management group and of a quality assurance program is to properly preplan work so that unplanned events are prevented or can be promptly responded to. This is accomplished by providing proper resources of personnel, training, procedures, equipment and supplies for the completion of a mission. However, one must recognize that just as we don't live in a perfect society, neither is a plant perfect. A pump fails to start or the wrong valve is opened are but two examples of a multitude of unplanned events that can happen. Once such an event occurs, the Commission is primarily interested in understanding what happened and why, so that corrective action can be taken to prevent recurrence. This same interest in preventing unplanned events is cultivated by management and is required by Appendix B, Criterion XVI, Corrective Action. Philosophy of management, both corporate and plant, is seen in the operation of a plant. The supply and utilization of resources is a management decision and beyond the scope of this presentation. Yet to assure a proper return on their investment, management needs to be involved sufficiently to verify that programs authorized are doing the job intended. For instance, a plant manager is not expected to run his job successfully from his desk. Periodically, he has to see for himself that conditions are as planned or reported. Similarly preplanned approaches of expected action or plant manipulations are documented in plant procedures. This is also required by Criterion V, Instructions, Procedures, and Drawings. Unless proper precautions are taken, unplanned events are likely to occur. One could say that an effective QA program understands Murphy's Law, i.e., if things can go wrong they will. Thus, the concept of a well run management is seen to inherently incorporate quality assurance principles. The quality assurance organization and program are resources which may and should be utilized to provide the confidence in construction or operations demanded by management. It is a systematic means to achieve the standards required. One of the criticisms often heard of a QA program concerns the cost of such a program in terms of manpower and time. It is true that a regulator is not directly affected by this cost, yet the author firmly believe that a specific preplanned effort will survive the long term objectives of safe reliable units whereas adhoc decisions and performance cannot give this same guarantee. One must look to management to make the Quality Assurance Organization an effective and useful arm of management. It is not intended to be a feather bedding exercise. The author is convinced that the QA program required by NRC regulations, when properly utilized and backed by management can bring a proper return in terms of safety and reliability. This recognizes that the advertised cost of plant downtime is thousands of dollars per day.

The safety record at nuclear power plants has been outstanding. No accidents have been experienced at nuclear power plants which have resulted in overexposure to a member of the public. These remarks are not intended to distract from this record. However, we should be mindful that there have been a number of occurrences involving safety related equipment which had the potential for compromising the health and safety of plant personnel and the general public if it were not for the "defense in depth" design concept required by the NRC. The NRC believes the best insurance policy in avoiding such occurrences is to rely on quality assurance in the design, construction, and operation of a nuclear plant. Nuclear quality requirements are mandatory and enforceable under Federal Regulations. The NRC inspection program treats this subject accordingly.

INSPECTION PROGRAM

The inspection program for a typical reactor begins at the time an applicant files an application for a construction permit and continues throughout the authorized lifetime of the facility. The review of this application is a Licensing function. The inspection program is directed principally toward verifying that a licensee is meeting the responsibilities he accepts when he receives the license for which he has applied. Inspections are normally unannounced and are conducted periodically at frequencies which vary depending on the significance of the activity. One to four inspectors participate in an inspection lasting perhaps three to four days. Inspection plans utilize sampling techniques based principally on judgement to determine the scope of inspection in a specific area. Note that the program is a sampling. It does not relieve the licensee of the responsibility for assuring that his activities comply with the requirements of his license. This recognition of the licensee's responsibility for safety and a belief that there should be no encroachment on this responsibility are fundamental bases for the inspection program. The inspections performed during the lifetime of a plant may be divided into several phases depending upon the activities in progress. These phases in chronological order (See Figure 2) are preconstruction, construction, preoperational testing and operational preparedness, and startup testing and operational. Each of these will be discussed in more detail.

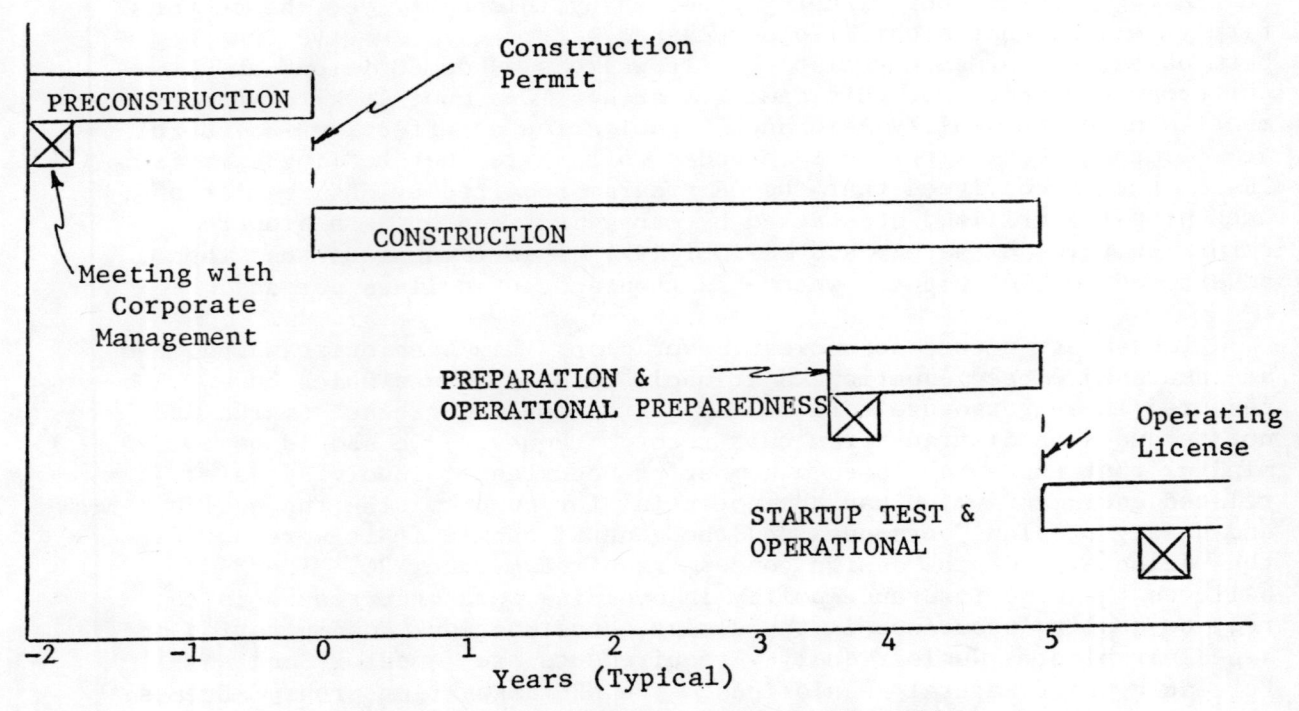

Fig. 2. Inspection Activity Timetable

Figure 2 also shows three meetings with corporate management. In order to communicate the inspection program to applicants/licensees, these meetings have been prescheduled at critical points. Corporate officer(s) responsible for the activities involved should be present. These and other meetings are scheduled as needed based on the inspection history and a utility's knowledge of the inspection program. Periodic corporate meetings are considered essential to assure an understanding of regulatory and licensee responsibilities. Typical topics covered (as appropriate) are: inspector-applicant/licensee relationship, establishment of channels of communication, methods of IE communication (exit interviews, inspection reports, bulletins, etc.), explanation of the inspection program, findings of previous inspections including enforcement history, requirements of future programs (Predocketing QA Inspection, Preoperational Testing Program, routine operations, etc.), inspection deadlines, licensee organization and responsibility, and plant schedules.

Pre-Construction Permit Phase

The pre-construction permit phase of the inspection program for a particular project begins approximately two years before the construction permit is issued. This phase of the program is primarily designed to verify that the applicant has developed an adequate quality assurance program for the design and procurement activities during the early stages of the nuclear project. Design and procurement activities encompass 10 CFR 50 Appendix B, Criteria I-VII and XVI-XVIII requirements. Prior to docketing, inspectors conduct corporate management meetings, perform an in-office review of the quality assurance program and procedures and forward substantive comments to Licensing. Unless major issues remain to be resolved, inspection at the corporate office is initiated. The inspectors hold discussions with utility employees and perform selected record reviews to ascertain the effectiveness of the applicant's quality assurance program related to design and procurement activities. A positive finding must be verified prior to docketing of the application by Licensing. As time progresses the inspection program verifies implementation of the QA program in preparation for construction activities. A principal objective of the program during this period is to verify that adequate controls are being imposed by applicants on their construction contractors and subcontractors to assure that work accomplished later can be shown to meet design requirements.

Construction Phase

The construction phase of the inspection program begins with issuance of the Construction Permit or a Limited Work Authorization and extends to issuance of the operating license. During this phase, inspections are planned to verify that the licensee and his contractors have developed and are implementing the approved quality assurance program and that it is effective in assuring that construction of the facility is consistent with the description contained in the application. These inspections include activities such as a selected review of quality control procedures prior to commencement of related construction work, observation of actual work in progress, discussions with personnel conducting licensee audits and inspection, and examination of work performance records. The particular construction activities which will be inspected within this phase of the program are shown in Table II.

TABLE II. CONSTRUCTION PHASE ACTIVITIES

Site Preparation	Reactor Coolant Pressure Boundary Piping
Environmental (For LWA-1)	Safety Related Piping
Foundations	Reactor Vessel Installation
Lakes, Dams, Canals	Safety Related Components & Supports
Containment and Safety Related Structures	Electrical & Instrumentation Components & Systems
Structural Concrete	Cables & Terminations
Prestressing (As Applicable)	Baseline (Inservice) Inspection
Steel Structures & Supports	
Structural Steel Welding	
Penetrations	

During this phase of the program, there is a shift in emphasis. No longer is an inspection largely a documentation review but now incorporates actual work inspection to verify that the work activities selected for inspection are being properly performed. For inspection of an activity such as the reactor coolant pressure boundary piping, the inspector will review QA implementing procedures prior to the start of the activity and periodically during the erection he will observe work in progress such as material handling, pipe fitup, welding, and non-destructive testing. The principal objective of this phase of the inspection program is to provide a basis for making a finding that the plant has or has not been built in accordance with the application. Such a finding must be made and found acceptable before an operating license can be issued.

Closely related to the construction inspection program is the program for inspection of the nuclear steam system suppliers, architect engineers, and certain independent suppliers or nuclear products and services. This program initiated in 1974 is identified as the Licensee Contractor Vendor Inspection Program (LCVIP) and is being carried out by a unit established in the Region IV office. The program looks at the implementation of the quality assurance program of a specific organization and is intended to minimize the impact of a multitude of individual organizations reviewing the same QA program. When findings are acceptable, a letter is issued to the vendor stating that acceptance of his QA program is contingent upon continued satisfactory performance. It is not designed to perform product acceptance type inspections. This responsibility remains with the licensee or his agent.

Preoperational Testing and Operational Preparedness Phase

Commencing about 18 months prior to scheduled issuance of the operating license, inspections are initiated to examine the licensee's preparations and procedures for conduct of the preoperational testing program and the quality assurance program related to preoperational testing activities. These inspections will proceed in parallel with the construction inspections just described. Inspector personnel changes occur here. Inspectors with operational oriented backgrounds are assigned. In addition, most utilities begin to shift responsibility from construction to their operations personnel for activities at this phase. The objectives of the inspection program for this phase of plant construction is designed to verify that the licensee has developed tests for all systems, subsystems and components related to safety and that the results of such tests demonstrate that the plant is ready for operation. The testing program described in the Safety Analysis Report forms the bases against which the plant is inspected. Regulatory Guide 1.68, Preoperational and Initial Startup Test Programs for Water-cooled Power Reactors (9), describes an acceptable method for complying with the Commission's regulations with regard to preoperational and initial startup testing programs. Appendix B of RG 1.68 discusses the Inspection and Enforcement inspection program and Appendix C covers the preparation of procedures. It is common for Regulatory Guides to be incorporated by reference in the Safety Analysis Report and thus become the bases against which the plant is inspected. This portion of the inspection program includes a review of the overall preoperational testing program, examination of selected procedures for technical content and adequacy, witnessing of selected tests, evaluation of certain test results, and ascertain that the licensee has evaluated all of the test results. Additionally, the applicant's QA organization is inspected to ascertain the development of internal controls.

The inspection effort regarding verification of operational preparedness is, to a large extent, conducted by the same inspectors that are involved in the preoperational testing inspections. This affords opportunities for the inspectors to develop a good understanding of the licensee's corporate and site organizational composition, as well as a basis for identifying any lack of technical competence or organizational weakness that may exist. Inspectors are expected to be well informed on the makeup and operations of the particular organization which they have responsibility for inspecting.

The inspection effort in the area of operational preparedness commences approximately 9 months prior to scheduled issuance of the operating license. Inspection activities in this area include verifying implementation of the operational quality assurance program, the environmental monitoring program, the staffing plan, the training program, the emergency plan, the physical security plan and the development of plant procedures. The latter includes examining the licensee's preparations and procedures for conducting initial core loading, and startup and power ascension testing. It also includes examining a sampling of administrative, plant and system operation, emergency, surveillance, and maintenance procedures. Appendix A of Regulatory Guide 1.33, Quality Assurance Program Requirements, and ANSI N18.7-1972, Administrative Controls for Nuclear Power Plants, and of course Regulatory Guide 1.68 previously mentioned, provides guidance for procedure requirements.

An in-depth review of the Operational Quality Assurance Program is conducted. The purpose of this inspection is to assure that an applicant has implemented his QA program for operations described in the SAR with implementing procedures and that activities are being conducted in accordance with these procedures. It is intended the program should be functioning routinely prior to fuel loading. Program areas inspected involve each of the QA criteria. The inspectors review the implementing procedures to assure that matters in the program are covered, determine that responsibility has been assigned to carry out the instruction, and test the system to confirm activities are being performed as specified. A team of 3-4 inspectors may spend 4-5 days at the plant performing this review. Program areas requiring corrective action will be reinspected and resolved prior to the receipt of the operating license. Examples of actual findings which required corrective action: responsibility to review proposed plant changes had not been assigned; the corrective maintenance program had not been implemented; and maintenance procedures did not specify preparation of the system for maintenance or retesting criteria subsequent to maintenance.

Startup Testing and Operational Phase

Upon issuance of an operating license, the operations phase of inspection activity is initiated. Here the licensee will be performing initial fuel loadings and operating for the first time under the requirements of his facility license and technical specification. The operations phase inspection program is subdivided into two distinct but overlapping inspection efforts of following the startup test program and the ongoing program of routine operations. The general inspection plan for the startup test program requires the inspector to perform examinations similiar to that described for the preoperational testing phase. Specifically, the test program is monitored to assure that test objectives are met and that plant response is as expected. Review of data and approval by plant personnel is expected prior to proceeding to the next higher power plateau. Typically, the startup test program is in progress from 3 to 6 months and would include one or two inspections each month.

Finally, the ongoing inspection program for operations consists of periodic inspections throughout the plant's lifetime to review such areas as plant operations, organization and administration, quality assurance program, training, procedures, maintenance, surveillance, calibration, in-service inspection, refuelings, emergency planning, radiation protection, radioactive waste management, material control and accounting, and physical protection. With regard to refueling, activities include observation of fuel movements, review of testing performed, review of outage activities including maintenance and restoration of systems, and the return to operation. One to three inspections can be expected each three months during this period.

Inspection Findings

Following an NRC inspection of a reactor facility, the inspection findings are discussed with licensee management prior to leaving the site. This is a basic requirement. If any timely corrective action is to result, issues must be communicated to the licensee. Findings are subsequently documented in an inspection report and formally transmitted to the licensee. Included in this correspondence is any proposed enforcement action for failure to comply with regulatory requirements disclosed during the inspection. Enforcement will be discussed in a moment, but for now let's say that the licensee is required to respond and provide the corrective action taken to prevent recurrence. Also included in the correspondence transmitting the inspection report are any identified safety matters or deviations from licensee commitments made to the Commission. The Regional Office acknowledges receipt of a licensee's response. The inspection report and correspondence are subsequently placed in the Public Document Rooms located near each facility.

Enforcement

The NRC expects licensees to manage and operate their facility within the requirements of Federal Regulations, orders, licenses, technical specifications, and permits issued pursuant to the Atomic Energy Act. When such requirements have not been complied with the following options are available in terms of increasing severity:

1. Notice of Violation (10 CFR 2.201) [10]

    Notice of violations are written notices to licensees, citing the apparent instances of failure to comply with regulatory requirements (violations but more commonly called noncompliance) which for purposes of categorization have been classified violations, infractions and deficiencies. Violations are the more serious and deficiencies the least serious. The vast majority of enforcement has fallen in the Notice of Violation option. Licensee's are required to respond in writing and provide corrective steps taken and results achieved; corrective steps taken to avoid further noncompliance; and the date when full compliance will be achieved. Corrective steps taken is subsequently reinspected. Unsatisfactory responses may be forwarded to Headquarters for evaluation if resolution cannot be obtained on the local level.

2. Civil Monetary Penalties (10 CFR 2.205) (10)

The Commission may levy civil monetary penalties against licensees for violations, infractions, or deficiencies with respect to the above mentioned regulatory requirements. These penalties are issued for more significant or repeated matters. The Commission is required to issue a "notice of violation" to the person charged before instituting proceedings to impose a civil penalty. The licensee has the option to pay the penalty or appeal. Civil penalties are less frequent and since their inception in 1971, fourteen cases involving reactor facilities have been issued. Seven of these cases involved failure to properly implement the Security Plan. Corrective steps taken is required in writing, similar to that described for the notice of violation.

3. Orders (10 CFR 2.202 and 2.204) (10)

The Commission has authority to issue orders to "cease and desist" and orders to suspend, modify, or revoke licenses. Such orders are ordinarily preceded by certain procedural requirements, including a written "notice of violation" to the licensee providing him with an opportunity to respond as to the corrective measures being taken. Experience has not required the use of this option in the nuclear power program. Licensee's have responded to either the lower level enforcement options or have agreed to voluntarily take any required actions.

INSPECTION AND ENFORCEMENT BULLETIN

One of the means available to communicate potential problems to industry or a group of licensees and obtain a specific corrective action is to issue an IE Bulletin. When a potential design, system, or component deficiency is identified, the matter is evaluated for potential generic significance. Bulletins are issued when an event meets the following criteria: event must have potential for occurrence at other facilities; event must involve safety equipment or affect safety; event must require action by the licensee, and the safety significance must warrant immediate dissemination.

These bulletins provide information on the potential deficiency and requests licensee's to investigate the matter, take required corrective action, and report their findings. Such responses are evaluated at the local and Headquarters level. A subsequent followup inspection confirms action reported. Examples of such bulletins are the: Cable Fire at Browns Ferry Nuclear Plant; Operability of Category I Hydraulic Shock and Sway Suppressors; and Defective Westinghouse Type OT-2 Control Switches.

CONCLUSION

This chapter gives a description of the Office of Inspection and Enforcement's inspection program and its relation to quality assurance for a typical power reactor facility. As was mentioned, the inspection program recognizes and insists that the licensee's organization be responsible for complying with the provisions of his license. The NRC believes this can be effectively achieved with an active quality assurance program that has the support of management.

REFERENCES

1. "Quality Assurance Criteria for Nuclear Power Plants," Code of Federal Regulations, Title 10, Part 50, Appendix B.

2. "Administrative Controls and Quality Assurance for the Operational Phase of Nuclear Power Plants," ANSI N18.7 - 1972, Rev. 1 (Draft No. 5) (February 12, 1975).

3. "Quality Assurance Program Requirements," USAEC Regulatory Guide 1.33, (Safety Guide 33) (November 3, 1972).

4. "Administrative Controls for Nuclear Power Plants," ANSI N18.7 - 1972, (December 20, 1972).

5. "Quality Assurance Program Requirements for Nuclear Power Plants," ANSI N45.2 - 1971, (October 20, 1971).

6. "Guidance on Quality Assurance Requirements During Design and Procurement Phase of Nuclear Power Plants - Revision 1," USAEC, WASH 1283, (May 24, 1974).

7. "Guidance on Quality Assurance Requirements During The Construction Phase of Nuclear Power Plants," USAEC, WASH 1309, (May 10, 1974).

8. "Guidance on Quality Assurance Requirements During The Operations Phase of Nuclear Power Plants," (USAEC Orange Book) (October 26, 1973).

9. "Preoperational and Initial Startup Test Programs for Water Cooled Reactors," USAEC Regulatory Guide 1.68, (November, 1973).

10. "Procedure for Imposing Requirements by Order, or For Modification, Suspension, or Revocation of a License, or for Imposing Civil Penalties," Code of Federal Regulations, Title 10, Part 2, Subpart B.

RELEASE OF RADIOACTIVE MATERIALS FROM REACTORS

Karl Z. Morgan
School of Nuclear Engineering
Georgia Institute of Technology
Atlanta, Georgia 30332

INTRODUCTION

Figure 1 gives the familiar double humped thermal neutron yield curves (1) for ^{235}U, ^{233}U and ^{239}Pu. The U-yields are similar in the two cases except from A = 80 to 95 and A = 108 to 125 where yields of ^{233}U are greater than yields of ^{235}U. The yields of ^{235}U and ^{233}U differ appreciably from those of ^{239}Pu. As a result there can be some important differences in fission product inventories in reactors of the same power and operating history but using ^{235}U, ^{233}U or ^{239}Pu as primary fuel or with the buildup of transplutonium radionuclides in the LWR's (especially if reprocessed fuel were used). It is noted from these curves that, for what are probably the most significant radionuclides--health wise, the yields of ^{90}Sr, ^{89}Sr and ^{85}Kr are about twice as large for ^{235}U and ^{233}U as for ^{239}Pu and that for ^{137}Cs, ^{131,132,129}I and ^{133}Xe the yields are about the same for all three types of fuel.

Radionuclides produced in a reactor are principally of three types: fission products, neutron activated radionuclides and actinide radionuclides. Table I lists some of the typical and more important radionuclides of these three types in terms of risk per µCi with respect to the values of maximum permissible concentration, MPC, in relation to the risk from ^{3}H. It should be emphasized that curie-for-curie some radionuclides are far more of a risk to man, for as indicated here, ^{239}Pu for example when released into the air is 3.3×10^6 times more of a hazard to man than ^{3}H. Some radionuclides like ^{239}Pu and ^{90}Sr are hazardous primarily because of their long effective half life T given in the simple case of single biological exponential elimination by the equation

$$T = \frac{T_r T_b}{T_r + T_b} \tag{1}$$

in which T_r = radioactive half life and T_b = biological half life. In the case of ^{90}Sr, $T \simeq T_b = 17.4$ y and in the case of ^{239}Pu, $T \simeq T_b = 200$ y. Also, the ^{239}Pu is a much greater risk than ^{90}Sr in part because of its higher energy per disintegration and in this case the quality factor Q_α (relating to linear energy transfer) has a value of 10 and a biological quality factor of N = 5 compared with $Q_\beta N_\beta = 1$ for the beta radiation from ^{90}Sr and its daughter ^{90}Y. Fortunately the fractional uptake by ingestion of ^{239}Pu in its usual insoluble forms is very low and is considered to be only about 3×10^{-5}. Radionuclides such as ^{131,132,129}I are hazardous primarily because of their large uptake and concentration in a small body organ, the thyroid (adult

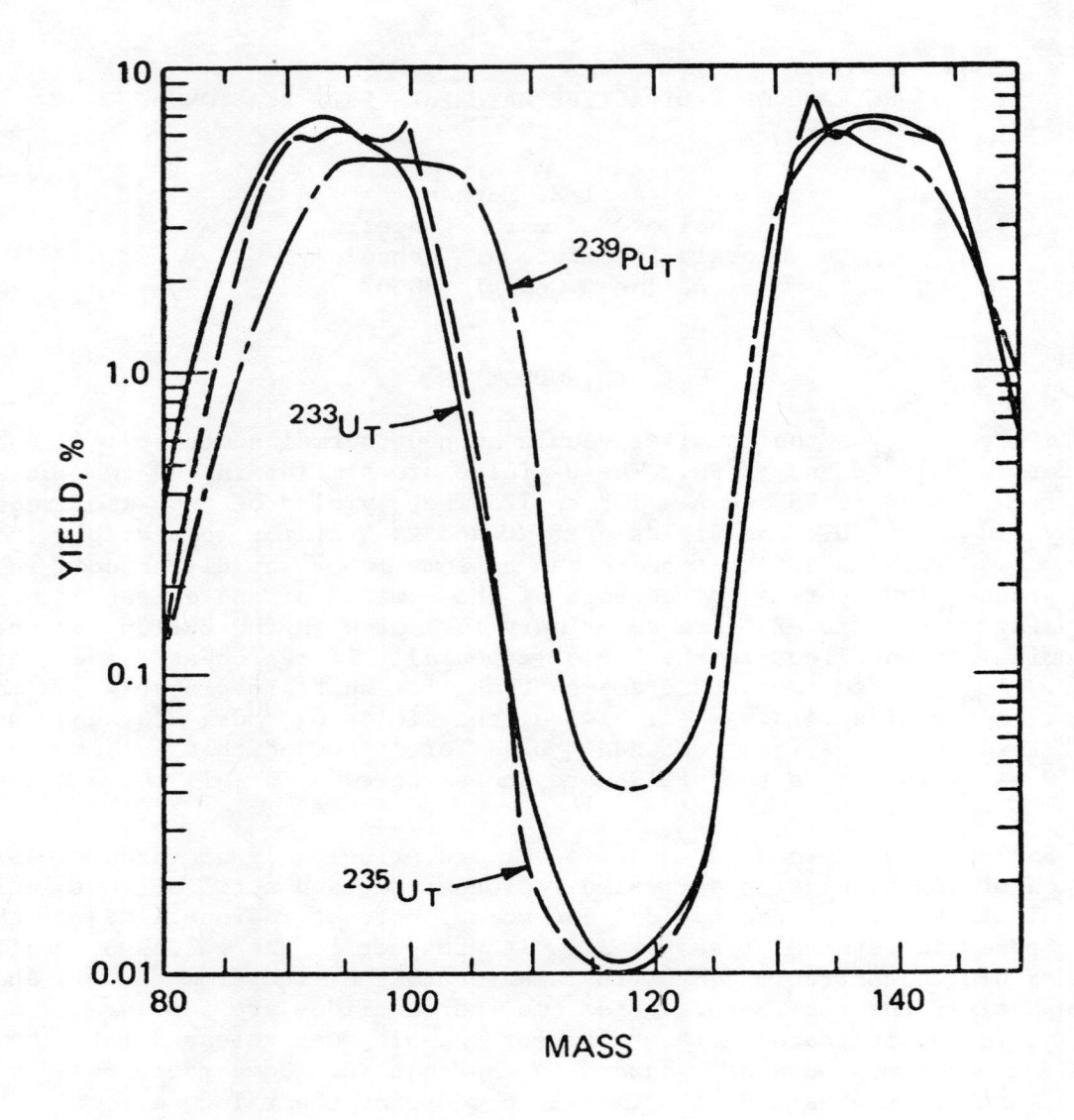

Fig. 1. Mass-yield Curves for Thermal-neutron Fission of ^{233}U, ^{235}U and ^{239}Pu (1)

Table I. Relative Risk of Radionuclides in Relation to Environmental Contamination from Nuclear Plant Operations with Respect to Maximum Permissible Concentrations (prepared by K. Z. Morgan using references 2 and 3)

Radionuclide	MPC in Water µc/cc	Relative Risk	MPC in Air µc/cc	Relative Risk
^{3}H	0.03	1	2×10^{-6}	1
^{51}Cr	0.02	1.5	8×10^{-7}	2.5
^{55}Fe	8×10^{-3}	3.7	3×10^{-7}	6.7
^{58}Co	10^{-3}	3.0	3×10^{-7}	200
^{54}Mn	10^{-3}	30	10^{-8}	200
^{95}Zr	6×10^{-4}	50	10^{-8}	200
^{59}Fe	5×10^{-4}	60	2×10^{-8}	100
^{63}Ni	3×10^{-4}	100	2×10^{-8}	100
^{60}Co	3×10^{-4}	100	3×10^{-9}	667
^{140}Ba	2×10^{-4}	150	10^{-8}	200
^{137}Cs	2×10^{-4}	150	2×10^{-8}	100
^{32}P	2×10^{-4}	150	2×10^{-8}	100
^{144}Ce	10^{-4}	300	2×10^{-9}	10^3
^{89}Sr	10^{-4}	300	10^{-8}	200
^{134}Cs	9×10^{-5}	333	4×10^{-9}	500
^{131}I	2×10^{-5}	1500	3×10^{-9}	667
^{90}Sr	4×10^{-6}	7500	4×10^{-10}	5×10^3
^{133m}Xe			4×10^{-6}	0.5
^{85}Kr			3×10^{-6}	0.67
^{133}Xe			3×10^{-6}	0.67
^{85m}Kr			10^{-6}	2
^{135}Xe			10^{-6}	2
^{14}C (CO_2)			10^{-6}	2
^{87}Kr			2×10^{-7}	10
^{233}U	4×10^{-5}	750	4×10^{-11}	5×10^4
^{239}Pu	5×10^{-5}	600	6×10^{-13}	3.3×10^6
^{226}Ra	10^{-7}	3×10^5	10^{-11}	2×10^5

103

thyroid 20 g) and because of their concentration by the grass-cow-milk-child food chain. The activity yield (Ci) of ^{129}I (in contrast to the mass yield) is much less than that of ^{131,132}I because of its long radioactive half life (1.7×10^7 y) but because of this long half life ^{129}I like ^{239}Pu (24,390 y), ^{238}Pu (86.4 y), ^{240}Pu (6580 y), ^{241}Am (458 y), ^{243}Am (7.95×10^3 y) etc. becomes a long range environmental problem.

The ^{137}Cs, because of its long radioactive half life (30 y) and large uptake into the soft tissues of the body ($f_w = 1.0$), including the gonads, would be one of the most dangerous of the radionuclides except for the fact that it has a short biological half life $T_b = 70$ days (compared with $T_b = 50$ y for ^{90}Sr). The noble gases such as ^{133}Xe and ^{85}Kr are a problem because of their high yield and the ease with which they escape into the environment, but fortunately they have no significant uptake or concentration in the body organs. The ^{3}H has a very low fission yield as indicated by Fig. 1 but a very high yield in some reactors due to neutron reactions with boron, lithium, nitrogen, helium, and deuterium*; as a consequence, even with its low relative hazard per curie, it can be an important health problem simply because of its high discharge rate at the reactor and/or during fuel reprocessing. ^{3}H (especially as the gas ^{3}H^{1}H or ^{3}H$_2$ rather than as the oxide) is perhaps the least radiotoxic of any of the radionuclides because: 1) its average β-energy per disintegration is only 0.005 MeV, 2) its $T_b = $ 12 days, 3) it is not concentrated in a food chain (for example, ^{32}P is concentrated in the Columbia River food chain by $> 10^5$), 4) it is not localized in a small body organ, and 5) most of all in this case, we have isotopic dilution eventually by all the water in our environment. Radionuclides such as ^{59}Fe, ^{63}Ni and ^{60}Co can be a problem due to activation if their precursor metals are exposed to a high neutron flux and can be eroded into the reactor coolant. Thus ^{60}Co has become the principal source of occupational exposure in the light water cooled reactors. The ^{85}Kr like ^{129}I is a special problem because of its long radioactive half life (10.76 y) and relatively easy access to the environment. It should be kept in mind that the risk relative to T_r acts in both directions so that other factors being equal it is radionuclides with intermediate T_r (i.e. $T_r = $ 10 to 50,000 years) that present the greater risk to man. In other words, if $T_r \to 0$, very little of the radionuclide may enter man's environment or into his body, and if $T_r \to \infty$ the radionuclide becomes stable and as is the case with ^{238}U, the chemical hazard may exceed that from radiation.

The values of relative risk per μc intake in Table I were obtained by dividing the $(MPC)_{w,a}$ for ^{3}H in water and air, respectively, by the corresponding values of $(MPC)_{w,a}$ of the radionuclide. In this way proper account in evaluating the risk is taken of: 1) retention function, 2) biological, and 3) radiological half lives and 4) fraction initially retained in the body following ingestion or inhalation (or skin penetration in the case of ^{3}H where the intake by skin respiration equals intake by inhalation), 5) submersion radiation (in the case of gases or swimming in water), 6) the fraction of that taken into the blood that is deposited in the critical

---

^{10}B + n_f → ^{11}B → ^{3}H + ^{8}Be* → ^{3}H + 2^{4}He; ^{6}Li + n_t → ^{3}H + ^{4}He;

^{7}Li + n_f → ^{3}H + ^{4}He + n, ^{14}N + n_f → ^{3}H + ^{12}C; ^{3}He + n_t → ^{3}H + ^{1}H; ^{2}H + n_t → ^{3}H.

organ, 7) the mass of critical organ, 8) the cross radiation from organ to organ, 9) the average energy per disintegration of the parent radionuclide and its daughter products, 10) the frequency of each type of radiation, 11) the quality factor Q relating to linear energy transfer (LET), 12) the biological quality factor N, 13) the fraction of the energy retained in the critical body organ, 14) the energy of the recoil atoms, internal x-rays, K-capture, internal conversion, bremsstrahlung, etc., 15) the ingestion and inhalation rates, and 16) the radiosensitivity of the critical tissue by setting dose limits for members of the public: MPD = 0.5 rem/y to total body, gonads and red bone marrow, MPD = 3.0 rem/y for bone, thyroid and skin (MPD = 1.5 rem/y for thyroid of child), and MPD = 1.5 rem/y for all other body organs.

It should be emphasized that the values of relative hazard given in Table I apply only to the concentrations in air, water and food of man and do not take into account reactor inventories, probability of release from a reactor, meteorology and hydrology of the site, environmental ecology and the food webs leading to man, eating and recreational habits of man, etc. These factors determine the concentrations of the radionuclides to which man is exposed and differ from reactor to reactor. However, they must be taken into account for each reactor in order to determine the risk to the local population from the radionuclides discharged to the environment.

Table II gives the critical body organ and values of T_r, T_b and T for several of the typical and more important (health wise) reactor radionuclides. The critical body organ is that organ receiving the radionuclide that results in the greatest body damage. It usually is the organ with the greatest concentration of the radionuclide but this is not always the case. For example, the erythrocytes receive the highest concentration of $^{55,59}Fe$ but they do not comprise the critical body tissue or critical organ because they are body cells that no longer undergo cell division and they can tolerate much higher doses than normal dividing cells without impairing their function.

Table III indicates the importance of the length of time before fuel is changed in a 1000 KW_t reactor (the values would be rather similar but about 3000 larger in a 1000 MW_e BWR or PWR). As would be expected, increasing the residence time from 100 days to 5 years does not increase the reactor inventory of short lived radionuclides such as ^{133}Xe (5.27 d), ^{131}I (8.05 d), ^{132}I (2.26 hr), ^{143}Pr (13.59 d), and ^{147}Nd (11.06 d) while for long-lived radionuclides such as ^{90}Sr (27.7 y) and ^{137}Cs (30.0 y) the activity is almost directly proportional to the residence time of the fuel in the reactor for the first few years.

Time is an important safety factor also in terms of the delay times before releasing to the environment radionuclides that escaped the reactor, especially in the case of short lived noble gases as shown in Tables IV, V and VI. Here it is noted that increasing the delay time from 30 minutes to 1 day to 60 days reduces the released Kr activity from 50,000 µCi/sec to 226 µCi/sec to 10 µCi/sec and reducing the delay time from 30 minutes to 15 days to 60 days reduces the Xe activity from 72,000 µCi/sec to 888 µCi/sec to 2.4 µCi/sec or a tenfold increase in delay time decreases the total noble gas discharged to the environment from 22,000 Ci/y to 3,700 Ci/y; further delay

Table II. Half-lives of Radionuclides in Target Organs (2,3)

Radionuclide	Critical Organ	Physical (Tr)		Biological (T_b)		Effective (T)	
Hydrogen-3[b,f,i] (Tritium)	whole body	12.6	y	12	d	11.97	d
Iodine-131[f]	thyroid	8	d	138	d	7.6	d
Strontium-90[f]	bone	28	y	50	y	18	y
Plutonium-239[c]	bone	24,400	y	200	y	198	y
	lung	24,400	y	500	d	500	d
Cobalt-60[i]	whole body	5.3	y	99.5	d	9.5	d
Iron-55[i]	spleen	2.7	y	600	d	388	d
Iron-59[i]	spleen	45.1	d	600	d	41.9	d
Manganese-54[i]	liver	303	d	25	d	23	d
Cesium-137[f]	whole body	30	y	70	d	70	d

[a] d = days, y = years
[b] Mixed in body water as tritiated water
[c] Actinide
[f] Fission product
[i] Induced

Table III. Important Fission Products in a Reactor (4)

Fission Product	Activity (curies)[a] after Selected Periods of Continuous Operation of a Reactor at a Power Level of 1000 kW		
	100 Days	1 Year	5 Years
Kr-85	53	191	818
Rb-86	0.25	0.26	0.26
Sr-89	28,200	38,200	38,500
Sr-90	402	1,430	6,700
Y-90[b]	402	1,430	6,700
Y-91	34,800	48,900	49,500
Zr-95	32,900	49,200	50,300
Nb-95(90 H)[b]	446	687	704
Nb-95(35 D)[b]	20,900	48,200	50,500
Ru-103	25,100	30,900	31,000
Rh-103[b]	25,100	30,900	31,000
Ru-106	753	2,180	4,220
Rh-106[b]	753	2,180	4,220
Ag-111	151	151	151
Cd-115	4.8	5.9	5.9
Sn-117	83	84	84
Sn-119	<24	<24	<100
Sn-123	4	9	10
Sn-125	100	101	101
Sb-125[b]	12	43	139
Te-125[b]	5	34	136
Sb-127	787	787	787
Te-127(90 D)[b]	146	260	277
Te-127(9.3 H)[b]	808	922	939
Te-129(32 D)	1,410	1,590	1,590
Te-129(70 M)[b]	1,410	1,590	1,590
I-131	25,200	25,200	25,200
Xe-131[b]	250	252	252
Te-132	36,900	36,900	36,900
I-132[b]	36,900	36,900	36,900
Xe-133	55,300	55,300	55,300
Cs-136	52	52	52
Cs-137	300	1,080	5,170
Ba-137[b]	285	1,030	4,910
Ba-140	51,500	51,700	51,700
La-140[b]	51,300	51,700	51,700
Ce-141	43,000	47,800	47,800
Pr-143	45,000	45,300	45,300
Ce-144	9,860	26,700	44,000
Pr-144[b]	9,860	26,700	44,000
Nd-147	21,800	21,800	21,800
Pm-147[b]	1,290	4,900	16,000
Sm-151	9	37	175

Table III. (Continued)

Fission Product	Activity (curies)[a] after Selected Periods of Continuous Operation of a Reactor at a Power Level of 1000 kW		
	100 Days	1 Year	5 Years
Eu-155	23	74	207
Eu-156	108	109	109
Total	563,691	693,573	767,547

[a] Calculated using fission product yields. From *Radiological Health Handbook*, 1960

[b] Daughter product.

Table IV. Estimated Gaseous Effluent Activity from 1000 MWe Light Water Reactors (5)

Nuclide	Half-Life	Release Rate (µCi/sec)[a]		
		30 Min Delay[b]	1 Day Delay[c]	60 Days Delay[d]
^{83m}Kr	1.86 h	3,600	0.5	0
^{85}Kr	10.76 y	10	10	10
^{85m}Kr	4.4 h	6,400	156	0
^{87}Kr	1.3 h	19,200	0.06	0
^{88}Kr	2.8 h	20,400	60	0
^{89}Kr	3.2 min	440	0	0
Total Kr Activity (µCi/sec)		~50,000	226	10
		30 Min Delay[b]	15 Days Delay[c]	60 Days Delay[d]
^{131m}Xe	11.9 d	15	6	0.4
^{133}Xe	5.27 d	6,400	880	2
^{133m}Xe	2.3 d	240	2	0
^{135}Xe	9.2 h	21,600	0	0
^{135m}Xe	15.6 min	10,400	0	0
^{137}Xe	3.8 min	1,120	0	0
^{135}Xe	14 min	31,200	0	0
Total Xe Activity (µCi/sec)		~72,000	888	2.4
Total Xe and Kr Activity (µCi/sec)		~122,000	1,100	12
Total Xe and Kr Activity (Ci/yr)		~3,850,000	35,000	380

[a] Assumes operation with 0.2% clad defects.
[b] Typical hold-up time in BWR's built to date (1973).
[c] Typical hold-up times for BWR's using charcoal beds.
[d] Typical hold-up time in PWR's.

Table V. Reactor Off-Gas Percent Composition at Various Decay Times (6)

Isotope	Half-Life	30 min	1 hr	3 hr	1 d	3 d	10 d	30 d	90 d	150 d
^{89}Kr	3.2 min	0.3								
^{137}Xe	3.8 min	0.9								
^{135m}Xe	15 min	8.0	3.0							
^{138}Xe	17 min	26.7	11.8							
^{87}Kr	1.3 h	15.7	18.0	1.4						
^{83m}Kr	1.9 h	2.5	3.1	0.8						
^{88}Kr	2.8 h	17.4	22.9	13.6	0.7					
^{85m}Kr	4.4 h	5.8	8.1	8.9	1.0					
^{135}Xe	9.2 h	17.6	25.2	50.0	38.5	2.3				
^{133m}Xe	2.3 d	0.2	0.3	0.9	1.8	2.1	0.5			
^{133}Xe	5.3 d	5.0	7.5	24.3	57.0	95.0	98.4	91.2	0.4	
^{131m}Xe	12.0 d			0.1	0.2	0.4	0.6	1.8	0.8	0.03
^{85}Kr	10.7 y				0.1	0.2	0.5	7.0	98.8	99.97

Table VI. Effect of Tenfold Increase in Delay Time on BWR Noble Gas Releases - Noble Gas Releases (Curies/y) at 0.8 Plant Factor for 1000 MW$_e$ LWR ([5])

Isotope	Base Case[a]	Tenfold Increase (in delay)
Kr-85m	140.	0
Kr-85	3,700.	3,700.
Kr-87	.35	0
Kr-88	76.	0
Xe-131m	120.	3.1
Xe-133m	88.	.006
Xe-133	18,000.	4.7
Xe-135	.054	0
Total	22,124	3,708

[a] 15 hour delay for krypton, 7 day delay for xenon.

would not be effective unless it were for decades since now essentially all that remains is ^{85}Kr (10.76 y). Thus, in the interest of maintaining population exposures as low as practicable it would seem prudent to continue development and improvement in techniques to increase this delay time in the reactors to about 90 days to remove essentially all the short-lived noble gases and to make serious plans to store the ^{85}Kr removed at the fuel reprocessing plants.

Before concluding this introduction, it might be helpful to glance at Fig. 2 as a reminder that although this chapter is limited to a discussion of radioactive materials from reactors there are many other stages in the nuclear fuel cycle from which radioactive contamination may escape to the environment and these stages are interrelated. For example, it may make little difference in terms of population exposure what fraction of the reactor-produced ^{85}Kr and ^{3}H escapes to the environment from the reactor instead of from the fuel reprocessing plant. This is because of the long half lives of these radionuclides and because what does not escape from the reactors will be permitted to escape later from the fuel reprocessing plant--that is, unless in the case of ^{85}Kr methods are developed and applied to recover the ^{85}Kr for long range storage. No technique is available to recover the ^{3}H when using aqueous solvent extraction of the fuel elements. In this case the probable solution of the problem would seem to be to produce less ^{3}H. This could be accomplished for the most part by not putting boron, lithium and nitrogen in the LWR's.

Also, the solid waste materials that are shipped from the reactor should be considered as "Release of Radioactive Materials from Reactors" but this subject is covered in another chapter in this series. This solid waste ranges from 9 to a high of 60 shipments per year among 10 power plants for which records are available. Typically these shipments might consist of 55 gallon steel drums into which was placed contaminated clothing, "hot" filters, contamination from sumps and drains, samples, etc. mixed with concrete.

POPULATION DOSE FROM VARIOUS MAN-MADE SOURCES OF IONIZING RADIATION

The population dose from ionizing radiation as a consequence of routine operation of nuclear power plants and their release of radioactive materials into the environment is considered to be very small. As estimated in Table VII it comprised in 1970 only 0.004% of the dose to the United States population from man-made sources of radiation. This dose undoubtedly will increase as more and larger nuclear power plants go into operation and as some of them approach retirement from operation, but I believe it is quite possible that by the year 2000 this dose will be less than an average of 1 mrem/y to the neighboring populations. Although not the subject of this lecture, it should be pointed out in the data of Table VII that the principal radiation problem in the U.S. is medical exposure, i.e. > 90%. I have shown in many publications and congressional hearings that most of this medical exposure in the U.S. is unnecessary and that with education, training and proper motivation of members of the medical profession and certification requirements this medical exposure could be reduced easily to less than 10% of its present value while at the same time providing better medical radiography. Although medical radiography saves tens of thousands of lives each year, this is no excuse for the careless use of ionizing radiation by members of the medical profes-

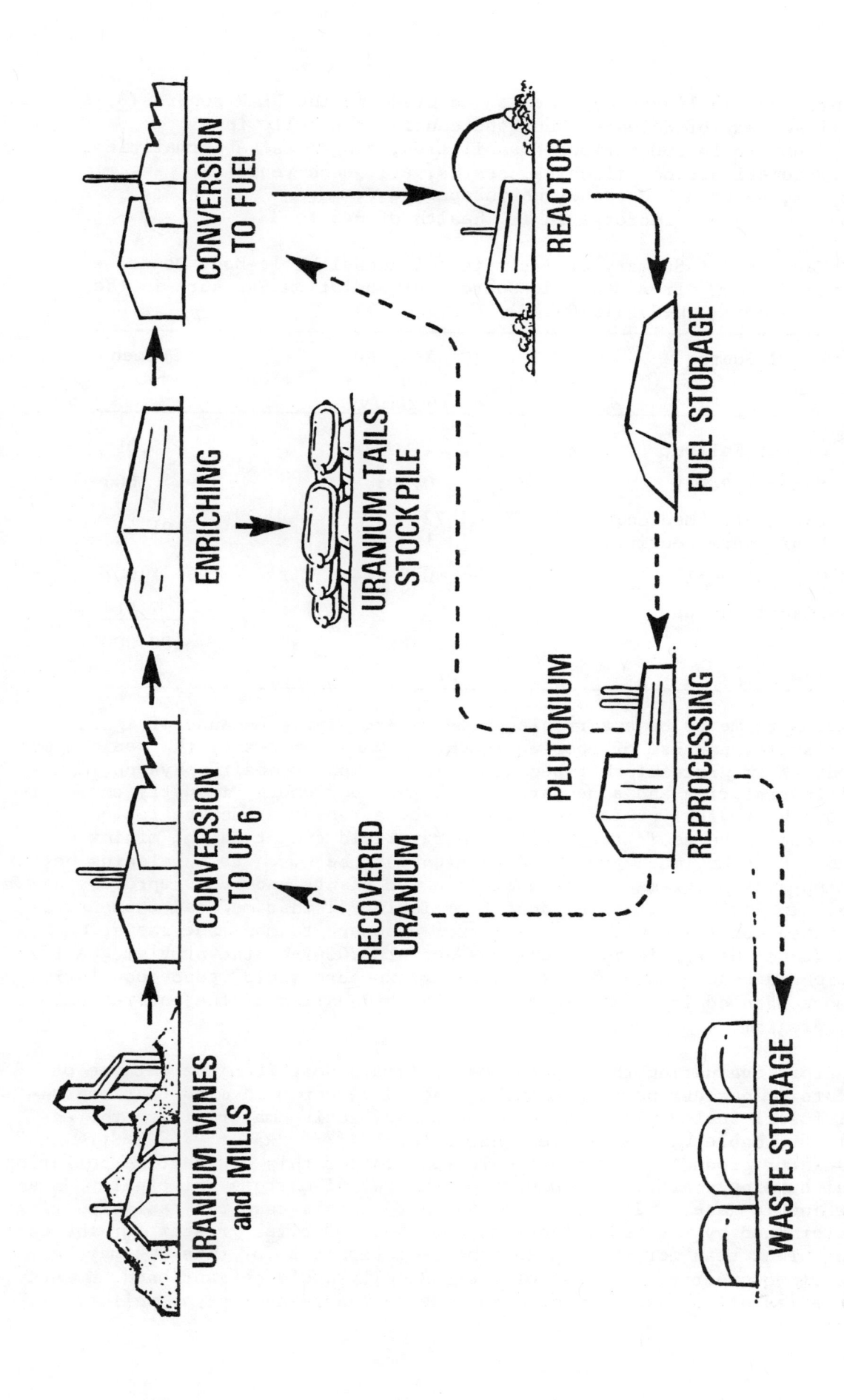

Fig. 2. The Light Water Reactor Nuclear Fuel Cycle (7)

sion which on the linear hypothesis (as given in the BEIR report (8) of the National Academy of Science) would be causing the following:
1) 600 to 14,000 serious disabilities, congenital abnormalities, constitutional diseases, deaths, etc. each year;
2) 1,600 to 8,000 cancer deaths per year;
3) an overall increase in ill health of 0.3 to 3%.

Table VII. Summary of Estimates of Annual Whole-Body Doses from Man-Made Sources of Radiation Exposure in the U.S. (1970) (8)

Source	Average Dose Rate (mrem/y)	Percent
Global Fallout	4	5.01
Nuclear Power	0.003	.004
Diagnostic Medical	72 ⎫ 73	91.48
Radiopharmaceuticals	1 ⎭	
Occupational	0.8	1.00
Miscellaneous	2	2.51
	79.803	100.000

This seems to me to be a terrible price we are paying because of ignorance, lack of skill, and use of poor equipment by many members of the medical profession. I am grateful that because of attention to health physics in national laboratories and some parts of the nuclear energy industry these mistakes of the medical profession have not been repeated such as to lead to serious consequences in the nuclear energy field except in the mining of uranium. However, much greater attention must be given to developing better health physics programs in the nuclear power plants and fuel reprocessing plants. In any case it is evident from Table VII that medical exposure is where those concerned with the harm from exposure to man-made radiation should focus their primary attention--not at 0.004% of the problem. A 1% reduction in unnecessary diagnostic x-ray exposure would reduce population dose more than could be attained by a discontinuation of the entire nuclear energy program.

Before evaluating the population radiation dose from the release of effluents of nuclear power plants, it should be borne in mind that conventional fossil fueled plants also release radioactive materials to the environment. Table VIII indicates that natural ^{226,228}Ra, ^{228,230,232}Th and the daughter products are responsible for most of this radioactive pollution by coal burning plants. By comparison the PWR discharges principally ^{3}H and the BWR noble gases. It will be noted that in this case the radiation risk (as determined by the MPC values) may be over 400 times greater for the coal burning plant than for the PWR and the radiation risk of this BWR may be 70,000 times greater than that of the PWR. It should be cautioned, however, that the radiation risks of the more modern plants--nuclear as well as fossil

Table VIII. Comparison of Modern Nuclear Plants with Modern Coal-Fired Power Plants[†] (9,10[**])

	Coal Plant	PWR	BWR
Stack discharges,[*] Ci/y/MW(e)			
^{226,228}Ra + ^{228,230,232}Th	4.8×10^{-5}		
Noble gases	4.8×10^{-5}	8×10^{-3}	1200[†]
Liquid discharges, Ci/y/MW(e)			
Fission products		8.2×10^{-3}	3×10^{-2}
^{3}H		3.8	1.5×10^{-2}
Critical organ	Bone	Total body	Total body
Limiting dose rate, mrem/y	3000	500	500
Dose rate, mrem/y	0.3(24)	0.005	150
Fraction ICRP MPE/MW(e)	11×10^{-8} (8.6×10^{-6})	2.1×10^{-8}	1.5×10^{-3}
Relative risk	5(410)	1	73,000

[†] Some of the more modern (second generation) BWR's are considered capable of operating at much lower levels of environmental radiation than this Dresden 1 plant.

[*] Coal plants are assumed to be similar to Willow Creek and are operating at about 1000 MW(e) with efficient air cleaning (97.5% efficiency) and tall stacks (800 ft). The PWR is the Connecticut Yankee plant operating at 200 MW(e). Ground-level release is assumed for the nuclear plants. Measurements were those at 1.1 to 1.7 miles downwind from the plants.

[**] Estimates by K. Z. Morgan using data of J. E. Martin, E. D. Harward and D. T. Oakley.

The values in parentheses are based on a comparison for 300 ft stacks that would not take credit for the 800 ft stack dilution since the larger dilution might simply deliver the same man-rem dose but involves more exposed persons.

fuel--may be much less than shown for the older plants in Table VIII. Also it should be noted that the 1600 y ^{226}Ra is far more dangerous than ^{3}H and the noble gases because it can remain in man's immediate environment much longer.

RELEASE OF RADIOACTIVE MATERIALS IN LIQUID EFFLUENTS BY NUCLEAR POWER PLANTS

Tables IX through XV summarize the releases of radioactive materials in liquid effluents from the various nuclear power plants. Some of these data such as given in Table IX are based only on gross β and γ counting. Perhaps in the early period when nuclear power plant facilities were limited and much of the program was still on an experimental basis such crude estimates by the nuclear power companies were acceptable, but it is generally believed that such data are no longer sufficient and that much more sophistication is not only warranted but required.

The findings of the Five Man Fact-Finding Committee of the Governor of Pennsylvania should serve as a lesson and underscore the need for more meaningful and relevant data on environmental releases of radioactive pollutants into the environment. This Committee was established to look into allegations of Dr. Sternglass that there was an increase in infant mortality, malignancies, heart diseases, and other causes of death due to radioactive effluents discharged by the Shippingport reactor and heard testimony at the Aliquippa Hearings on July 31 and August 1, 1973. I believe the most important observations and conclusions made by this Committee can be summarized as follows:

1) The radiation dosimetry and health physics supervision provided by the Shippingport operation and its contractors were very inadequate during its past operations.

2) There were abnormally high radiation levels in the vicinity of the Shippingport reactor during the period 1957-1971.

3) There was an increase in cancer incidence, infant mortality, and heart disease in populations living in the vicinity of the Shippingport reactor during the period 1957-1971, but this increase probably was the result of changes in the composition and characteristics of the population.

4) Because of lack of adequate data it is impossible to state with certainty that radiation levels resulting from the Shippingport effluents were or were not in part responsible for increased cancer incidence, infant mortality, and heart disease in the populations living in the vicinity of the Shippingport reactor during the period 1957-1971.

How much better it would have been if an adequate health physics program had been in operation by Shippingport! Its position would have been much stronger if it had had in its employ a qualified health physicist. Hopefully the NRC Regulatory Guides 8.8 ('73) and 8.10 ('74) and the revisions of these guides by the NRC will in time awaken the nuclear power companies to their responsibilities and correct this serious shortcoming of many nuclear plants. Guide 8.8 stated, "A specific individual, i.e. the health physics chief or manager, should be given explicit responsibility and authority to ensure that exposures are ALAP. He should be directly responsible to someone at high

Table IX. Annual Discharge Rates (Ci/year) of Gross Beta and Gamma Radioactivity (excluding tritium) in Liquid Effluents of Light-water Nuclear Power Stations[a] (11)

Nuclear Power Station	1959	1960	1961	1962	1963	1964	1965	1966	1967	1968	1969	1970	1971	1972	Average Discharge (Ci/year)	Geometric Mean Discharge (Ci/year)
PWR's																
Shippingport	0.083	0.021	0.129	0.09	0.19	0.53	0.14	0.06	0.07	0.08	0.208[b]	0.07[b]	c	c	0.14	0.10
Yankee Rowe			0.008	0.008	0.003	0.002	0.029	0.036	0.009	0.009	0.019	0.034	0.0115	0.0206	0.016	0.011
Indian Point 1				0.130	0.164	13.0	26.3	43.7	28.0	34.6	28.0	7.8	81.12	25.4	26.2	10.3
San Onofre									0.32	1.5	8.0	7.6	1.54	30.30	8.20	3.33
Conn. Yankee									0.39	3.9	12.0	29.5	5.88	4.78	9.40	4.97
Saxton									0.02	0.009	0.01	0.012	0.01	c	0.012	0.012
Ginna										c	0.02	10	0.96	3.75	3.68	0.92
H. B. Robinson													0.736	c	0.736	0.736
Point Beach 1													0.15	1.53	0.84	0.48
Maine Yankee														0.0169	0.017	0.017
Palisades														6.81	6.81	6.81
Surry 1														0.0251	0.025	0.025
All PWR's															4.68	0.34
BWR's																
Dresden 1		0.770	2.095	2.61	2.78	3.82	8.7	11.5	4.3	6.1	9.5	8.2	6.15	6.76	5.63	4.57
Big Rock Point				0.2	0.63	6.22	2.80	6.12	10.1	7.9	12.0	4.7	3.46	1.09	5.01	3.07
Humboldt Bay					0.397	0.664	1.89	2.10	3.13	3.2	7.5	2.4	1.84	1.40	1.85	1.58
LaCrosse									<0.005	0.074	8.5	6.4	17.10	48.5	13.43	1.60
Elk River			0.0012[d]	0.001[e]	0.0046[e]	0.026[e]	0.016[e]	0.013[e]	c	c	c	c	c	c	0.01	0.004
Nine Mile Point											0.9	28.0	32.2	34.6	23.92	12.94
Oyster Creek											0.48	18.5	12.1	10.0	10.26	5.72
Dresden 2,3												13	23.2	22.1	19.43	18.81
Millstone Point													19.65	51.5	35.57	31.81
Monticello													0.014	2.90 x 10⁻⁶	0.007	0.007
Pilgrim														1.45	1.45	1.45
Quad Cities 1,2														2.41	2.41	2.41
All BWR's															9.91	1.52
All PWR's + BWR's															7.30	0.71

[a] Data taken from references 15,20,27,28 and 29 of report ORNL-TM-3801 unless specified otherwise.
[b] These values from "Radioactive Waste Discharges to the Environment from Nuclear Power Facilities," by Joe E. Logsdon, Surveillance and Inspection Division, Office of Radiation Programs, U.S. Environmental Protection Agency, Rad. Data and Reports, Vol. 13, No. 2, pp. 117-129 (1972).
[c] Data not available.
[d] October 1, 1969 to March 31, 1972.
[e] April 1 of previous year to March 31 of year indicated.

Table X. Radionuclides That Should Be Considered When Evaluating the Environmental Impact of Radioactivity Released in the Liquid Effluent of Existing Light-Water Power Reactors (11)

Radionuclide	Radioactive Half-Life	Important Daughter Radionuclide	Measured Primary Coolant Concentration[a] $\div (MPC)_w$ PWR	BWR
H-3	12.3 y		10^3	10
C-14	5730 y		10^{-2}	
Na-22	2.58 y			
Na-24	15.0 h		10-200	110
P-32	14.3 d		3	
S-35	88 d		10^{-2}	
Sc-46	84 d			
Cr-51	27.7 d		0.5	0.01-1.0
Mn-54	313 d		5	10
Fe-55	2.7 y		0.5	0.1
Fe-59	45 d		3	<0.01
Co-57	276 d		10^{-2}	0.02
Co-58	71 d		10	1-100
Co-60	5.26 y		2	0.2-400
Ni-63	92 y		10	
Cu-64	12.8 h		10	10-30
Zn-65	244 d		$<10^{-2}$	0.03-1.0
Zn-69m	14 h	Zn-69		0.1
Rb-86	18.7 d			
Sr-89	50.5 d		3	30
Sr-90	28.5 y	Y-90	1	30
Y-90	64.2 h			
Sr-91	9.7 h	Y-91m, Y-91	1-5	5-350
Y-91	59 d			
Y-93	10 h		<0.03	200
Zr-95	63 d	Nb-95	1	150
Nb-95	35 d		0.5	5
Zr-97	17 h	Nb-97m, Nb-97		
Mo-99	66 h	Tc-99m	1	10
Ru-103	41 d	Rh-103m	<1	10
Rh-105	35.5 h			4
Ru-106	1.0 y	Rh-106	<0.1	10
Ag-110m	270 d	Ag-110	1	0.003
Sb-122	2.8 d		10	
Sb-124	60.2 d		2	0.02
Sn-125	9.5 d	Sb-125, Te-125m		
Sb-125	2.7 y	Te-125m		
Te-125m	58 d			
Sb-127	93 h	Te-127m, Te-127		

Table X. Continued

Radionuclide	Radioactive Half-Life	Important Daughter Radionuclide	Measured Primary Coolant Concentration[a] ÷ $(MPC)_w$ PWR	BWR
Te-127m	109 d	Te-127		
Te-127	9.3 h			
Te-129m	33 d	Te-129		
I-130	12.6 h			
Te-131m	30 h	Te-131 I-131		
I-131	8.05 d		100-300	100-30,000
Te-132	77 h	I-132	<0.1	25
I-133	20.9 h		800-1000	200-60,000
Cs-134	2.1 y		3	0.01-1
I-135	6.7 h		10	10^5
Cs-136	13 d		0.01	0.3
Cs-137	30.0 y	Ba-137m	1	1
Ba-140	12.8 d	La-140	1	0.02-50
La-140	40.2 h			
Ce-141	32 d		<0.1	10
Ce-143	32 h	Pr-143	<0.1	100
Pr-143	13.7 d			
Ce-144	290 d	Pr-144	<0.1	30
Nd-147	11.3 d	Pm-147		50
Pm-147	2.6 y			
Ta-182	115 d		1	0.5
W-185	74 d		0.1	
W-187	24 h		5	3
Np-239	2.3 d	Pu-239	<0.01	200

[a] $(MPC)_w$ values are given in 10 CFR 20, Appendix B, Table II, Column 2.
[b] Reader should refer to ORNL-TM 3801 (11) for more detail and references.

Table XI. Liquid Effluent Summary 1972 - Mixed Fission and Activation Products (12)

Facility	Curies Released	Average Concentration (μCi/ml)	Percent of Limit
Boiling Water Reactors			
Oyster Creek	10.0	8.62×10^{-9}	32.3
Nine Mile Point	34.6	7.80×10^{-8}	3.18
Millstone 1	51.5	8.35×10^{-8}	7.04
Dresden 1	6.75	2.32×10^{-8}	23.2
Dresden 2,3	22.1	1.49×10^{-8}	14.9
LaCrosse	48.5	1.46×10^{-7}	6.39
Monticello	2.90×10^{-6}	2.78×10^{-16}	2.05×10^{-8}
Big Rock Point	1.09	1.06×10^{-8}	0.88
Humboldt Bay	1.40	7.78×10^{-9}	0.11
Pilgrim*	1.45	5.58×10^{-8}	7.98
Quad Cities 1,2	2.41	1.73×10^{-9}	1.73
Vermont Yankee*	\multicolumn{3}{No liquid discharges for the report period.}
Pressurized Water Reactors			
Maine Yankee*	0.0169	1.75×10^{-10}	0.0919
Palisades	6.81	8.50×10^{-9}	0.005
Yankee	0.0206	1.28×10^{-10}	0.0249
Indian Point 1	25.4	5.12×10^{-8}	2.88
R. E. Ginna	0.375	5.18×10^{-10}	0.00234
Connecticut Yankee	4.78	6.20×10^{-9}	0.233
H. B. Robinson	0.826	3.82×10^{-9}	3.82
San Onofre	30.3	5.22×10^{-8}	1.49
Point Beach 1,2	1.53	3.11×10^{-9}	0.0052
Surry 1*	0.0252	1.83×10^{-10}	0.00395
Nonwater Reactors			
Peach Bottom 1	0.0209	1.57×10^{-9}	1.57
Fermi	0.222	9.53×10^{-9}	7.53

*Plants operated less than 1 year.

Table XII. Annual Discharge Rates (Ci/y) of Identified Radionuclides (excluding tritium) in Liquid Effluents of Light-water Nuclear Power Stations Measured During 1972[a] (11)

Radio-nuclide	Measured BWR Releases										Measured PWR Releases						
	Humboldt	Nine Mile Point	Oyster Creek	LaCrosse	Monti-cello	Big Rock Point	Mill-stone Point	Pilgrim	Quad Cities 1,2	Yankee Rowe	Indian Point 1	San Onofre	Ginna	Connect-icut Yankee	Point Beach 1	Palisades	Maine Yankee	Surry 1
C-14										1.71(-2)						1.58(-2)		3.31(-1)
Na-24								3.08(-2)	1.41(-1)		1.32(+0)					2.77(+0)		
Cr-51	7.4 (-2)	5.6 (-2)	1.18(-1)	1.30(+0)			8.54(-1)	7.15(-1)	1.27(-1)	4.3 (-4)		8.0 (-2)	3.1 (-3)		1.0 (-3)	9.19(-2)	8.06(-4)	2.2 (-4)
Mn-54		4.67(+0)	6.3 (-1)	2.39(-1)			1.09(+0)		1.07(-3)	5.08(-4)	2.36(+0)		2.5 (-3)			6.7 (-2)		
Fe-59		5.4 (-2)	2.0 (-2)				1.09(-1)				1.39(-1)					3.9 (-3)		
Co-57																		
Co-58	1.1 (-3)	7.25(-1)	1.53(-1)	2.71(+1)	7.9 (-7)	1.7(-2)	1.35(+0)	1.57(-1)	1.08(-1)	2.67(-4)	3.5 (-1)	1.52(+0)	7.9 (-3)	9.71(-1)	1.2 (-2)	3.17(+0)	5.48(-4)	6.05(-2)
Co-60	8.4 (-2)	9.1 (-1)	1.68(+0)	1.86(+0)		8.5(-2)	2.83(+0)	8.58(-4)		3.3 (-4)	1.24(+0)	1.45(-1)	4.21(-2)	1.15(+0)	1.0 (-3)	7.18(-2)	2.0 (-4)	1.02(-2)
Cu-64								5.78(-3)	6.56(-2)		8.2 (-3)							
Zn-65	1.74(-1)			3.55(-1)		2.3 (-2)										3.9 (-3)		
Sr-89	5.5 (-2)	3.94(-1)	2.28(-1)	3.17(-2)		4.6(-3)	1.03(+0)	7.8 (-4)	2.55(-2)		2.06(-2)							4.15(-5)
Sr-90	3.3 (-3)	2.5 (-2)		1.87(-1)	1.0(-8)	3.1(-3)	2.16(-1)	7.8 (-4)	1.04(-3)	1.10(-5)	1.29(-3)				1.0 (-3)	4.0 (-5)		4.66(-6)
Y-90							2.16(-1)											
Zr-95													1.17(-1)			3.19(-3)		
Nb-95			2.15(-1)													3.84(-2)		
Mo-99													3.18(-2)	8.96(-2)			1.0 (-5)	
Ru-103													8.9 (-3)	3.04(-1)				
Ru-106													3.3 (-3)					
Ag-110m																		
Sb-124			3.0 (-3)															
I-131	6.0 (-2)		4.5 (-1)	6.08(+0)	6.1(-7)	5.2(-2)	1.14(+1)	4.19(-1)	7.50(-1)	1.56(-3)	1.95(+0)	1.95(+0)	4.52(-3)		7.38(-1)	9.2 (-3)	4.54(-3)	1.29(-3)
I-133									1.86(-1)		3.55(+0)	7.87(-1)			1.0 (-1)	2.82(-2)		
Cs-134	2.37(-1)	3.90(+0)	2.06(+0)	2.86(+0)		2.9(-1)	8.86(+0)			1.64(-4)	4.48(+0)	8.66(+0)	4.39(-2)	5.75(-1)	2.6 (-2)	6.0 (-4)		
I-135											2.46(+0)				5.0 (-3)			
Cs-137	5.76(-1)	1.03(-1)	3.05(+0)	5.33(+0)	8.3(-7)	3.0(-1)	1.64(+1)			2.33(-4)	6.44(+0)		7.01(-2)	7.06(-1)	3.8 (-2)	1.21(-2)		
Ba-140	5.06(-2)		6.7 (-2)	3.67(-1)							4.08(-2)					3.1 (-3)		
Ce-144	2.5 (-2)									2.3 (-5)			3.81(-2)	2.27(-1)		6.0 (-5)		1.46(-4)
W-187									7.55(-3)							8.0 (-4)		
Np-239			6.83(-1)					3.98(-2)										

[a] Reported in Tables 6 through 29 of "Report on Releases of Radioactivity in Effluents and Solid Waste from Nuclear Power Plants for 1972," Directorate of Regulatory Operations, U.S. Atomic Energy Commission, Washington, D.C.

Table XIII. Summary of Annual Discharge Rates (Ci/year) of Radionuclides in Liquid Effluents of Light-water Nuclear Power Stations (11)

Radionuclide	Number of Entries in Table 4 of Ref. 11	Mean Release of all Entries (X_m)	Geometric Mean Release of All Entries (X_R)	Range Factor for Radionuclides with Five or More Entries (X_r)
C-14	4	8.98(-2)	1.61(-2)	-
Na-22	1	2.0 (-2)	2.0 (-2)	-
Na-24	11	8.97(-1)	1.51(-1)	44
P-32	4	5.91(-3)	5.73(-4)	-
S-35	1	6.0 (-4)	6.0 (-4)	-
Sc-46	1	1.0 (-2)	1.0 (-2)	-
Cr-51	29	1.18(+0)	6.22(-2)	199
Mn-54	31	1.38(+0)	5.79(-2)	264
Fe-55	3	3.15(-2)	1.20(-2)	-
Fe-59	17	5.50(-1)	6.51(-3)	1196
Co-57	4	1.81(-3)	7.70(-4)	-
Co-58	42	1.87(+0)	1.14(-1)	166
Co-60	42	9.33(-1)	8.05(-2)	108
Ni-63	1	1.0 (-3)	1.0 (-3)	-
Cu-64	7	3.19(-2)	9.20(-3)	21
Zn-65	13	2.86(-1)	2.94(-2)	85
Zn-69m	2	2.55(-4)	6.96(-5)	-
Rb-86	1	1.0 (-3)	1.0 (-3)	-
Sr-89	22	1.63(-1)	1.53(-2)	72
Sr-90	25	3.23(-2)	7.15(-4)	563
Y-90	3	7.55(-2)	1.03(-2)	-
Sr-91	3	3.13(-2)	2.43(-2)	-
Y-91	3	2.93(-2)	2.02(-2)	-
Y-93	2	1.55(-3)	1.14(-3)	-
Zr-95	5	2.01(-1)	3.09(-3)	876
Nb-95	6	3.60(-1)	9.02(-3)	498
Zr-97	2	2.52(-4)	5.05(-5)	-
Mo-99	13	2.13(-1)	7.87(-3)	304
Ru-103	5	7.92(-2)	4.86(-3)	166
Rh-105	2	7.05(-3)	1.18(-3)	-
Ru-106	3	1.00(-3)	5.65(-4)	-
Ag-110m	8	4.19(-2)	3.09(-3)	129
Sb-122	1	5.0 (-3)	5.0 (-3)	-
Sb-124	4	3.57(-3)	1.53(-3)	-
Sn-125	1	1.0 (-3)	1.0 (-3)	-
Sb-125	1	6.5 (-4)	6.5 (-4)	-
Te-125m	1	1.5 (-4)	1.5 (-4)	-
Sb-127	1	1.0 (-4)	1.0 (-4)	-
Te-127m	1	1.5 (-4)	1.5 (-4)	-
Te-127	1	1.5 (-4)	1.5 (-4)	-
Te-129m	2	2.40(-4)	2.22(-4)	-

Table XIII. (Continued)

Radionuclide	Number of Entries in Table 4 of Ref. 11	Mean Release of all Entries (X_m)	Geometric Mean Release of All Entries (X_R)	Range Factor for Radionuclides with Five or More Entries (X_r)
I-130	1	1.0 (-4)	1.0 (-4)	-
Te-131m	1	1.5 (-4)	1.5 (-4)	-
I-131	41	1.51(+0)	8.98(-2)	167
Te-132	2	4.36(-2)	4.17(-3)	-
I-133	17	5.97(-1)	1.44(-1)	26
Cs-134	26	2.15(+0)	5.55(-2)	473
I-135	3	8.51(-1)	1.03(-1)	-
Cs-136	3	1.44(-3)	8.93(-4)	-
Cs-137	35	2.06(+0)	5.96(-2)	411
Ba-140	18	1.24(-1)	1.40(-2)	76
La-140	5	1.35(-1)	1.57(-3)	1266
Ce-141	4	1.25(-1)	1.29(-3)	-
Ce-143	2	2.58(-4)	9.22(-5)	-
Pr-143	2	2.93(-4)	2.08(-4)	-
Ce-144	10	3.66(-2)	2.67(-3)	134
Nd-147	3	2.20(-4)	1.25(-4)	-
Pm-147	1	5.0 (-4)	5.0 (-4)	-
Ta-182	3	1.90(-2)	4.60(-3)	-
W-185	2	1.21(-3)	9.16(-4)	-
W-187	5	2.23(-2)	3.63(-3)	277
Np-239	8	2.42(-1)	2.87(-2)	72

Table XIV. Annual Discharge Rates (Ci/y) of Tritium in Liquid Effluents of Light-water Nuclear Power Stations[a] (11)

Nuclear Power Station	1959	1960	1961	1962	1963	1964	1965	1966	1967	1968	1969	1970	1971	1972	Average Discharge (Ci/y)	Geometric Mean Discharge (Ci/y)
PWR's																
Shippingport	64.0	99.0	13.2	1.33	2.17	1.39	3.04	27.3	34.8	35.2	20[b]	1.71[b]	c	c	25.26	10.13
Yankee Rowe			c	c	c	c	1300	1920	1690	1170	1200	1500	1685	803	1408	1365
Indian Point 1			c	c	c	c	491[d]	125	297	787	1100	410	725	574	564	477
San Onofre									c	2350	3500	4800	4570	3480	3740	3627
Conn. Yankee									221	1740	5200	7400	5830	5890	4380	2825
Saxton											<1	10	4.14	c	5.05	3.5
Ginna											1.26[b]	110	154	199	116	45
H. B. Robinson													118.3	c	118.3	118.3
Point Beach I													266	558.0	412	385
Maine Yankee														9.22	9.22	9.22
Palisades														120.0	120	120
Surry I														5.03	5.03	5.03
All PWR's															908.6	108.6
BWR's																
Dresden I	Average of 5 to 10 Ci/y from 10/1/59 to 12/31/66									2.9	~6	5	8.7	98.8	24.3	9.4
Big Rock Point	Average of 20 Ci/y from 9/1/62 to 5/1/67									34	28	54	10.3	10.4	27.3	22.3
Humboldt Bay	Average of 20 to 100 Ci/y from 1963 through 1968					<54	<60	<166	<6.6	<5	<7	<6.8	13.0	63.2	27.3
Nine Mile Point											<1	20	12.4	27.8	15.3	9.1
Oyster Creek										c	5	22	21.5	61.6	27.5	19.5
LaCrosse										c	25	20	91.4	120.0	64.1	48.4
Elk River	Average of 1 to 10 Ci/y from 9/1/61 to 4/1/67									c	c	c	c	c	~5	~5
Dresden 2,3												31	38.5	25.9	31.8	31.4
Millstone Point													12.7	20.9	16.8	16.3
Monticello													0.59	7.6 x 10^{-5}	0.3	0.3
Pilgrim														4.18	4.18	4.18
Quad Cities 1,2														4.70	4.70	4.70
All BWR's															23.7	9.8

[a] Data taken from Refs. 15, 20, 27 and 28 of Report ORNL-TM-3801 unless specified otherwise
[b] Values from "Radioactive Waste Discharges to the Environment from Nuclear Power Facilities," by Joe E. Logsdon, Surveillance and Inspection Division, Office of Radiation Programs, U.S. Environmental Protection Agency, Rad. Data and Reports, Vol. 13, No. 2, pp. 117-129, 1972
[c] Data not available.
[d] Feb. 1, 1965 to Jan. 31, 1966

Table XV. Liquid Effluent Summary 1972 - Tritium (12)

Facility	Curies Released	Average Concentration (μCi/ml)	Percent of Limit
Boiling Water Reactors			
Oyster Creek	61.6	5.31×10^{-8}	0.00177
Nine Mile Point	27.8	6.28×10^{-8}	0.00209
Millstone 1	20.9	3.39×10^{-8}	0.00113
Dresden 1	43.3	1.48×10^{-7}	0.00493
Dresden 2,3	25.9	1.81×10^{-8}	0.0006
LaCrosse	120.0	3.61×10^{-7}	0.012
Monticello	7.60×10^{-5}	7.31×10^{-15}	2.44×10^{-10}
Big Rock Point	10.4	1.07×10^{-7}	0.0034
Humboldt Bay	13.0	7.22×10^{-8}	0.00241
Pilgrim*	4.18	1.61×10^{-7}	1.61
Quad Cities 1,2	4.70	3.38×10^{-9}	0.000113
Vermont Yankee*	No liquid discharge during the report period.		
Pressurized Water Reactors			
Maine Yankee*	9.22	9.57×10^{-8}	1.91
Palisades	208.0	2.60×10^{-7}	0.430
Yankee	803.0	4.97×10^{-6}	0.166
Indian Point 1	574.0	1.16×10^{-6}	0.040
R. E. Ginna	119.0	1.64×10^{-7}	0.00547
Connecticut Yankee	5890.0	7.64×10^{-6}	0.255
H. B. Robinson	4.05×10^{2}	1.88×10^{-7}	0.000625
San Onofre	3480.0	5.99×10^{-6}	0.2
Point Beach 1,2	563.0	1.13×10^{-6}	0.037
Surry 1*	5.03	3.64×10^{-8}	0.091
Nonwater Reactors			
Peach Bottom 1	1.7	1.28×10^{-7}	0.0427
Fermi	None		

*Plants operated less than 1 year.

management level. The health physics group should not be a part of operations or production oriented divisions . . .. The individual responsible for recommending and implementing the radiation control program, i.e. the health physics chief or manager, should be a professional of recognized competence in this field, preferably with power reactor experience. Where this individual does not have qualifications equivalent to those required for certification by the American Board of Health Physics, he should be supported by and have available immediate access to one or more consultant and/or staff member who is qualified and who is in the facility at least once a month."

Unfortunately, the revisions of these guides by the NRC are somewhat watered down and weaker than the above but I consider that unless the goals expressed in Guides 8.8 and 8.10 are carried out, the nuclear power industry cannot expect to merit the confidence and acceptance of the public.

It is not my assignment here to review what I consider would have been adequate instrumentation by the Shippingport operation but if I did, I would begin my list of equipment today, for example, with a Ge(Li) detector which can have an energy resolution of 2 keV at 1 MeV and a sensitivity of 0.02 pCi/g of sample. With such modern equipment identification of each of the radioactive contaminants and its origin is possible.

Table X is a list of 63 radionuclides prepared by R. S. Booth (11) which he indicates might include most of those that should be identified and measured in liquid releases from the LWR's. Perhaps a few others and especially the radionuclides of Pu, Am and Cm should be added to the list. The use of such a list will obviate the necessity of looking for the several hundred radionuclides that are released; most of them in extremely low concentration and for which the ratio of primary coolant concentration to the $(MPC)_w$ is very small. It is to be noted in Table X that this ratio is large for $^{131,133}I$, $^{58,60}Co$, ^{239}Np, $^{89,90,91}Sr$, ^{93}Y, ^{3}H, ^{24}Na, ^{95}Zr and ^{143}Ce.

Referring again to Table IX and the Shippingport reactor it is noted that for the period 1959-1970 the reported gross β plus γ discharges were consistently less than those from most of the other plants. It would be interesting to know whether or not such data have much comparative value or meaning. Some of the plants shown in Table IX have conspicuously high discharge of radioactive liquid effluents, for example Nine Mile Point in 1970, 1971 and 1972; Millstone 1 in 1971 and 1972; LaCrosse in 1971 and 1972; Indian Point 1 in 1965-1969 and 1971 and 1972; and San Onofre in 1972. Table IX indicates that, if we exclude ^{3}H, BWR's had liquid effluent radioactive discharges that were about twice those of the PWR's and Table XI indicates the BWR's in 1972 reached a higher percent of the permissible limit than did the PWR's and that Oyster Creek and Dresden 1 had the worst records in this respect.

Table XII lists about half of the more important radionuclides (as given in Table X) indicating the amounts discharged in 1972. Perhaps the most significant feature of this table is the absence of information in many cases. Table XIII is a summary indicating the mean releases of radionuclides as obtained from available data provided by the various LWR's. Booth points out from these data the wide range of values for the various reactors and that the same reactor for subsequent years can differ in these discharges by orders of magnitude (11).

Tables XIV and XV give comparative data on the discharge of ^{3}H from various power reactors. As would be expected, in most cases the ^{3}H discharges were much greater from the PWR's than from the BWR's. Some of the PWR's such as Connecticut Yankee and San Onofre had consistently high discharges of ^{3}H. Figure 3 indicates that the projected build up of ^{3}H in the environment from nuclear power reactors is expected to reach the natural equilibrium level in about 1980 (13).

RELEASE OF RADIOACTIVE MATERIALS IN GASEOUS EFFLUENT BY NUCLEAR POWER PLANTS

Table XVI indicates, as would be expected, the amounts of noble gases released in liquid waste are very low. Table XVII shows the expected high levels of noble gases that are airborne from the BWR's (e.g. 866,000 Ci from Oyster Creek in 1972) and indicates that Oyster Creek and Humboldt Bay were releasing radioactive noble gases in the air at 10.4 and 27.3%, respectively, of the limit.

Table XVIII indicates the airborne releases of halogens and particulates. Oyster Creek, Dresden 1,2,3 and Humboldt Bay had the highest halogen and particulate releases in 1972 but La Crosse and Humboldt Bay reached the highest percent of their limit. It should be noted that the design limit varies considerably from reactor to reactor.

DOSE ESTIMATES RESULTING FROM NUCLEAR POWER OPERATION

Figure 4 is a typical food web showing the pathways of radionuclides from nuclear power plants leading to man.

Table XIX indicates estimated doses to the neighboring populations from various pathways of the airborne effluents, ^{131}I, ^{89}Sr and ^{137}Cs, emitted from the Dresden 1 plant. Table XX indicates the average individual dose (mrem) and the population dose (man-rem) within 50 miles of the reactors during 1971. Dresden 1, 2 and 3 and Humboldt Bay scored worst in this regard. It is to be noted also that in these two cases the man-rem doses probably exceeded any reasonable or acceptable ALAP or ALARA limits. Table XXI indicates that Dresden and Humboldt Bay exceeded the 5 mrem/y limiting dose to an individual if measured at the boundary of the plant control area. Table XXII gives the percent of the annual dose to various body organs from the several methods of intake of the radionuclides discharged by nuclear power reactors. Table XXIII indicates the number of deaths from radioactive effluents resulting to the year 2020 from past to present and from future operation of nuclear power reactors based on the linear hypothesis. These numbers in terms of percent of the U.S. population are small but certainly are not insignificant. Here again is emphasized the risk of malignancies that may develop in a man many years after exposure to plutonium and the other actinide elements. Figure 5 provides some comparison of risks of radiation, SO_2 and NO_2. Table XXIV indicates the accumulation of radioactive wastes from the nuclear power industry to year 2000. Here it can be predicted that ^{244}Cm, ^{243,241}Am and ^{238,240}Pu probably will be a much greater reactor waste problem than ^{239}Pu.

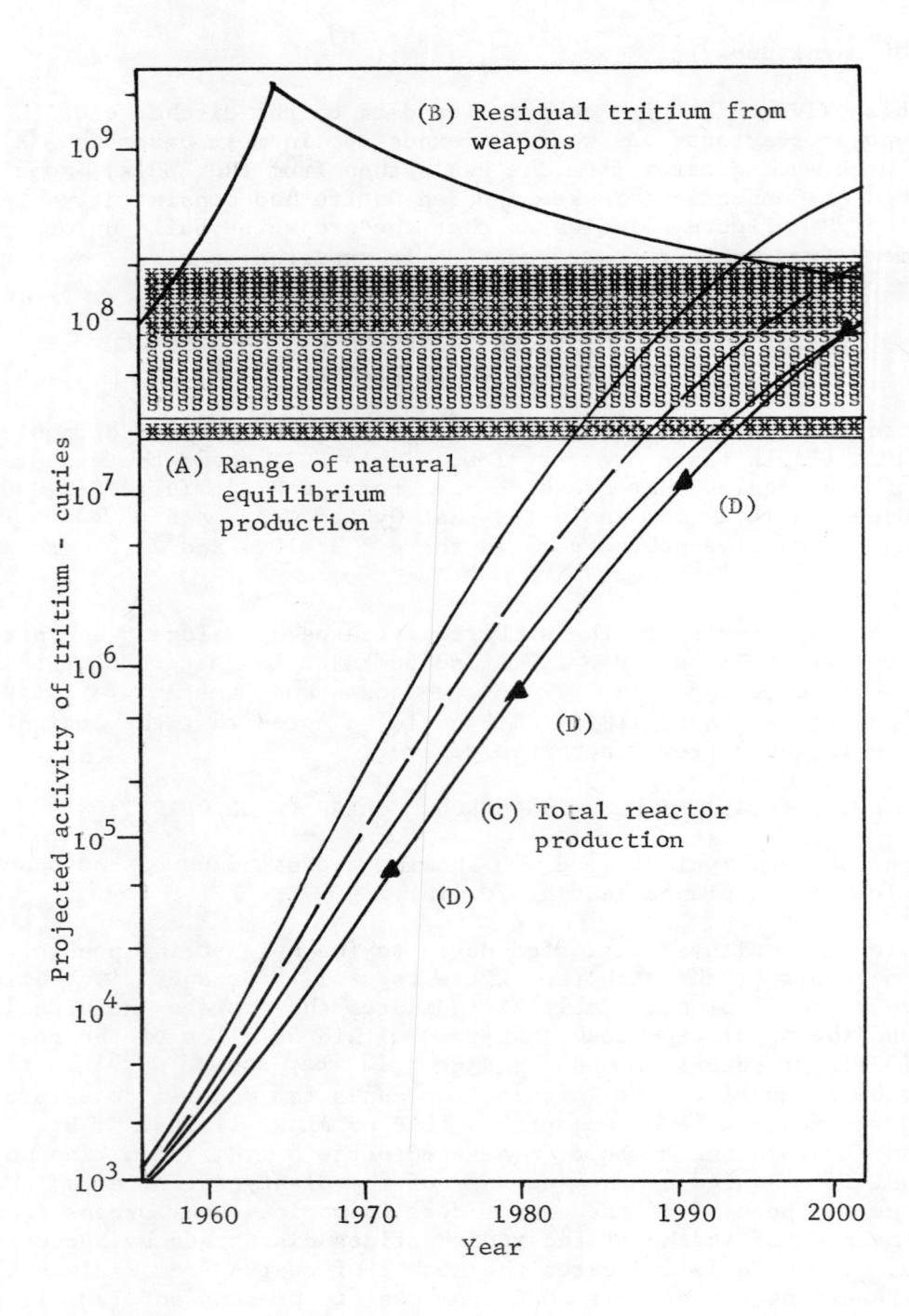

Fig. 3. Comparison of Tritium Activity (13) from: (A) Natural Production (wide range = all estimates; narrow range = most probable estimate), (B) Residual Weapons Fallout, (C) U.S. Reactor Production and (D) Reactor Production (▲) as Estimated by Cowser et al. (14) Using Earlier Data

Table XVI. Gaseous Activity in Liquid Waste in 1972 (12)

Facility	Xe-133 (Ci)	Xe-135 (Ci)
Boiling Water Reactors		
Oyster Creek	7.84×10^{-1}	2.48
Nine Mile Point	NR*	NR*
Millstone 1	$<2.42 \times 10^{-1}$	$<2.80 \times 10^{-1}$
Dresden 1	NR	
Dresden 2,3	NR	
LaCrosse	$<8.7 \times 10^{-1}$	$<8.02 \times 10^{-2}$
Monticello	NR	NR
Big Rock Point	NR	NR
Humboldt Bay	NR	NR
Pilgrim	NR	NR
Quad Cities 1,2	2.83×10^{-3}	4.70×10^{-4}
Vermont Yankee	No Liquid Releases	
Pressurized Water Reactors		
Maine Yankee	1.44×10^{-3}	NR
Palisades	1.79	4.0×10^{-2}
Yankee	2.50×10^{-2}	6.36×10^{-4}
Indian Point 1	NR	NR
R. E. Ginna	NR	NR
Connecticut Yankee	7.35	1.73×10^{-1}
H. B. Robinson	--	--
San Onofre	5.43	5.71×10^{-2}
Point Beach 1,2	5.96×10^{-1}	$<1.0 \times 10^{-3}$
Surry 1	NR	NR
Nonwater Reactors		
Peach Bottom 1	NR	NR
Fermi	NR	NR

*None reported.

Table XVII. Airborne Effluent Comparison by Year - Noble Gases (12)

Facility	Curies/Year 1970 (X1000)	Curies/Year 1971 (X1000)	Curies/Year 1972 (X1000)	% of Limit 1972
Boiling Water Reactors				
Oyster Creek	110	516	866	10.4
Nine Mile Point	9.5	253	517	2.01
Millstone 1	--	276	726	2.91
Dresden 1	900	753	877	5.30
Dresden 2,3	--	580	429	1.51
LaCrosse	0.95	0.53	30.6	5.13
Monticello	--	75.8	751	8.8
Big Rock Point	280	284	258	8.19
Humboldt Bay	540	514	430	27.3
Pilgrim*	--	--	18.1	0.343
Quad Cities 1,2	--	--	132	3.80
Vermont Yankee*	--	--	55.2	2.38
Pressurized Water Reactors				
Maine Yankee*	--	--	0.002	4.19
Palisades	--	--	0.505	0.0208
Yankee	0.017	0.0128	0.0183	0.0263
Indian Point 1	1.7	0.36	0.543	0.00337
R. E. Ginna	10	31.8	11.8	0.678
Connecticut Yankee	0.7	3.25	0.645	0.252
H. B. Robinson	--	0.018	0.257	0.54
San Onofre	0.42	7.67	19.1	1.09
Point Beach 1,2	--	0.838	2.81	0.0874
Surry 1*	--	--	(0.0126)†	1.67×10^{-6}
Nonwater Reactors				
Peach Bottom 1	0.006	0.122	0.058	0.031
Fermi	--	0.18	0.184	19.5

*Plants operated less than 1 year.

†Actual value, X1000 does not apply; all other values of Curies in table are to be multiplied by 10^3.

Table XVIII. Airborne Effluent Comparison by Year - Halogens and Particulates (12) (half-life greater than eight days)

Facility	Curies 1970	1971	1972	% of Limit 1972
Boiling Water Reactors				
Oyster Creek	0.32	2.14	6.48	5.13
Nine Mile Point	<0.001	<0.06	0.969	2.01
Millstone 1	--	4.0	1.32	1.37
Dresden 1	3.3	<0.67	2.75	3.65
Dresden 2,3	1.6	8.68	5.89	3.60
LaCrosse	<0.06	<0.001	<0.712	43.4
Monticello	--	0.052	0.589	1.6
Big Rock Point	0.13	0.61	0.148	0.416
Humboldt Bay	0.35	0.3	1.78	31.4
Pilgrim*	--	--	0.0319	5.71×10^{-5}
Quad Cities 1,2	--	--	0.747	0.454
Vermont Yankee*	--	--	0.171	1.63
Pressurized Water Reactors				
Maine Yankee*	--	--	3.71×10^{-6}	0.189
Palisades	--	--	9.7×10^{-3}	0.291
Yankee	<0.001	<0.0001	7.77×10^{-4}	0.0607
Indian Point 1	0.08	0.21	0.928	1.25
R. E. Ginna	0.05	0.17	0.035	3.80
Connecticut Yankee	0.002	0.03	0.0181	8.71
H. B. Robinson	--	None detected	0.0268	{ 11.04** 3.6†
San Onofre	<0.001	<0.0001	4.74×10^{-4}	0.000108
Point Beach 1,2	--	--	1.75×10^{-4}	0.09
Nonwater Reactors				
Peach Bottom 1	<0.001	<0.003	None	--
Fermi	--	<0.001	0.001	1.84

*Operated less than 1 year.
**Halogens
†Particulates

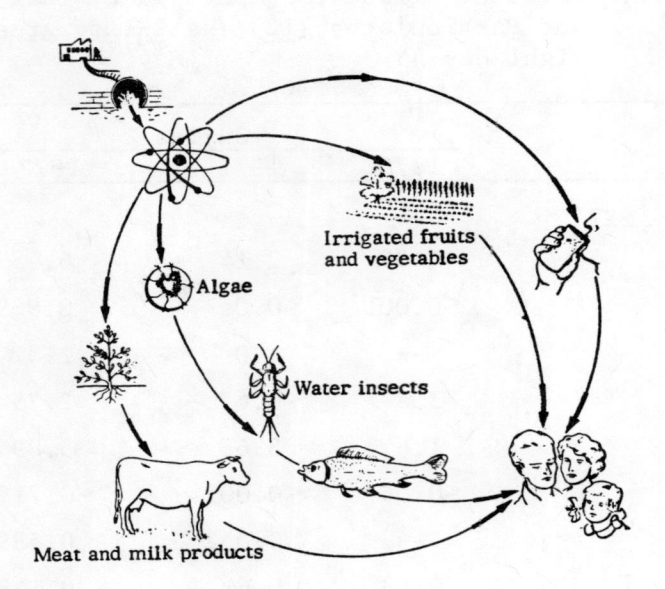

Figure 4. Paths of Radioactive Substances to Man via Water (from Tsivoglou, Harward, and Ingram (15)

Nuclear power plants and reprocessing facilities routinely release various radioactive isotopes into the air and water. These become more concentrated as they move up the food chain, as animals eat plants and smaller animals. For example, a study of emissions from the Hanford reactor in Washington (16) shows the following concentration factors in the Columbia River:
1. Radioisotope concentrations in water below the NRC "permissible levels"
2. Plankton have 2,000 times more radioactive phosphorus-32 per gram than water
3. Ducks have 40,000 times more radioactive phosphorus-32 than water
4. Fish have 150,000 times more than water
5. Swallows have 500,000 times more than water
6. Egg yolks of birds have 1,500,000 times more

Source: "Principles of Radiation Protection Engineering," Dr. L. Dresner; McGraw-Hill Book Co., New York (1965)

Yet, the NRC's new "interim" standards suppose a reconcentration of only 1,000 and are given only in terms of effect on humans.
- - - Are NRC radioactive emissions standards sufficiently conservative?

Table XIX. Estimated Doses Along Various Pathways Near Dresden BWR Unit 1 in 1968 and 1970 (17,18)

Pathway	Nuclides	Organ	Distance (km)	Estimated Dose Rates (mrem/y) 1968	1970
Cloud immersion	Gaseous fission products	Whole body	1.2	14	32
Inhalation	I-131	Thyroid	---	0.0002	0.013
Ingestion					
Milk	I-131	Thyroid	3.4	0.28	31
Drinking	I-131	Thyroid	170	0.35	3.2
Fish	I-131	Thyroid	3.8	0.063	0.4
Leafy vegetables	I-131	Thyroid	65	0.043	7.2
Beef	Sr-89 Cs-137	Bone Whole body	2.3	0.00037	0.033

Table XX. Curies of Noble Gases Released, Boundary and Average Individual Doses and Population Doses (Man-Rem) ([19],[20][*])

Type Facility	Curies Released	Boundary Dose (mrem)	Within 50 Miles Average Individual Dose (mrem)	Within 50 Miles Population Dose (man-rem)
Pressurized Water Reactors				
Indian Point	360	.035	.00005	.77
Yankee Rowe	13	.3	.0003	.41
San Onofre	7,670	2.2	.002	6.3
Conn Yankee	3,250	5.6	.003	11
Ginna	31,800	5.0	.004	4.5
H. B. Robinson	18	.05	.00002	.015
Point Beach	838	.2	.0008	.15
Boiling Water Reactors				
Oyster Creek	516,000	31.	.013	46.
Nine Mile Point	253,000	4.8	.009	8.2
Dresden (1,2,3)	1,330,000	32.	.057	420.
Humboldt Bay	514,000	160.	.54	61.
Big Rock Point	234,000	4.6	.026	3.1
Millstone	276,000	5.5	.0056	15.
Monticello	76,000	4.4	.0036	4.4

[*]Releases are for 1971.

Table XXI. Calculated Annual Radiation Exposures to Unshielded Individuals and Populations in the Vicinity of Nuclear Power Plants Based on Gaseous Emissions for 1969*

Reactor Site	Type	Max. at Boundary D_m (mrem)	Within Circle of 50 Mile Radius P_{50} (thousands)	D_{50} (man-rem)	D_{50} (mrem)
Dresden	BWR	18	5,715	360	.063
Humboldt Bay	BWR	155	101	107	1.06
Nine Mile Point	BWR	.005	533	.012	.000023
Big Rock Point	BWR	3.25	100	3.64	.036
Oyster Creek	BWR	.375	1,158	.606	.00052
San Onofre	PWR	.23	2,696	1.02	.00037
Saxton	PWR	.030	837	.05	.00006
Indian Point	PWR	.055	13,324	1.94	.000145
Connecticut Yankee	PWR	5	2,682	15.56	.0058
R. E. Ginna	PWR	.005	953	.0077	.000008
LaCrosse	PWR	.5	328	.301	.00092
Yankee Rowe	PWR	.11	1,209	.70	.00059
Peach Bottom	HTGR	.19	4,405	1.79	.00041
All (Total or Average)		14.1	33,841	492.6	.0145

*Extracted from Gammertsfelder (1971) (21).

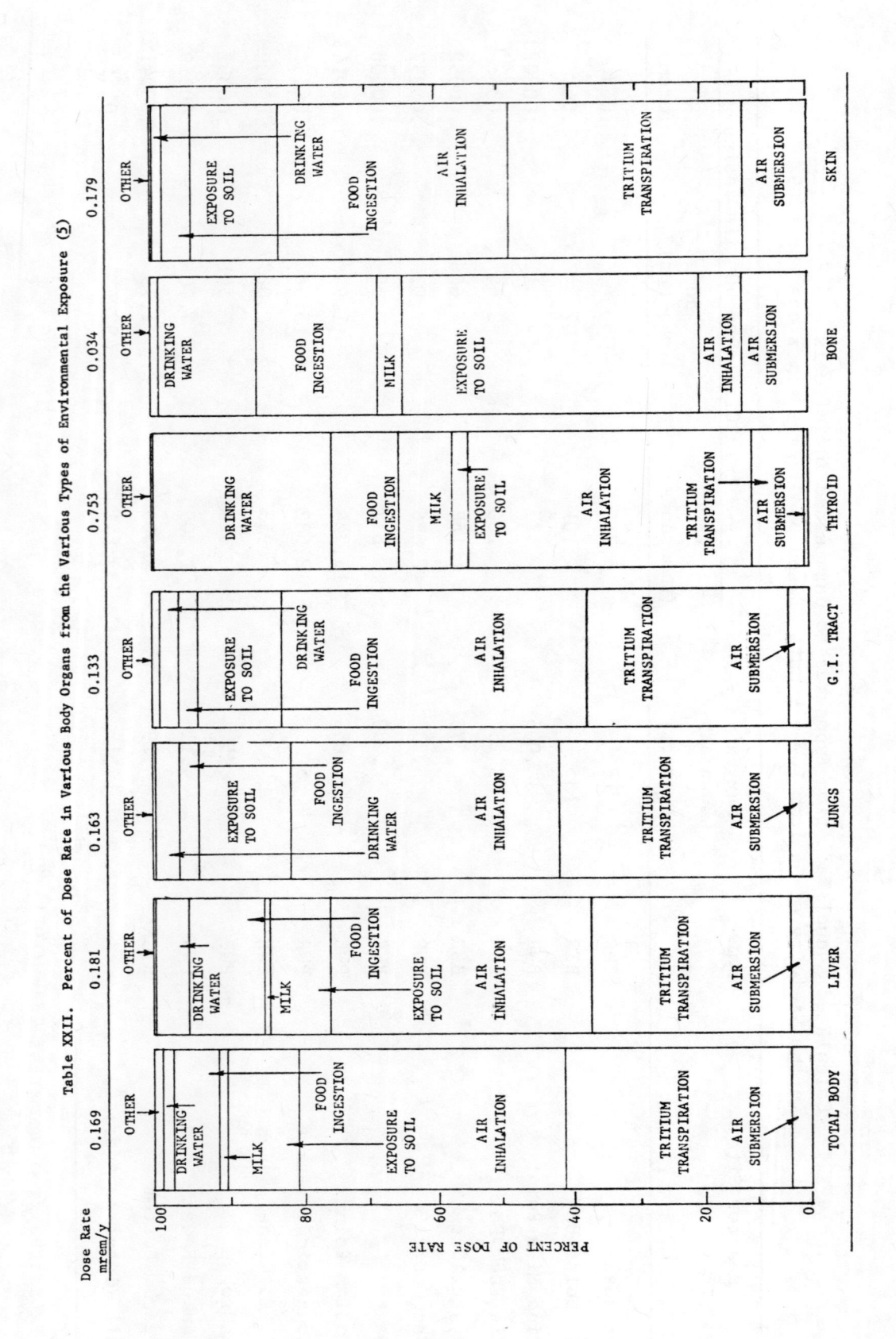

Table XXII. Percent of Dose Rate in Various Body Organs from the Various Types of Environmental Exposure (5)

Dose Rate mrem/y	0.169	0.181	0.163	0.133	0.753	0.034	0.179
Organ	TOTAL BODY	LIVER	LUNGS	G.I. TRACT	THYROID	BONE	SKIN

Table XXIII. Estimated Cumulative Numbers of Potential Health Effects Committed By Operation of the Nuclear Power Industry (22)

Year	Iodine-129 Past-Present	Iodine-129 Future	Tritium Past-Present	Tritium Future	Krypton-85 Past-Present	Krypton-85 Future	Actinides Past-Present	Actinides Future
1970	0	0	0	0	0	0	0	0
1975	0	0	2	0.5	0.3	5	2	26
1980	0	1	11	3	3	26	12	140
1985	1	4	35	6	14	79	96	440
1990	3	9	88	21	42	190	96	1,100
1995	6	17	190	43	110	410	210	2,200
2000	11	32	360	81	230	760	400	3,900
2005	21	53	630	140	460	1,300	720	6,500
2010	34	82	1,000	230	830	2,100	1,200	10,000
2015	53	120	1,600	340	1,400	3,200	1,900	15,000
2020	78	170	2,300	500	2,300	4,600	2,800	21,000
	about 25% fatal		about 64% fatal		about 61% fatal		about 100% fatal	

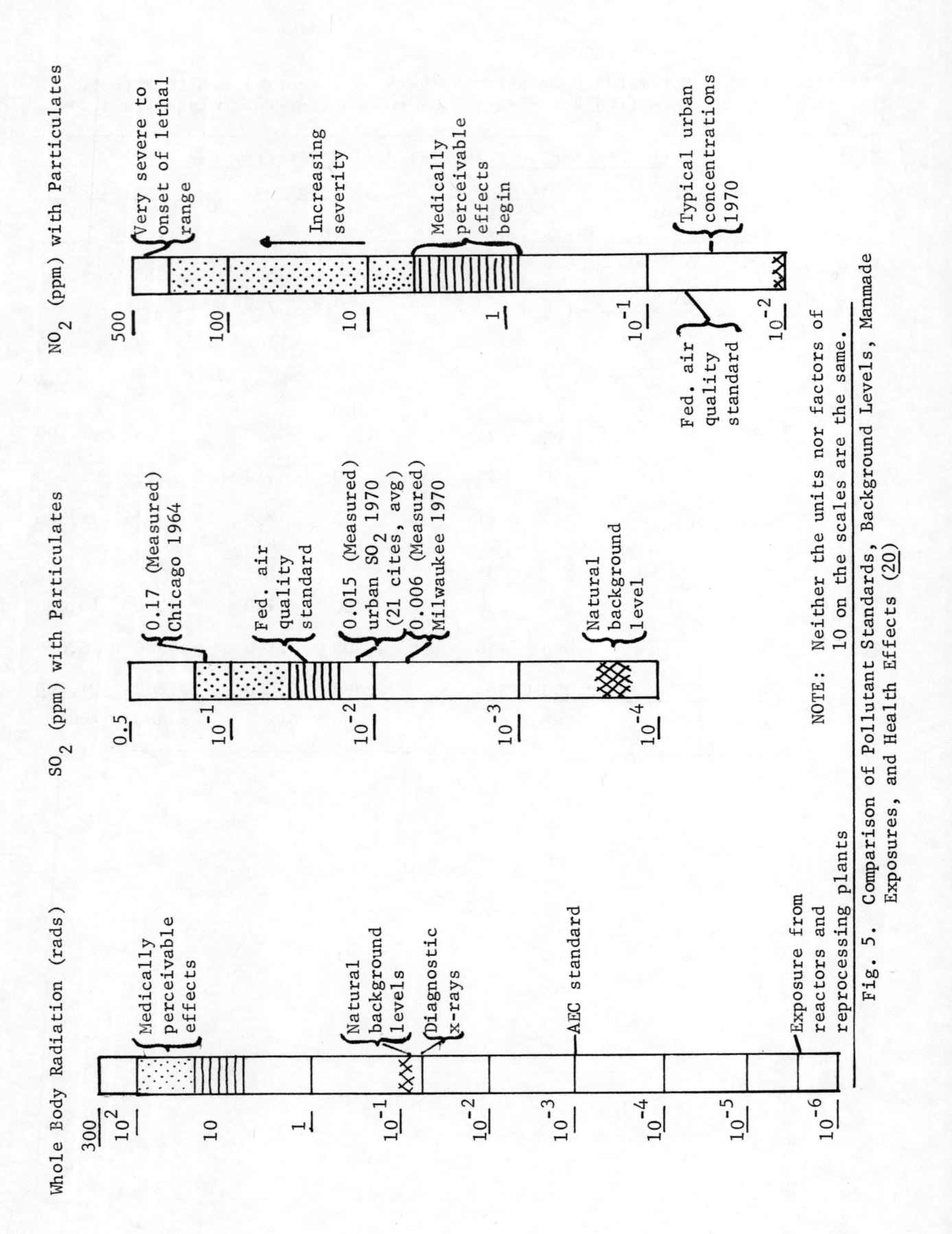

Fig. 5. Comparison of Pollutant Standards, Background Levels, Manmade Exposures, and Health Effects (20)

NOTE: Neither the units nor factors of 10 on the scales are the same.

Table XXIV. Projected Fuel Processing Requirements and High-Level Waste Conditions for the Civilian Nuclear Power Program (24)

Requirements	Half-Life (yrs)	1970	1980	1990	2000
Installed capacity, MW(e)[a]		14,000	153,000	368,000	735,000
Electricity generated, 10^9 kWhr/yr[a]		71	1,000	2,410	4,420
Spent fuel shipping					
Number of casks shipped annually		30	1,200	6,800	9,500
Number of loaded casks in transit		1	14	60	85
Spent-fuel processed, metric tons/yr[a]		94	3,500	13,500	15,000
Volume of high-level liquid waste generated[b,c]					
Annually, 10^6 gal/yr		0.017	0.97	2.69	4.60
Accumulated, 10^6 gal		0.017	4.40	23.8	60.1
Volume of high-level waste, if solidified[b,d]					
Annually, 10^3 ft^3/yr		0.17	9.73	26.9	46.0
Accumulated, 10^3 ft^3		0.17	44.0	238	601
Solidified Waste Shipping[e]					
Number of casks shipped annually		0	3	172	477
Number of loaded casks in transit[f]		0	1	4	10
Significant radioisotopes in waste[g,h]					
Total accumulated weight, metric tons		1.8	450	2,400	6,200
Total accumulated beta activity (MCi)		210	18,900	85,000	209,000
Total heat-generation rate (MW)		0.9	80	340	810
^{90}Sr generated annually (MCi)	27.7	4.0	230	560	770
^{90}Sr accumulated (MCi)		4.0	960	4,600	10,000
^{137}Cs generated annually (MCi)	30	5.6	320	880	1,500
^{137}Cs accumulated (MCi)		5.6	1,300	6,500	15,600
^{129}I generated annually (Ci)	1.7 x 10^7	2.0	110	440	670
^{129}I accumulated (Ci)		2.0	480	2,700	7,600
^{85}Kr generated annually (MCi)	10.76	0.6	33	90	150
^{85}Kr accumulated (MCi)		0.6	124	570	1,200
^{3}H generated annually (MCi)	12.3	0.04	2.1	6.2	12
^{3}H accumulated (MCi)		0.04	7.3	36	90
^{238}Pu generated annually (MCi)	86.4	0.0007	0.041	0.2	0.6
^{238}Pu accumulated (MCi)		0.0007	1.20	8.3	31

140

Table XXIV. Concluded

Requirements	Half-Life (yrs)	1970	1980	1990	2000
Significant radioisotopes in waste (cont.)					
^{239}Pu generated annually (MCi)		0.00009	0.005	0.05	0.2
^{239}Pu accumulated (MCi)	24,390	0.00009	0.02	0.24	1.3
^{240}Pu generated annually (MCi)		0.00012	0.007	0.06	0.21
^{240}Pu accumulated (MCi)	6,580	0.00012	0.04	0.4	1.9
^{241}Am generated annually (MCi)		0.009	0.5	4.4	15
^{241}Am accumulated (MCi)	458	0.009	2.3	23	120
^{243}Am generated annually (MCi)		0.00021	0.01	0.1	0.5
^{243}Am accumulated (MCi)	7950	0.00021	0.23	1.5	5.2
^{244}Cm generated annually (MCi)		0.13	7.4	18	23
^{244}Cm accumulated (MCi)	17.6	0.13	30	140	260
Volume of cladding hulls generated[i]					
Annually, 10^3 ft^3		0.3	8	40	90
Accumulated, 10^3 ft^3		0.3	40	320	1,030

[a] Data from Phase 3, Case 42, Systems Analysis Task Force (April 11, 1968).
[b] Based on an average fuel exposure of 33,000 MWd/ton and a delay of 2 yr between power generation and fuel processing.
[c] Assumes wastes concentrated to 100 gal/10,000 MWd (thermal).
[d] Assumes 1 ft^3 of solidified waste per 10,000 MWd (thermal).
[e] Assumes 10-year-old wastes, shipped in 36 6-in.-diameter cylinders per shipment cask.
[f] One-way transit time is 7 days.
[g] Assumes light-water reactor (LWR) fuel continuously irradiated at 30 MW/ton to 33,000 MWd/ton and fuel processing 90 days after discharge from reactor; liquid metal fast breeder reactor (LMFBR) core continuously irradiated to 80,000 MWd/ton at 148 MW/ton, axial blanket to 2,500 MWd/ton at 4.6 MW/ton, radial blanket to 8,100 MW/ton, and fuel processing 30 days after discharge.
[h] Assumed 0.5 percent of Pu in spent fuel is lost to waste.
[i] Based on 2.1 ft^3 of cladding hulls per ton of LWR fuel processed and 8.7 ft^3 of cladding hardware/ton of LMFBR mixed core and blankets processed.

Figure 6 shows the build up of ^{85}Kr from the U.S. nuclear power industry to the year 2060. The nuclear industry release of ^{85}Kr in the U.S. is expected to result in doses to the skin of about 10 mrem/y and doses to the total body (from bremsstrahlung) of about 0.1 mrem/y by the year 2000. These figures are expected to double due to the release by other countries. It seems advisable that long before then the ^{85}Kr released from the fuel elements at the reprocessing plant should be recovered and stored because otherwise this released ^{85}Kr might be expected to result in about 10,000 cancer deaths per year on a worldwide basis by the year 2000. Table XXV indicates the dose to various body organs from an infinite cloud of ^{85}Kr per μCi/m^3 of ^{85}Kr.

SAFETY OF NUCLEAR POWER PLANT OPERATIONS IN THE FUTURE

Figure 7 indicates how the ^{131}I release may change from month to month and as indicated from past experience this radionuclide can be expected to account for a major fraction of environmental exposure from the operation of nuclear power plants. In this case the level at Oyster Creek plant varied by two orders of magnitude over a two month period. Table XXVI indicates the wide variation in concentration of ^{131}I in vegetation and in its transfer to milk as observed at the Monticello plant and thus the need of collecting frequent samples (of air, grass, milk, thyroids from slaughter houses, etc.) in order to assure adequate environmental vigilance in satisfying the public that the internal dose from ^{131}I is maintained at an acceptable level. These data emphasize the necessity of maintaining continuous air monitoring in the vicinity of a nuclear power plant and providing immediate and appropriate health physics response in case high levels of radioactive pollution are indicated at any time.

Finally, I would like to remind us that the occupational exposure in a nuclear power plant as well as environmental exposure of members of the public must be kept to a minimum. In this case we are concerned with minimizing exposure from all in-plant sources and operations such as fuel handling, steam generator inspection and repair, inservice inspections, liquid waste treatment, cooling pond operation, filter changes, cleanup of spills, etc. Table XXVII indicates the distribution of occupational exposure in the PWR and BWR power plants. Also, in this case the facility must be designed to provide sufficient shielding for personnel protection from the short lived radionuclides such as excited ^{16}O* (10^{-11} sec) from the radioactive decay of ^{16}N (7.3 sec) which emits very penetrating gamma radiation of 5 to 7.1 MeV. I believe the nuclear power industry is somewhat negligent in minimizing occupational exposure from the in-plant sources of radiation that are summarized in Table XXVII. Sometimes I have heard officials of nuclear power operations claim that because of greater attention to reducing environmental exposure of members of the public, the in-plant exposures of necessity have increased. I believe there is no acceptable reason why occupational exposure (in man-rem) should be greater than 10% of the population dose. Both occupational workers and members of the public should be made to understand and appreciate the consequences of excessive radiation exposure and see that appropriate measures are taken to keep these exposures ALAP or ALARA. Power plant employees as well as the public must not fear ionizing radiation but they must learn to respect it and avoid all unnecessary exposure.

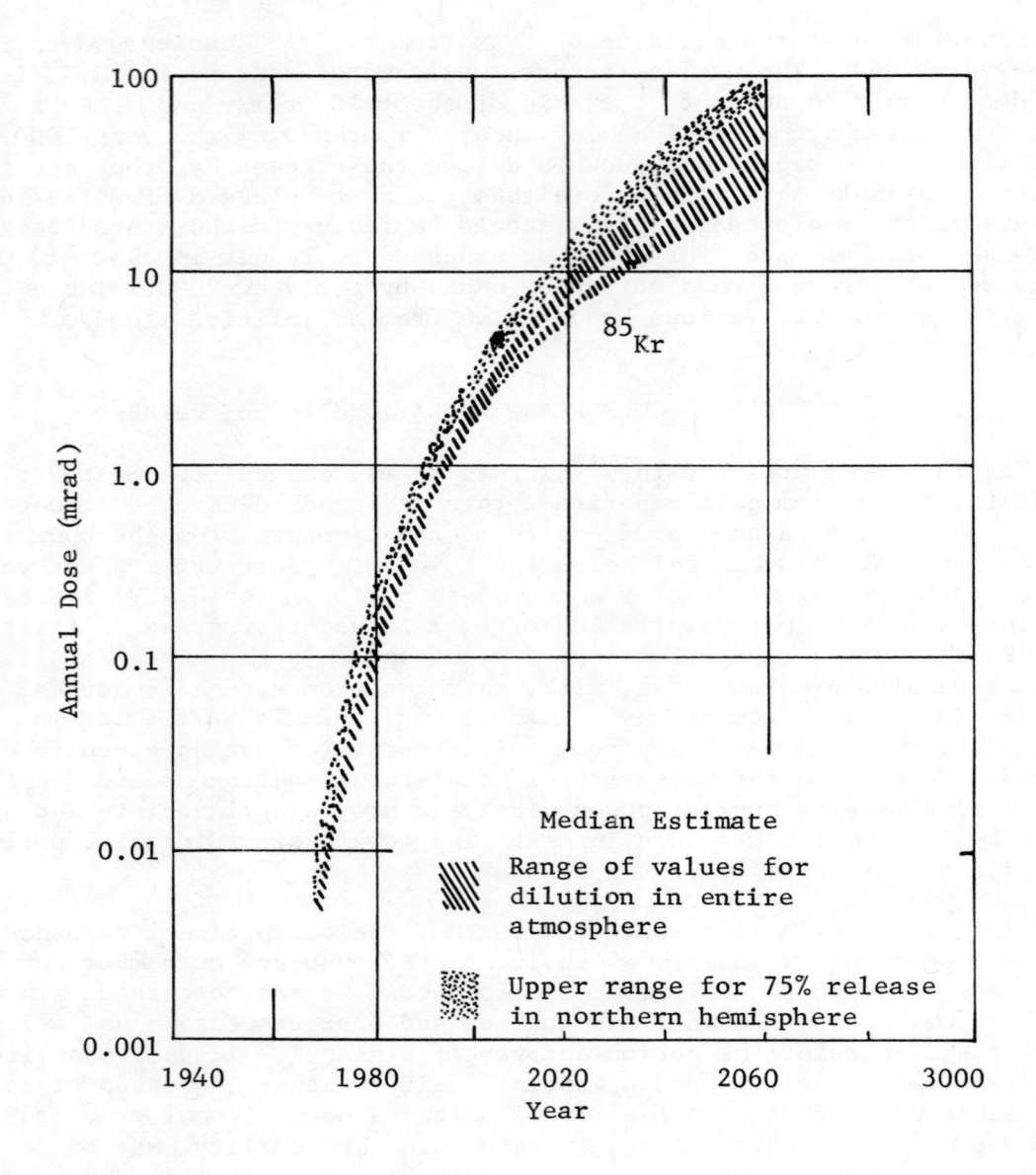

Fig. 6. Estimated Annual Dose from Krypton-85 -- 1970-2060
(14, 25, 26)

Table XXV. Total Dose Rate to Organs of the Body from an Infinite Cloud of ^{85}Kr (29) (rads/yr)/(μCi/m^3)

Organ	Dose Rate
Skin	1.8
Adipose tissue	1.5×10^{-2}
Lungs	3.2×10^{-2}
Red bone marrow	1.8×10^{-2}
Skeleton	1.9×10^{-2}
Ovaries	6.2×10^{-3}
Testes	1.6×10^{-2}
Lenses of the eyes	1.8×10^{-2}

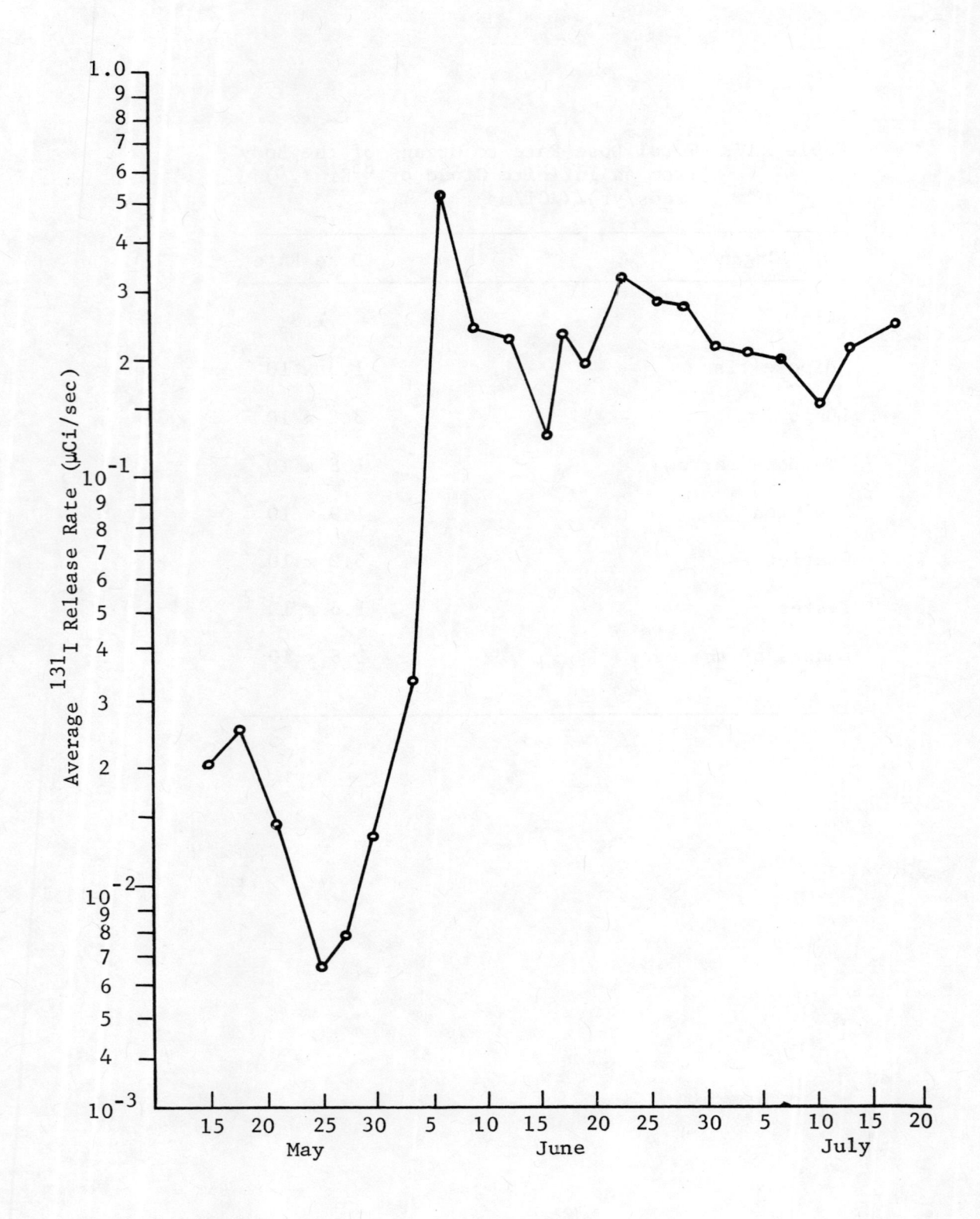

Fig. 7. Oyster Creek Plant Release Rates in 1973 (27)

Table XXVI. Milk-to-Vegetation Transfer Parameters, Monticello Pasture (28)

Vegetation Data			Milk Data		Transfer Parameters	
1973	C_v (pCi/m^2)	K_v (pCi/kg)[a]	1973 Date	C_m (pCi/liter)	C_m/C_v (m^2/liter)	100 C_m/I (%/liter)
6-17	<3	<25	6-19	0.21 ± .05	>0.07	>0.08
6-17	<3	<35	6-21	0.17 ± .05	>0.05	>0.05
6-25	<3	<19	6-26	0.24 ± .05	>0.08	>0.13
6-25	<3	<19	6-27	0.26 ± .05	>0.08	>0.14
6-27	<3	<21	6-28	0.26 ± .1	>0.08	>0.12
6-29	<4	<34	6-30	0.31 ± .05	>0.07	>0.09
7-1	<5	<38	7-2	0.32 ± .05	>0.06	>0.08
7-1	<5	<34	7-3	0.23 ± .05	>0.04	>0.07
7-5	<16	<125	7-7	22.6 ± .3	>1.4	>1.8
7-6	111 ± 15	650 ± 88	7-8	58.2 ± .5	0.52 ± .07	0.90 ± .12
7-7	82 ± 6	530 ± 40	7-9	52.8 ± .5	0.64 ± .05	1.00 ± .07
7-8	51 ± 3	750 ± 45	7-10	91 ± .5	1.78 ± .10	1.21 ± .07
7-9	40 ± 3	550 ± 38	7-11	79.7 ± .5	1.99 ± .15	1.45 ± .10
7-10	60 ± 2	540 ± 20	7-12	73.0 ± .4	1.22 ± .04	1.36 ± .05
7-11	161 ± 6	990 ± 35	7-13	71.0 ± .4	0.44 ± .02	0.72 ± .02
7-12	85 ± 6	400 ± 25	7-14	77.0 ± .4	0.91 ± .06	1.92 ± .12
7-13	39 ± 3	470 ± 29	7-15	64.2 ± .4	1.65 ± .13	1.37 ± .08
7-14	59 ± 5	400 ± 30	7-16	54.1 ± .3	0.92 ± .08	1.36 ± .10
7-15	32 ± 3	320 ± 50	7-17	45.1 ± .3	1.41 ± .13	1.40 ± .13
7-16	28 ± 3	430 ± 30	7-18	26.2 ± .2	0.94 ± .10	0.61 ± .05
	Mean of 11 Positive Values				1.13	1.21
	Variation about the Mean (1σ)				0.52	0.38

[a] Dry weight of vegetation; determined by drying to constant weight at 110°C.

Table XXVII. Average Fraction[1] (in Percent) of Annual Plant Exposure for 18 Activity Categories (30)

Activity Category	PWR Av	PWR High	PWR Low	PWR (N)[1]	BWR Av	BWR High	BWR Low	BWR (N)[1]
Liquid waste treatment	4.1	17	0.8	(8)	5.6	11	1.9	(4)
Solid waste handling	2.5	6.8	0.4	(10)	3.3	9.0	0.7	(6)
Gaseous waste systems	0.4	–	–	(1)	2.7	5.5	0.9	(3)
Head removal and installation	6.5	12	1.8	(8)	1.4	3.2	0.4	(12)
Fuel handling	3.6	15	0.2	(10)	5.5	17	1.0	(12)
Instrumentation work, including calibration	1.3	2.8	0.3	(10)	3.0	11	0.6	(8)
Inservice inspection	5.6	8.3	0.6	(4)	4.9	9.6	1.4	(11)
Control rod drive work	minimal				3.2	9.4	0.5	(12)
Major equipment failures	not included				not included			
Recirculation pumps, including cleanup systems	not applicable				7.8	29	0.6	(13)
Steam generator inspection and repair	27	88	5.1	(14)	not applicable			
Reactor coolant pumps	2.8	5.5	0.6	(7)	not applicable			
Main coolant loops	5.1	11	0.8	(7)	not applicable			
Charging pumps	1.4	4.2	0.2	(8)	not applicable			
Valves	2.6	4.1	0.7	(5)	5.7	16	0.5	(8)
Turbine and auxiliary equip.	minimal				2.7	10	0.6	(7)
Fuel pool including cleanup system	0.3	0.7	0.1	(6)	0.5	1.3	0.1	(6)
Condensate demineralizers	not applicable				1.2	3.9	0.1	(4)
Total	63.2%				47.5%			

[1] Number of annual fractions used to compute average.

One of the most serious problems we face in the nuclear energy industry is the potential population exposure from plutonium and the other actinide elements. This situation will become even more serious if and when mixed oxide (MOX) fuels, i.e. UO_2 + PuO_2, are used in light water cooled (LWR) reactors and when the LMFBR's come into operation. In Table XXVIII data are given indicating the inventory of the actinides and a few other elements in the BWR using once through fuel in contrast with the inventory of the BWR using MOX fuel as presently contemplated. Here it is noted that there is a buildup of radionuclides which are many times more hazardous than ^{239}Pu and which unlike ^{239}Pu emit high energy gamma radiation and neutrons. For example, the 2.04×10^4 g of ^{244}Cm in the MOX fuel corresponds to a neutron source of 2.2 Ci and I have estimated ([31]) the relative hazards of ^{239}Pu, ^{238}Pu, ^{241}Am, ^{242m}Am, ^{243}Am, ^{243}Cm and ^{244}Cm are 1, 150, 16, 50, 1, 45 and 32, respectively.

Tamplin and Cochran have emphasized the potential seriousness of the "hot particle problem," i.e. small particles of ^{239}Pu which deliver tens of rems/day when they are deposited in the lungs, and have called for a reduction of the present 0.015 µCi permissible occupational lung burden by a factor of 10^5. If this should have to be done, it probably would be the end of the BWR, PWR and LMFBR but probably would not affect the HWBR or the LWBR. I do not believe we have enough biological information to evaluate the hot particle problem or to set a satisfactory permissible body burden for ^{239}Pu and the other actinide elements based on the hot particle at the present time. I have shown ([32]), however, that there are experimental data which strongly suggest the present maximum permissible bone burden for occupational workers should be reduced from its present value of 0.04 µCi by a factor of about 240. I believe nuclear power could live with this tightening of radiation standards. In any case, I believe it is clear that the release of plutonium and the transplutonium radionuclides from a nuclear power reactor must be kept very low and probably less than 1 mCi/year. Figure 8 indicates that in New York the fallout of ^{239}Pu from past weapons testing has resulted in a cumulative deposit of about 2.7 mCi/km^2. Thus any ^{239}Pu discharged from a nuclear power operation must account for any addition to the ^{239}Pu already in the environment.

The irony of this situation is that we probably know much more about the effects of ^{239}Pu and other actinide elements on man than we know about the effects of many of the chemical pollutants produced by a fossil fueled plant. However, the more we know the more we realize we need to know. I believe if the nuclear power plants strengthen their health physics programs as provided in the Radiation Guides of NRC and as I have suggested here, it will be possible to keep the environmental discharges of ^{239}Pu and all the actinide radionuclides to a very low level in the years ahead, and barring serious accidents with reactors or fuel reprocessing plants, we can keep this environmental contamination to an acceptable level. I hope before the end of this century we will be able to evaluate adequately the hot particle problem and set more satisfactory permissible levels for occupational and non-occupational exposure to the actinide radionuclides. I believe when all the necessary information is available, it will be evident that the environmental risks from a properly operated nuclear energy industry can be less than those from a well run fossil fueled industry.

Table XXVIII. Comparison of BWR With MOX Fuel - 172 Spent Fuel Assemblies (Grams) (1)

Nuclides	BWR at Discharge	120 Day Decay	1.15 SGR at Discharge	120 Day Decay*
^{4}He	9.05×10^{0}	9.9×10^{0}	1.23×10^{2}	1.40×10^{2}
^{90}Zr	4.12×10^{6}	NC	4.12×10^{6}	NC
^{91}Zr	8.92×10^{5}	NC	8.92×10^{5}	NC
^{92}Zr	1.38×10^{6}	NC	1.38×10^{6}	NC
^{93}Zr	1.03×10^{3}	NC	1.03×10^{3}	NC
^{94}Zr	1.36×10^{6}	NC	1.39×10^{6}	NC
^{95}Zr	2.53×10^{1}	NC	2.53×10^{1}	7.03×10
^{96}Zr	2.24×10^{5}	NC	2.24×10^{5}	NC
^{230}Th	3.8×10^{-1}	4.15×10^{-1}	2.36×10^{-1}	NC
^{232}Th			6.56×10^{-3}	NC
^{231}Pa	1.74×10^{-1}	NC	1.07×10^{-1}	NC
^{232}U	5.35×10^{-2}	NC	3.42×10^{-2}	3.52×10^{-2}
^{233}U	9.2×10^{-1}	NC	5.85×10^{-1}	NC
^{234}U	3.75×10^{4}	NC	2.34×10^{4}	NC
^{235}U	2.55×10^{5}	NC	1.93×10^{5}	NC
^{236}U	1.32×10^{5}	NC	8.94×10^{4}	NC
^{237}U			1.36×10^{2}	3.29×10^{-3}
^{238}U	3.06×10^{7}	NC	3.03×10^{7}	NC
^{237}Np	1.33×10^{4}	1.35×10^{4}	9.54×10^{3}	9.68×10^{3}
^{239}Np			1.58×10^{3}	3.03×10^{-2}
^{236}Pu			1.37×10^{-2}	1.27×10^{-2}
^{238}Pu	4.46×10^{3}	4.58×10^{3}	1.94×10^{4}	2.00×10^{4}
^{239}Pu	1.67×10^{5}	NC	2.50×10^{5}	NC
^{240}Pu	6.60×10^{4}	NC	1.49×10^{5}	NC
^{241}Pu	2.94×10^{4}	2.90×10^{4}	9.11×10^{4}	8.97×10^{4}
^{242}Pu	8.95×10^{3}	NC	6.42×10^{4}	NC
^{241}Am	1.05×10^{3}	1.50×10^{3}	6.38×10^{3}	7.78×10^{3}
^{242m}Am	3.80×10^{1}	NC	4.01×10^{2}	NC
^{242}Am			9.43	4.80×10^{-3}
^{243}Am	2.12×10^{3}	NC	3.66×10^{4}	NC

Table XXVIII. Continued

Nuclides	BWR at Discharge	120 Day Decay	1.15 SGR at Discharge	120 Day Decay[*]
^{242}Cm	2.89×10^2	1.74×10^2	1.75×10^3	1.06×10^{-3}
^{243}Cm	2.14×10^0	NC	1.08×10	NC
^{244}Cm	5.91×10^2	NC	2.04×10^4	NC
^{245}Cm	3.50×10	NC	2.23×10^3	NC
^{246}Cm	3.50×10^0	NC	2.11×10^2	NC
^{247}Cm	4.0×10^{-2}	NC	3.51	NC
^{248}Cm			2.72×10^{-1}	NC

[*] For comparative decay. Absolute values are sensitive to the number of MO rods per bundle and amount of plutonium per rod.
NC - no change
MOX refers to Mixed Oxide Fuel ($UO_2 PuO_2$)
1.15 SGR refers to 115% of a Self Generating Reactor

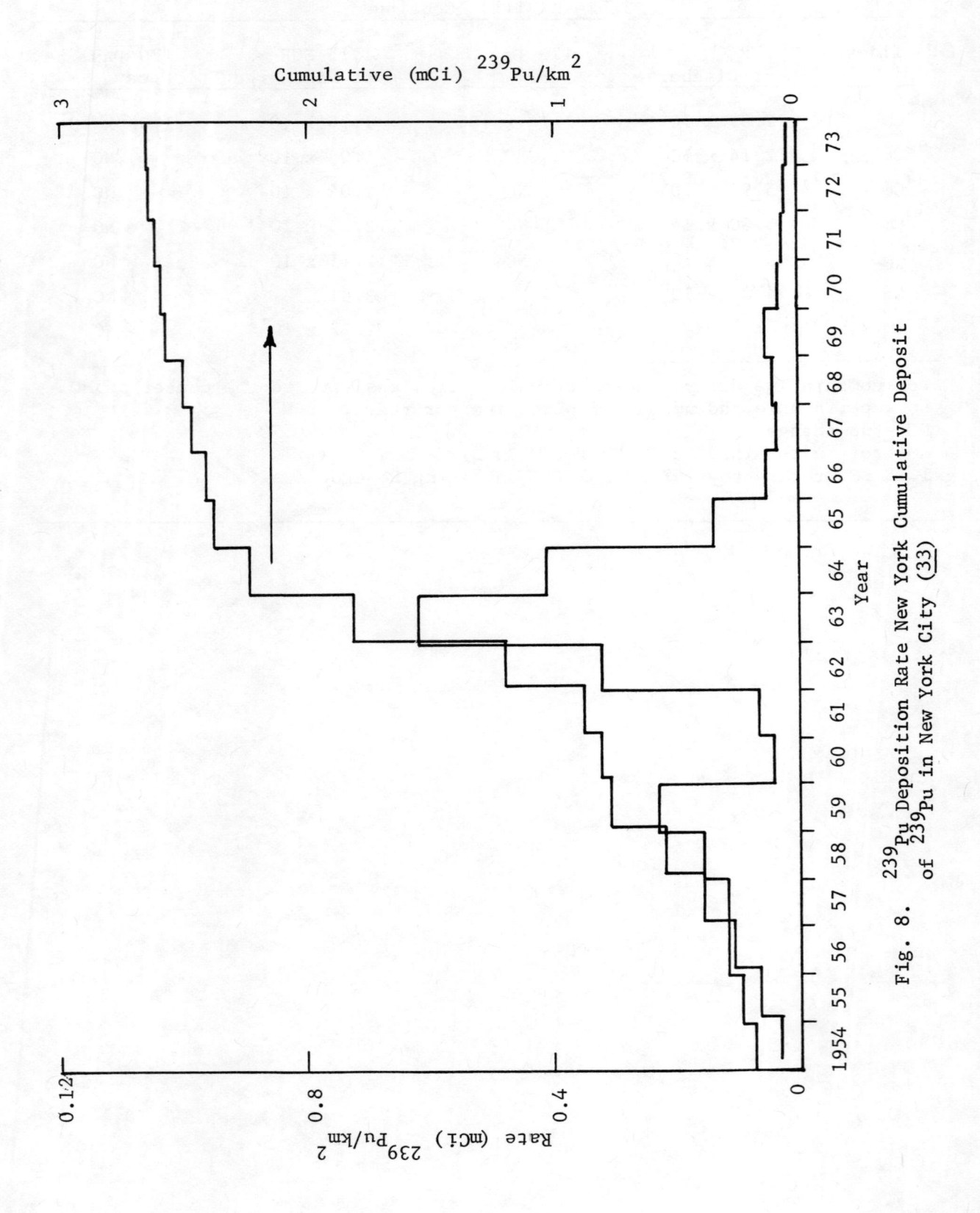

Fig. 8. ^{239}Pu Deposition Rate New York Cumulative Deposit of ^{239}Pu in New York City (33)

REFERENCES

1. "Generic Environmental Statement--Mixed Oxide Fuel," WASH-1327, Vol. 3 (August 1974).

2. International Commission on Radiological Protection, "A Report by the Internal Dose Committee II," ICRP Publication 2, Pergamon Press, New York (1959).

3. International Commission on Radiological Protection, "A Report by the Internal Dose Committee II, 1962," Supplement of Committee II on Permissible Dose for Internal Radiation, ICRP Publication 6, Pergamon Press, New York (1964).

4. H. CEMBER, Introduction to Health Physics, Pergamon Press, New York (1969).

5. "The Potential Radiological Implication of Nuclear Facilities in the Upper Mississippi River Basin in the Year 2000 (The Year 2000 Study," USAEC Rept., WASH-1209 (January 1973).

6. Report of the United Nations Scientific Committee on the Effects of Atomic Radiation, Report No. 25 (A-8-725), Vol. I, "Ionizing Radiation Levels and Effects," UNSCEAR Report 1972.

7. "Nuclear Fuel Cycle," A Report by the Fuel Cycle Task Force, USERDA-ERDA-33 (March 1975).

8. "Effects on Populations of Exposure to Low Levels of Ionizing Radiation," Report of Committee on Biological Effects of Ionizing Radiation - BEIR Report, NAS-NRC (1972).

9. J. D. MARTIN, E. D. HARWARD and D. T. OAKLEY, "Comparison of Radioactivity from Fossil Fuel and Nuclear Power Plants," Appendix 14 is Part I of hearings before the Joint Committee on Atomic Energy on "Environmental Effects of Producing Electric Power," U.S. Congress, Washington, D. C. (November 1969).

10. K. Z. MORGAN, "Adequacy of Present Radiation Standards," The Environment and Ecological Forum, TID-25857 (1972).

11. R. S. BOOTH, "A Compendium of Radionuclides Found in Liquid Effluents of Nuclear Power Stations, ORNL-TM-3801 (March 1975).

12. "Report on Release of Radioactivity in Effluents and Solid Waste from Nuclear Power Plants for 1972," U.S. Atomic Energy Commission (August 1973).

13. H. T. PETERSON, JR., J. E. MARTIN, C. L. WEAVER and E. D. HARWARD, "Environmental Tritium Contamination from Increasing Utilization of Nuclear Energy Sources," USPHS, HEW, BRH-SM-117/78 (March 1969).

14. K. E. COWSER, "^{85}Kr and ^{3}H in an Expanding World Nuclear Power Industry," Health Physics Annual Progress Report, ORNL-4007 (October 1966).

15. E. C. TSIVOGLOU, E. D. HARWARD and W. M. INGRAM, "Stream Surveys for Radioactive Waste Control," 2nd Nuclear Engineering and Science Conf., Philadelphia, Pa., Paper 57-NESC (March 21, 1957).

16. L. DRESNER, Translation of T. JAEZER'S, Principles of Radiation Protection Engineering, McGraw-Hill Book Co., Inc., New York (1965).

17. R. L. BLANCHARD et al., "Radiological Surveillance Studies at a BWR Nuclear Power Station--Estimated Dose Rates," Proc. Health Physics Soc. Fifth Annual Midyear Symp., Idaho Falls, Idaho (1970).

18. M. J. SHUMKLARSKY, "Environmental Radioactivity in Illinois, 1970," Radiation Data and Reports, 13, 589 (November 1972).

19. "Report on Releases of Radioactivity from Power Reactors in Effluents During 1971," USAEC Rept., WASH-1198 (1971).

20. "The Safety of Nuclear Power Reactors (Light Water-Cooled) and Related Facilities," USAEC Rept., WASH-1250 (July 1973).

21. C. C. GAMERTSFELDER, "Regulatory Experience and Projections for Future Design Criteria," Paper given at Southern Conference on Environmental Radiation Protection at Nuclear Power Plants, St. Petersburg, Florida (1971).

22. "Assessment of the Possible Environmental Dose Commitment Resulting from Release of Long-lived Radionuclides Produced by Operation of the Nuclear Energy Industry for the Next Fifty Years," Environmental Protection Agency (June 1973).

23. "Comparative Risk Cost-Benefit Study of Alternate Sources of Electrical Energy," USAEC Rept., WASH-1224 (1973).

24. "Radionuclides in Foods," prepared by Committee on Food Protection - Food and Nutrition Board, National Academy of Sciences, Washington, D.C. (1973).

25. J. R. COLEMAN and R. LIBERANCE, "Nuclear Power Production and Estimated Kr-85 Level," Radiological Health Data and Reports, USPHS, p. 615 (November 1966).

26. W. P. KIRK, "Krypton-85 -- A Review of the Literature and an Analysis of Radiation Hazards," Eastern Env. Rad. Lab, EPA (January 1972).

27. B. H. WEISS, P. G. VOILLEQUE, J. H. KELLER, B. KAHN, H. L. KRIEGER, A. MARTIN and C. R. PHILLIPS, "Detailed Measurement of ^{131}I in Air, Vegetation and Milk," Report of EPA and former AEC, NUREG-75/021 (March 1975).

28. Interim Report on "Results of Measurements of Iodine-131 in Air, Vegetation and Milk at Three Operating Reactor Sites," Directorate of Regulatory Operations, USAEC (October 1973).

29. W. S. SNYDER, L. T. DILLMAN, M. R. FORD and J. W. POSTON, "Populations of the Absorbed Dose to a Man Immersed in an Infinite Cloud of ^{85}Kr," Health Physics Division, Oak Ridge National Laboratory, Oak Ridge, Tenn. (1973).

30. C. A. PELLETIER, L. SIMMONS, M. BARBIER and J. H. KELLER, "Compilation of Analysis on Data on Occupational Radiation Exposure Experienced at Operating Nuclear Power Plants," National Environmental Studies Project-- Atomic Industrial Forum, New York, N. Y. (1974).

31. K. Z. MORGAN, W. S. SNYDER and M. R. FORD, "Relative Hazard of the Various Radioactive Materials," Health Physics, 10, 151 (1964).

32. K. Z. MORGAN, "Suggested Reduction of Permissible Exposure to Plutonium and Other Transuranium Elements," Amer. Ind. Hygiene Assn. J, 36(8), 567 (1975).

33. M. E. WRENN, "Environmental Levels of Plutonium and the Transplutonium Elements," AEC presentation at the EPA Plutonium Standards Hearings, Washington, D. C. (December 10-11, 1974).

WAYS OF REDUCING RADIATION EXPOSURE IN A FUTURE
NUCLEAR POWER ECONOMY

Karl Z. Morgan
School of Nuclear Engineering
Georgia Institute of Technology
Atlanta, Georgia 30332

INTRODUCTION

I believe exposures in connection with the early atomic energy programs and within the National Laboratories of the AEC (now ERDA) have been kept as low as practicable (ALAP) from their beginning primarily because of the high standards given us by the early directors of the Metallurgical Project (Drs. A. H. Compton, Director, and R. S. Stone, Associate Director) at the University of Chicago in 1942 and 1943 where Health Physics had its beginning. They emphasized to us that this new profession was being established by them in order not to permit a repetition of the sad experience of the radium dial painting industry where many young women had suffered and died as a consequence of unnecessary exposure to radium and its daughter product. Unfortunately, little if any of this admonition or guiding philosophy of Health Physics brushed off onto members of the medical profession (from whence more than 90% of unnecessary exposure to man-made sources of ionizing radiation in the U.S. derives) and I consider it a sad commentary that the nuclear power industry in many respects has failed to follow the good example set by these early fathers and the National Laboratories. I consider myself a strong supporter of the nuclear power industry but only when and if appropriate attention is given to radiation protection of the occupational workers and of members of the public.

WHY REDUCE RADIATION EXPOSURE IN A FUTURE NUCLEAR POWER ECONOMY?

Often we are asked why go to the expense of keeping radiation exposures in the nuclear energy industry ALAP and why attempt to further reduce these exposures when at the present time and into the foreseeable future they will be less than one percent of population exposure in the U.S. either from medical sources (diagnostic plus therapeutic) or from natural background radiations (terrestrial plus cosmic). This comparison with medical exposure is a proper and meaningful one because both medical x-rays and electrical power are of great benefit to mankind and because they should be reduced. For example medical exposure could be reduced to less than 10% of its present value while at the same time greatly increasing and enhancing the medical x-ray information and its benefits. I consider, however, that proponents or advocates of the nuclear power industry make a serious mistake when they compare exposures from the operation of a nuclear power facility with the dose received by the population from natural background radiation. This is because 1) natural background radiation is unavoidable for practical reasons, 2) natural background radiation is not beneficial and 3) natural background radiation on the best estimates (the linear hypothesis) accounts for a considerable toll of suffering and death (approximately 20,000 genetic and somatic deaths per year in the U.S.).

The exposure in a future energy economy must be reduced and kept ALAP or ALARA (as low as reasonably achievable is the most recent terminology used by the Nuclear Regulatory Commission, NRC) because at present there is a preponderance of evidence that there is no threshold dose of ionizing radiation so low that it is safe or such that the risk of damage (even serious damage such as leukemia) is zero. In the early period (before 1945) it was commonly believed that at very low doses and dose rates no radiation damage would manifest itself in the life of the individual. As a consequence we used the nomenclature "tolerance dose" and even expressed it in units of "threshold erythema." Today experimental evidence strongly suggests that the probability of most types of chronic radiation damage (with the exception of cataractogenesis) increases monotonically with the accumulated dose. For the past two decades the standards setting bodies (NCRP, ICRP, FRC, AEC, etc.) have been saying that in setting levels of maximum permissible exposure for the occupational worker or dose limits for members of the public we must make the "conservative assumption" of a linear relationship between dose and effect and some members of these organizations went out of their way (out on a limb) to emphasize that the linear hypothesis is very conservative (and perhaps too restrictive). The more recent report of the BEIR (1) Committee (Committee on the Biological Effects of Ionizing Radiations of the National Academy of Sciences) took a neutral position and indicated the linear hypothesis might be a conservative assumption but on the other hand it might be non-conservative. More recently some information has been accumulated (2) which suggests that at low doses and dose rates the linear hypothesis may in fact be non-conservative. This increased risk seems most strongly suggested for cancer production by high LET radiations.

In this struggle to assess the risk from exposure to ionizing radiation we find a wide range of opinions and it sometimes is difficult to determine whether these opinions are objective and without bias. Thus we have, for example, two orders of magnitude variation in estimates of the value of a man-rem saved, i.e. from "a few pounds sterling" (3) (~ $10) to $1,000. (4) Also, we have a wide range of opinion regarding the risks of ^{239}Pu. For example, Tamplin and Cochran (5) suggest the present maximum permissible level should be reduced by a factor of 10^5 while Cohen's (6) publications suggest the present level may be too conservative. I have suggested (2) the present level (based on bone as the critical tissue) is too high by a factor of about 200 and that since there is contradictory evidence regarding the increased risk from Pu particles in the lung, no satisfactory level can be set for particulates in the lung until we have more experimental data. Until that time, I believe the nuclear power industry should take extreme precaution to keep the environmental release of Pu and all the transplutonium radionuclides to as near zero as practicable (e.g. less than 1 mCi/y from a 1000 MWe plant).

During the past 10 to 15 years new data have indicated the risk of radiation induced cancer in humans is 10 or more times what we considered it to be in 1960 and there is no evidence of the existence of a safe threshold dose.

For example in the early period the only chronic damage observed or predicted among the survivors of the atomic bombings of Hiroshima and Nagasaki was leukemia but now there appears to be an increased incidence of many other types of malignancies. Also, a decade ago Hempleman (7) observed that radiation induced thyroid carcinoma seemed to relate linearly to the radiation dose down to doses as low as 20 rem and in the three human populations he studied he found a risk of 2.5 cancers/10^6 children/rad/year for a mean dose of 229 rad. Recently Silverman and Hoffman (8) have evaluated a number of studies pointing out that the studies of Modan et al. (9) of 11,000 persons in Israel indicate a much higher risk of radiation induced thyroid carcinoma, i.e. 6.1 cancers/10^6 children/rad/year for a mean dose of only 6.5 rad. Thus the risk of thyroid cancer per rad seems to be greater at the lower doses than at higher doses and data of Stewart and Kneale (10) indicated earlier this same trend in the case of childhood leukemia, i.e. a greater risk of leukemia per rad at low in utero exposures (0.25 to 0.5 rad) than at higher doses to the fetus. With this increasing awareness regarding chronic radiation risks (i.e. the risks of low level exposure to ionizing radiation are much greater than was considered to be the case in the past) I consider it ironical that the ICRP at the present time is not giving serious consideration to decreasing but rather to increasing the radiation exposure limits.

Finally, before concluding this discussion on why reduce radiation exposure and summarizing ways by which this may be accomplished in the nuclear power economy, it should be made clear that in this discussion we are referring both to occupational exposure and exposure to members of the public; we are referring both to internal and external exposure, and the dose limits about which we are concerned are dose equivalent (rem/y) to an individual, population dose (man·rem), dose commitment both to the individual (rem from a year's commitment) and to the population (man·rem from a year's commitment). This dose commitment includes the dose a person or population receives from external exposure in a year plus internal exposure from radionuclides taken into the body in a year. This internal exposure from a year's intake is equal to the resulting integral of the internal dose rates over the person's (or persons') expected remaining life. As shown in Fig. 1 (11) there are many secondary, tertiary and quaternary permissible exposure levels and radiation protection guides but of these dose commitment is the most important and only it provides the necessary and sufficient radiation protection to the individual or members of the public.

HOW TO REDUCE RADIATION EXPOSURE IN A FUTURE NUCLEAR POWER ECONOMY

The radiation exposure which must be reduced and kept ALAP in an acceptable future nuclear power economy includes both exposure of the occupational worker and of members of the public and it must include exposure from all stages of the industry, i.e. U-mines, U-mills, processing plants, fuel conversion and enrichment, fuel fabrication, reactor operations, fuel reprocessing, radioactive waste disposal, shipping of fuel and wastes, etc. In each of the individual operations from the U-mines to the final resting place of the radioactive waste, exposures should be kept ALAP or ALARA and no exposures should be permitted if they do not result in human benefit and if they can reasonably be avoided. It would take thousands of pages to review in detail how exposures can be reduced in each of these operations so only some of the

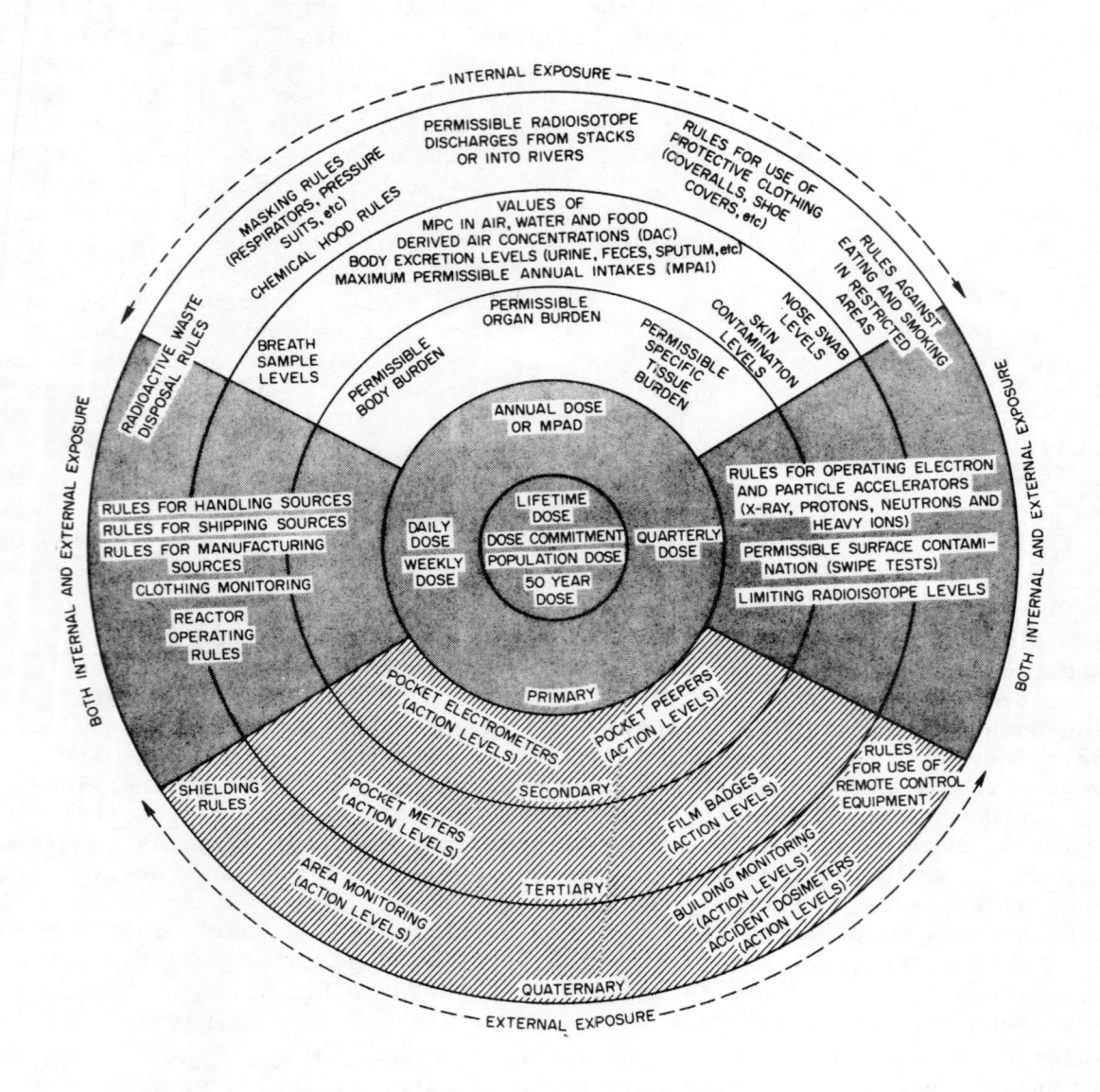

Fig. 1. Radiation Protection Guides, Permissible Exposure
Levels and Other Provisions Designed for Radiation
Protection of Radiation Workers and Members of the
Public (11)

general principles by which this objective may be accomplished and a few examples can be discussed here.

Exposures from U-mining and milling operations can be kept at an acceptable level only if no person is permitted exposure above the maximum permissible value of 4 working levels per year and if efforts as great as practicable (AGAP) are made to avoid all unnecessary exposure. Environmental exposures from mine and mill tailing and their ultimate disposition have in some cases greatly exceeded acceptable levels. These exposures must be substantially reduced in the future through careful control of the tailings, by limiting their public use, by taking effective action to prevent the release of ^{230}Th, ^{226}Ra, ^{210}Pb, and all the radioactive daughters of intermediate half life (e.g. ^{222}Rn and ^{210}Po) to man's environment or into his food chains, and where possible by disposing of the tailings permanently by such measures as returning them to spent mines. The use of U-tailings for ground fill or the leaching and draining of them into surface and groundwater systems must not be permitted.

In general, exposures at the uranium processing, fuel conversion, enrichment, and fabrication plants have been relatively low and the resulting environmental exposure has been minimal, but in some cases more care should be exercised in reducing occupational exposure to uranium dusts, for example in the preparation and processing of fuel oxides.

There are three areas that deserve special attention in reducing radiation exposures: 1) reactors, 2) fuel and waste shipping, and 3) fuel reprocessing and so reductions of radiation exposure in these areas are discussed in the following.

With few exceptions environmental exposures from the three above mentioned operations have been kept ALAP or ALARA and this record serves as an encouraging sign post of what we can expect and must demand in the future. I believe there is little doubt that ^{131}I is the radionuclide of principal concern for population exposure from routine operations of nuclear power reactors and perhaps it will prove to be one of the more serious sources of population exposure in the event of a reactor accident (either major or minor). This radionuclide can be a problem because of its concentration in drinking water or on leafy vegetables, but in most cases its potential for concentration in milk is the principal concern. For example, the estimated potential dose rate to the thyroids of persons using milk from cows grazing at 3.4 Km from the Dresden BWR Unit 1 in 1970 was 31 rem/y (12). The release of ^{131}I from a power plant can vary by orders of magnitude from day to day or month to month; for example at the Oyster Creek plant releases reached a low of 6 x 10^{-2} μCi/sec in May and a peak of 1.5 μCi/sec in June of 1973 (13). A number of improvements in the design and operation of more recently constructed reactors have lessened the amount of ^{131}I release from reactors, but there is still room for improvement. For example, the calculated milk to thyroid dose from ^{131}I to persons living in the house nearest the H. B. Robinson PWR Unit No. 2 (14) in 1974 was 41 mrem (0.27 mile distance) and this dose was orders of magnitude greater than the calculated dose to members of the public from other modes of exposure to ^{131}I or for exposure to other radionuclides. The next largest population dose (calculated) was 0.4 mrem/y to the total body from ^{3}H exposure. In order to minimize the consequences of release of iodine radionuclides in routine or accident situations, I believe serious consideration should be given to the development and use of equipment which would automatically release at all times to the discharge stacks enough stable iodine to provide adequate thyroid blockage. Also, KI tablets should be provided for immediate dispensing to plant employees and the neighboring public in case of a serious emergency in which large amounts of radioiodine isotopes are released.

Releases of radioactive noble gas is the principal source of population exposure from routine operation of BWR's. For example the AEC (15,16) estimated population doses from noble gases out to 50 miles of 420 man·rem for Dresden 1, 2, and 3 (BWR's) and 40 man·rem for Oyster Creek (BWR) power reactors in 1971. Gamertsfelder (17) estimated a potential boundry dose of 155 mrem in 1969 for the Humbolt Bay (BWR) reactor. Much has been accomplished to reduce the population dose from noble gas release by increasing the hold up time in the operations of some of the newer power reactors and much more deserves to be accomplished in this direction, especially in reducing the dose from the short lived radionuclides. For example, it has been shown (17) that the principal contributor to noble gas radioactivity is ^{138}Xe for 30 minutes hold up, ^{135}Xe for 1 to 8 hours hold up, ^{133}Xe for hold up of 1 to 30 days, and ^{85}Kr for hold ups of more than a few months.

Much has been said regarding the risk from total release of ^{85}Kr from the nuclear power industry. Since the half life (10.7y) of ^{85}Kr is so large, it makes little difference whether it leaks to the environment from the power plant or from the fuel reprocessing plant. Snyder et al. (18) have calculated the dose in (rads/y)/(μCi/m^3) to be 1.8 to skin and about 1% of this value to critical internal organs. A number of writers have shown that for uniform dispersion in the troposphere this would extrapolate to about 20 mrem/y to the skin (and 1% of this to other critical organs) of all persons on earth by the year 2000. Many methods have been developed by which the ^{85}Kr can be removed at the fuel reprocessing plants and be stored and I believe there is no question but that this must be done in the future.

Another gaseous release from power reactors that must be watched is ^{14}C. Kunz et al. (19) have estimated it could deliver a dose of 1 mrad/y to persons living 1 Km from a 1000 MWe reactor. Thus, it may be necessary to restrict the N_2 and O_2 from entering the neutron flux region of reactors whence the ^{14}C is produced.

As mentioned above ^{3}H can be an environmental problem especially for the PWR and CANDU reactors. Certainly extreme precautions must be continued with the CANDU reactors to prevent release of ^{3}H, and the sources of ^{3}H in the PWR (other than the fission), i.e. B and Li, should be replaced by some other form of automatic shim.

In the early period of nuclear power development there was much concern about ^{90}Sr. Perhaps this to some extent was warranted from the ^{89}Sr and ^{90}Sr fallout during the Windscale accident in England in 1957 but for routine operations ^{134,137}Cs is a problem that is greater by several orders of magnitude. Eisenbud (20) has indicated it is 10^4 times the problem of ^{90}Sr for nuclear power plant environmental releases.

With regard to environmental contamination and potential exposure from fuel reprocessing plants, we can say the Nuclear Fuel Services Co. of West Valley, New York has been the only commercial operating plant (out of operation in 1975) and it made about all the mistakes one could imagine in relation to poor health physics practices, of excessive releases of radioactive contamination, and storage and burial of radioactively contaminated materials. Future plants must not repeat this poor record of operations.

It is surprising to some persons to learn that in many cases the estimated population doses from the shipment of irradiated fuel and radioactive waste are greater than those from the operation of the reactor from which they are shipped. Furthermore, the dose would be less if shipments were made by rail rather than by truck and still less if shipments were by barge. For example, the dose per reactor year to people along the shipping route of irradiated fuel is calculated (21) to be 1, 0.2 and 0.03 man-rem by truck, rail and barge, respectively.

The above discussion has dealt with ALAP environmental exposure or dose commitment but it is of equal importance to maintain these exposures for the occupational workers ALAP. Although it is reported (22) that the average occupational exposures have remained relatively constant (i.e. 1 rem/y in 1969 and again in 1973), the average man·rem/plant year was 188 in 1969 but rose to 404 in 1973. Some have resorted to the excuse that occupational exposures of necessity rose because of pressure to apply the ALAP philosophy in reference to exposure of members of the public. I can only call this a most deplorable and unacceptable excuse. From my 30 year's experience in directing the health physics and environmental radiation programs at Oak Ridge National Laboratories, I found the reverse relationship could be made and was made to come to pass; namely as greater restrictions and lower exposure levels were set and enforced either for the occupational workers or for members of the public, both exposures tended to decline. Studies of the AEC (now NRC) also have failed to support the contention that application of the ALAP philosophy to members of the public is the cause of increased occupational exposure and concluded much of the occupational exposure in nuclear power plants is due for example to such things as improperly shielded tanks, lack of access for maintenance, lack of remote controls and failure to provide remote viewing equipment. In other words the fault is in the design of many of the present day power plants--they did not incorporate well proven principles of good health physics (which have long been used as basic requirements at AEC National Laboratories) into the plant design. Furthermore, adequate training programs from the lowest employee on the totum pole to the top executive have not developed and many of the employees have not been properly motivated by the long accepted and applied health physics philosophy of ALAP or ALARA. Webster et al. (14) have indicated the success the health physics programs have had at the H. B. Robinson (PWR) Unit No. 2 plant in reducing both environmental and inplant exposures and I believe this is the only type of program to which we should subscribe or which we can condone.

Martel (23) predicted there would be a sixfold increase in the nuclear plant occupational exposure (25,000 man·rem in 1975 to 150,000 man·rem in 1985) in the coming decade. I believe health physicists must do everything in their power to see that this prediction does not come to pass. On the linear hypothesis this 150,000 man·rem would result in a total of only about 150 dealths/year (both somatic and genetic) and some would say this is a negligible price to pay for the benefits from the nuclear energy industry. However, my reply is that we must lead the way in such reforms and make major efforts that are AGAP to avoid unnecessary exposure.

It seems to be generally agreed that ^{60}Co is the source of most of this large occupational exposure and that corrosion of pipes and tubes rather than wear of hard faced alloys is the principal source of ^{60}Co and ^{58}Co exposure in nuclear power plants. Specifically, the AIF study (24) points out that inspection and repair of the steam generator (containing accumulated ^{58,60}Co) is the major source (up to 88%) of occupational exposure with the PWR's; while operations about recirculation pumps, including the cleanup systems, are the major sources (up to 29%) of occupational exposure with the BWR's.

I would like to register a strong complaint regarding a practice that has developed in the nuclear energy industry of "burning" and even "burning out" temporary employees. By this we mean employing poorly instructed and untrained persons temporarily to carry out "hot" jobs. Because of limited training, poorly developed skills and a lack of appreciation of the risks of chronic exposure, such employees are much more likely to be involved in radiation accidents that could result in harm to themselves and others. I consider the practice of burning out employees to be highly immoral and unless the nuclear energy industry desists from such practices, I (and I am afraid many others) will cease to be strong supporters of this industry.

The Nuclear Fuels Services, Inc. of West Valley, New York (the only industrial fuel reprocessing plant that has been in operation) had an especially bad record regarding the employment of transient workers without sufficient specific training and indoctrination relative to safety problems and precautions regarding radiation hazards. The high average exposures of plant personnel of 2.74, 3.81, 6.76 and 7.15 rem in 1968 through 1971, respectively, provided strong evidence of the lack of an acceptable health physics program (25). AEC inspectors observed that there were levels of radiation exposure exceeding the in-plant design criteria and in addition they did not meet acceptable radiation standards outside this reprocessing plant in the unrestricted areas.

The record of exposure to outside workers (contract workers, transient workers, etc.) in the LWR power plants is very bad; for example in the period 1969-1974 the total occupational exposure was 5297 man·rem and of this 2853, 505 and 1969 man·rem were received by station employees, utility employees and outside workers, respectively. About half this occupational exposure was from routine maintenance.

Finally and in conclusion I list some of the steps that can be taken to reduce radiation exposure in the nuclear energy industry as follows:

1. Select and build only those reactors that can be expected to deliver the lowest radiation doses (occupational and public) from routine and accident situations. I believe this means discontinuance of the LMFBR program in favor of a safer and more economical breeder.

2. Abandon plans to reprocess nuclear fuel and to operate light water

reactors on mixed oxides (UO$_2$ plus PuO$_2$) until the questions of safer breeder reactors, permanent high level radioactive waste disposal and security of large amounts of purified plutonium are better resolved.

3. Ship spent fuel elements and high level radioactive waste primarily by barge and rail.

4. Completely revise and upgrade health physics programs in the nuclear industry. Perhaps the best example of an inadequate health physics program in a nuclear power operation was the Shippingport reactor facility. It was brought out at the hearings before the Fact-Finding Committee of the Governor of Pennsylvania in Aliquippa, Pa. in July 1973 that the health physics program was woefully inadequate. An outside contractor had furnished data indicating large releases of radioactive material to the environment and yet the organization just sat on the data and waited for Dr. E. J. Sternglass to develop allegations that this radioactive contamination was causing an increase in infant mortality and leukemia in the exposed populations. As a result of the hearings the allegations of Dr. Sternglass could not be confirmed but it was concluded that the health physics environmental data were unreliable, the radiation exposure to neighboring populations may or may not have been excessive and said exposure may or may not have caused an increase in infant mortality and leukemia. How much better it would have been if Shippingport had had a health physics program that had the answers!

A few of the requirements of an acceptable health physics program are:

a) Place the health physics program at each plant (i.e. nuclear power plant, fuel reprocessing plant, etc.) under the direction of a senior health physicist. He should be certified by the American Board of Health Physics or have equivalent qualifications. He should have a line of reporting to top management of the utility or company. Although this may be a "dotted line" of reporting, he should be urged and in fact required to use this channel of reporting whenever there is any question about proper attention to radiation protection at the local plant level.

b) The senior health physicist should be part of the team during the early planning stages. He should assist in selecting the plant site, choosing the type of reactor, checking the design drawings, and in determining the adequacy of the shielding, safety devices, fuel changing equipment, waste handling, filter system, etc.

c) The senior health physicist should assist in preparation of environmental impact statements.

d) He should order, place in use, and maintain all health physics instruments.

e) He should set up an adequate personnel monitoring program, including body fluid analyses, etc.

f) He should establish an adequate environmental monitoring program.

g) He must prepare the radiation protection manuals, maintain ex-

posure records, prepare reports to NRC, etc.
- h) He must employ the health physics manpower he needs.
- i) He must conduct radiation protection training programs for all employees.
- j) He must develop and maintain a proper interface with the community (i.e. with the fire department, police, public health agencies, hospitals, etc.).

5. See that the reactor incorporates all the recommended safety devices and that these are maintained in proper condition. Here we are considering such items as double or triple containment, ECCS, spray cooling system, avoidance of common mode failure systems, etc.

6. Avoid the use of such materials as B and Li in the primary cooling system (to reduce ^{3}H) of the PWR, and avoid use of alloys with large amounts of ^{58}Ni and ^{59}Co (the precursors of ^{58}Co and ^{60}Co).

7. Design reactors to reduce the carbon, nitrogen and oxygen reactions which lead to the production of ^{14}C in the fuel rods, coolant, etc.

8. Increase the hold up time of gases from the reactors (especially the BWR's) to reduce the radioactive noble gases released to the environment.

9. Remove and store the ^{85}Kr at the fuel reprocessing plants.

10. Develop and rehearse an adequate emergency evaluation plan.

11. Develop a system of chemical prophylaxis (e.g. use of KI tablets and automatic stack release of stable iodine) to reduce uptake of radionuclides of iodine and of other radioactive contamination released to the environment.

12. Use a fast flowing transport clean up system (26) to remove activation products.

13. Add extra shielding in major areas of high exposures.

14. Identify all the major dose producing operations and reduce exposures by redesign of equipment, addition of shielding, installation of remote control and TV equipment, etc. and by the rigid enforcement of better administrative control.

15. Provide better storage facilities for radioactively contaminated equipment.

16. Use such design concepts (27) as back-flushable filters, a centralized waste processing control room, separation of high maintenance equipment from low maintenance equipment, careful piping layout, better shielded tanks, better access for maintenance, extensive use of remote control and TV, remote charcoal change systems, reduction of waste volume and of waste handling, remote monitoring and maintenance, separation of radioactive and non radioactive equipment, pipes, etc.

17. Apply good housekeeping at all times.

18. Design and use special tools to reduce maintenance time, refueling time, waste handling time, etc.

19. Develop and use new techniques to correct the plugging tubes of the PWR steam generator.

20. Make appropriate use of administrative control to minimize all in-plant exposures.

REFERENCES

1. "The Effects of Populations of Exposure to Low Levels of Ionizing Radiation," Report of the Advisory Committee on the Biological Effects of Ionizing Radiations (BEIR Committee), Nat. Acad. of Sciences, National Research Coucil (1972).

2. KARL Z. MORGAN, "Suggested Reduction of Permissible Exposure to Plutonium and Other Transuranium Elements," Amer. Ind. Hygiene Journal, 36, 8, 567 (August 1975).

3. H. J. DUNSTER and A. M. MCLEAN, "The Use of Risk Estimates in Setting and Using Basic Radiation Protection Standards, "Health Physics, 19, 121 (1970).

4. ROBERT ALEXANDER, In a presentation before a subcommittee of the ACRS in Washington, D. C., the figure of $1,000 per man·rem was referred to as a guiding figure for some of the considerations of NRC (August 7, 1975).

5. ARTHUR R. TAMPLIN and T. B. COCHRAN, "Radiation Standards for Hot Particles," a publication of the Natural Resources Defense Council, Washington, D. C. (February 14, 1974).

6. BERNARD L. COHEN, "Hazards of Plutonium Dispersed," draft report widely circulated, Institute for Energy Analysis (November 11, 1975).

7. L. H. HEMPLEMAN, "Risk of Thyroid Neoplasms after Irradiation in Childhood," Science, 160, 159 (April 12, 1968).

8. CHARLOTTE SILVERMAN and D. A. HOFFMAN, "Thyroid Tumor Risk from Radiation During Childhood," Preventive Medicine, 4, 100 (1975).

9. B. MODAN, H. MART, H. BAIDATZ, R. STEINITZ and S. G. LEVIN, "Radiation-induced Head and Neck Tumors," Lancet, 1, 277 (1974).

10. ALICE STEWART and G. W. KNEALE, "Radiation Dose Effects in Relation to Obstetric X-Ray and Childhood Cancers," Lancet, 1, 1185 (1970).

11. KARL Z. MORGAN, "Proper Use of Information on Organ and Body Burdens of Radioactive Materials," Symposium in Stockholm, Sweden, November 22-26, 1971 of IAEA and WHO on Assessment of Radioactive Contamination in Man, IAEA, Vienna (1972).

12. R. L. BLANCHARD, H. L. KRIEGER, H. E. KOLDE and B. KAHN, "Radiological Surveillance Studies at a BWR Nuclear Power Station--Estimated Dose Rates," Proc. of Health Physics Society Midyear Symp., Vol. II, Idaho Falls, Idaho (1970).

13. "Results of Measurement of ^{131}I in Air Vegetation and Milk at Three Operating Reactor Sites," Interim Oyster Creek Report (October 1973).

14. J. A. PADGETT, B. H. WEBSTER and D. L. SWINDLE, "Distribution of Radionuclides and Radiation Dose in the Vicinity of an Operating Nuclear Power Plant," presented at Fourth National Symposium on Radioecology (May 12, 1975).

15. "Report on Releases of Radioactivity from Power Reactors in Effluents During 1971," USAEC Rept. WASH-1198, Washington, D. C. (1971).

16. "The Safety of Nuclear Power Reactors (Light Water-Cooled) and Related Facilities," USAEC Rept. WASH-1250, Washington, D. C. (July 1973).

17. "Report of the United Nations Scientific Committee on the Effects of Atomic Radiation," UNSCEAR Rept. (A-8-725) Vol. 1, Ionizing Radiation Levels and Effects (1972).

18. W. S. SNYDER, L. T. DILLMAN, M. R. FORD and J. W. POSTON, "Absorved Dose to a Man Immersed in an Infinite Cloud of ^{85}Kr, "Health Physics Division ORNL Rept. (1973).

19. C. KUNZ, W. E. MAHONEY and T. W. MILLER, "C-14 Gaseous Effluents from Pressurized Water Reactors," Proc. Midyear Symposium of the Health Physics Society, Knoxville, Tennessee (November 1974).

20. M. EISENBUD, data presented at the meeting of a committee of ACRS on health physics manpower needs in nuclear power plants, Boston, Mass. (September 4, 1975).

21. "Environmental Survey of Transportation of Radioactive Materials to and from Nuclear Power Plants," USAEC Rept. WASH-1238, Washington, D. C. (December 1972).

22. T. D. MURPHY, "Compilation of Occupational Radiation Exposure From Light Water-Cooled Nuclear Power Plants, 1969-1973," USAEC Rept. WASH-1311, Washington, D. C. (May 1974).

23. L. J. MARTEL, "Purpose of Decontamination Seminar," Seminar on Decontamination of Nuclear Plants, Columbus, Ohio (May 7-9, 1975).

24. C. A. PELLETIER, L. SIMMONS, M. BARBIER and J. H. KELLER, "National Environmental Studies Project," American Industrial Forum (1974).

25. L. D. DOW, "Noncompliance Items" given in AEC inspection report to Nuclear Fuel Services (March 16, 1972).

26. R. WILSON, G. A. VIVIAN, C. BIEBER, D. A. WATSON and G. G. LEGG, "Man·Rem Expenditure and Management in Ontario Hydro Nuclear Power Stations," presented at 20th annual meeting of Health Physics Society, Buffalo, N. Y. (July 13-17, 1975).

27. W. C. MCARTHUR, B. G. KNIAZEWYCZ and B. H. WEBSTER, "Design Methods for Reducing In-Plant Exposures," Carolina Power and Light Co., Unpublished paper (1975).

REACTOR CONTROL

Lynn E. Weaver
School of Nuclear Engineering
Georgia Institute of Technology
Atlanta, Georgia 30332

INTRODUCTION

The basic requirement of a power plant is to supply electrical energy to the power distribution system on demand. This demand represents a load to the plant turbine-generator. To meet this demand (load), the Power Conversion System must respond with the correct flow of preconditioned steam to the turbine. In a nuclear power plant, the correct steam conditions and flow are dependent on the reactor power level. Therefore, for power plant operation, the reactor must be responsive to changes in energy demand (turbine-generator load). At equilibrium conditions, there is a direct relationship between the electrical output and the neutron flux density. By controlling the neutron generation it is possible to obtain rates of heat generation proportional to the demand electrical power output. This is normally accomplished automatically by the reactor control system in conjunction with the load demand program.

In addition to normal operation the control system must be capable of controlling unexpected nonroutine situations, such that safety limits are not exceeded. The most severe cases occur with large transients associated with an accident. These generally are considered in the establishment of limiting conditions that affect component design. Further, the control system must assure stable operation at all power levels.

It is not the intent in this presentation to go into depth in the complex problems of reactor kinetics and control for this is the responsibility of the specialist. The discussion here is for the nonspecialist who is concerned mainly with conceptual relations and their interplay with other phases of reactor safety. These conceptual relationships are developed starting with the equations describing dynamic behavior.

EQUATIONS OF REACTOR DYNAMICS (1,2)

The basic equations describing the dynamic behavior of a nuclear reactor, neglecting spatial effects and dependence on neutron energy, are given below

$$\frac{dn}{dt} = k_{eff} (1-\beta) \frac{n}{\ell_o} + \sum_{i=1}^{6} \lambda_i c_i - \frac{n}{\ell_o} \qquad (1)$$

$$\frac{dc_i}{dt} = \frac{\beta_i k_{eff} n}{\ell_o} - \lambda_i c_i$$

where

$\frac{dn}{dt}$ = rate of change of neutron density or power

$k_{eff}(1-\beta)\frac{n}{\ell_o}$ = prompt neutron production

$\sum_{i=1}^{6} \lambda_i c_i$ = delayed neutron production

$\frac{n}{\ell_o}$ = neutrons lost/sec

and

$\frac{dc_i}{dt}$ = rate of change of delayed neutron precursor concentration

$\frac{\beta_i k_{eff} n}{\ell_o}$ = production of delayed neutron precursor concentration

$\lambda_i c_i$ = loss of delayed neutron precursor concentration

In the equations above; n = neutron density or power; β = delayed fraction of neutrons present in the system; β_i = that fraction of the delayed neutrons which belongs to the ith group (six groups of delayed neutron emitters are generally recognized in reactor dynamics); c_i = the ith precursor concentration; λ_i = the decay constant of the ith precursor; ℓ_o = mean neutron life time for a neutron in the reactor. In equations (1) k_{eff} is the effective multiplication constant of the reactor which depends on the materials and geometry of the system. It defines the state of the system in a broad sense in that when $k_{eff} = 1$, the reactor is in the steady state (critical); when $k_{eff} > 1$, the reactor power is increasing (super critical); and when $k_{eff} < 1$, the reactor power is decreasing (subcritical).

Since our interest is in the time dependency of the neutron density or power in relation to various feedback effects, the space-independent or point reactor model given by equations (1) is adequate for identifying important parameters.

Equations (1) can be put into the form

$$\frac{dn}{dt} = \left(\frac{\rho-\beta}{\ell}\right) n + \sum_{i=1}^{6} \lambda_i c_i$$

$$\frac{dc_i}{dt} = \frac{\beta_i}{\ell} n - \lambda_i c_i \qquad (2)$$

where ρ, called the reactivity, is given by

$$\rho = \frac{k_{eff} - 1}{k_{eff}} \qquad (3)$$

and ℓ, the generation time, is defined by

$$\ell = \frac{\ell_o}{k_{eff}} \qquad (4)$$

Obviously from equations (2), ρ is the forcing function to which the reactor power responds. When $\rho=0$, the reactor is at constant power $dn/dt = 0$ (critical) and accordingly, when ρ is greater or less than zero, the reactor power is increasing (supercritical) or decreasing (subcritical), respectively.

Reactivity changes can occur due to external input (control rod motion) or internal effects caused by changes in temperature, pressure, fission product poisoning, and void formations in the coolant depending on the particular reactor system. Reactivity is usually measured in terms of dollars and cents, a dollar's worth of reactivity is $\rho=\beta$. Accordingly, $(100/\beta)\rho$ gives reactivity in cents. When $\rho=\beta$, the reactor is critical without delayed neutrons and is referred to as prompt critical. For values of $\rho > \beta$ the reactor is said to be super prompt critical. $\beta = 0.0065$ for U-235 fueled thermal reactors.

It is easily shown that the solutions of equations (2) for constant reactivity ρ_o are linear combinations of exponentials $e^{\omega_j t}$ and can be written as

$$n(t) = \sum_{j=1}^{7} A_j e^{\omega_j t} \qquad (5)$$

The relationship between ρ_o and ω_j is

$$\rho_o = \ell \omega + \sum_{i=1}^{6} \frac{\beta_i \omega}{\omega + \lambda_i} \qquad (6)$$

where each ω_j satisfies the equation. Equation (6) is referred to as the in-hour equation. The in-hour equation has all negative real roots except one real root with the same sign as ρ_o. As time increases one term in equation (5) will dominate and

$$n(t) = A_1 e^{\omega_1 t} \qquad (7)$$

where ω_1, is the algebraically greatest ω_j. For positive reactivity, this term grows exponentially with time. A characteristic time referred to as the reactor period is given by $T = 1/\omega_1$, the time required for the reactor power to increase by a factor e.

For very large ω either positive or negative, equation (6) gives the relationship

$$\omega = \frac{\rho_o - \beta}{\ell} \qquad (8)$$

It is clear from equation (8) that when $\rho_o \gg \beta$ equation (7) becomes

$$n(t) = A_1 e^{\rho_o t/\ell} \tag{9}$$

Since $10^{-5} < \ell < 10^{-3}$ it is apparent that ρ_o should be kept less than β to avoid rapid increases in reactor power. Obviously, ρ doesn't remain a constant and will change as the temperature in the reactor increases. Power reactors are designed such that a negative reactivity feedback effect occurs when the temperature of the reactor increases, thus limiting power for various values of ρ_o. This behavior can be described by the block diagram in Fig. 1.

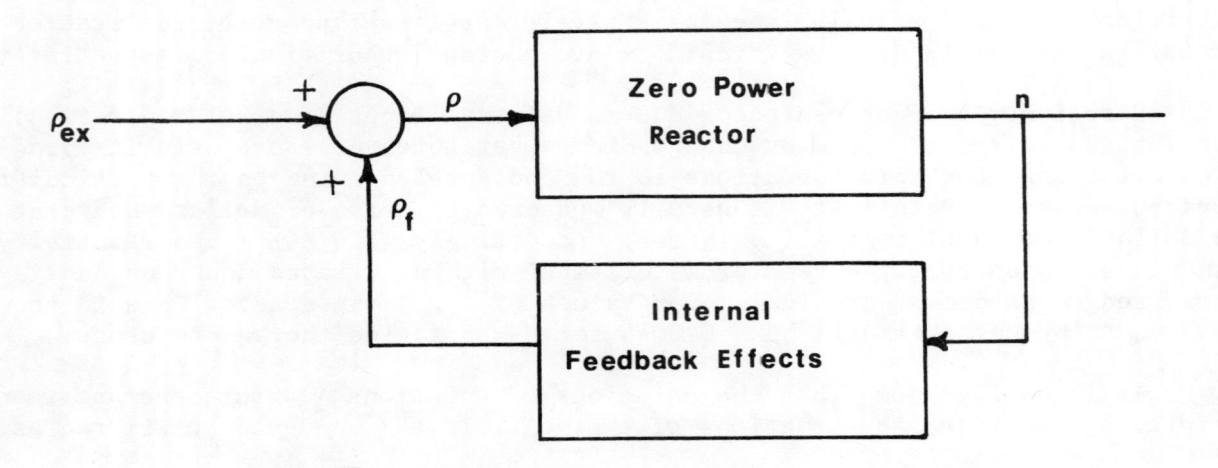

Fig. 1. Reactor Block Diagram

In Fig. 1 ρ is the total system reactivity and ρ_{ex} and ρ_f are the externally induced and internally generated feedback reactivity, respectively. Therefore

$$\rho = \rho_{ex} + \rho_f \tag{10}$$

Reactors are designed such that ρ_f is negative.

The reactor model in Fig. 1 can be described by three systems of equations:
  (1) The neutron-kinetics equation (equation (2))
  (2) A family of thermal-transport and hydrodynamic equations relating reactor power to temperature change within the core
  (3) The equations relating temperature change to reactivity.

TEMPERATURE COEFFICIENT OF REACTIVITY (2,3)

The temperature coefficient of reactivity α_T relating temperature change to reactivity is made up of prompt and delayed components. Prompt effects depend on the instantaneous state of the fuel (Doppler effect, thermal distortion of fuel elements, etc.), while delayed effects are mainly associated with the moderator or coolant (neutron temperature, thermal expansion of moderator material, etc.). In fast power excursions, delayed effects may be negligible; however, they might dominate in a quasi-static situation. The temperature coefficient of reactivity, for power reactors, associated with prompt effects α_p is on the order of 10^{-5} $\rho/°F$ while the temperature coefficient for delayed effects α_D is in the neighborhood of 10^{-4} $\rho/°F$.

An important prompt acting coefficient is the result of the Doppler effect, i.e. neutron absorption in heavy elements at epithermal energies is dominated by many resonance peaks in the cross section vs. neutron energy curve. As the temperature of the fuel increases, it is found that the curve of cross section vs. energy is broadened and flattened. The integral of the energy dependent absorption cross section over the energy region remains unchanged however and the number of absorptions would not change if the neutron spectrum were independent of energy (flat). Flux depression near the surface of a fuel pin in a heterogeneous reactor results in a reduced reaction rate in the fuel interior, effectively shielding the atom of resonance absorber inside the pin. As the resonance peak broadens the self-shielding effect for fuel pins decreases, thus increasing absorption further. The Doppler contribution to the temperature coefficient is prompt, because it is due to thermal agitation of fuel atoms, and is therefore important to reactor safety.

Delayed reactivity feedback effects are due to changes in the temperature of the core components, which are caused by shifts in the heat generation - heat removal balance. Since there is a time delay due to heat transfer, the equations relating power to delayed reactivity feedback are time dependent. Fortunately, these time dependent temperature effects, such as changes in moderator density and void formations in boiling water reactors tend to limit increase in power.

In a general sense, the reactor can be viewed as given by the simplified block diagram of Fig. 2.

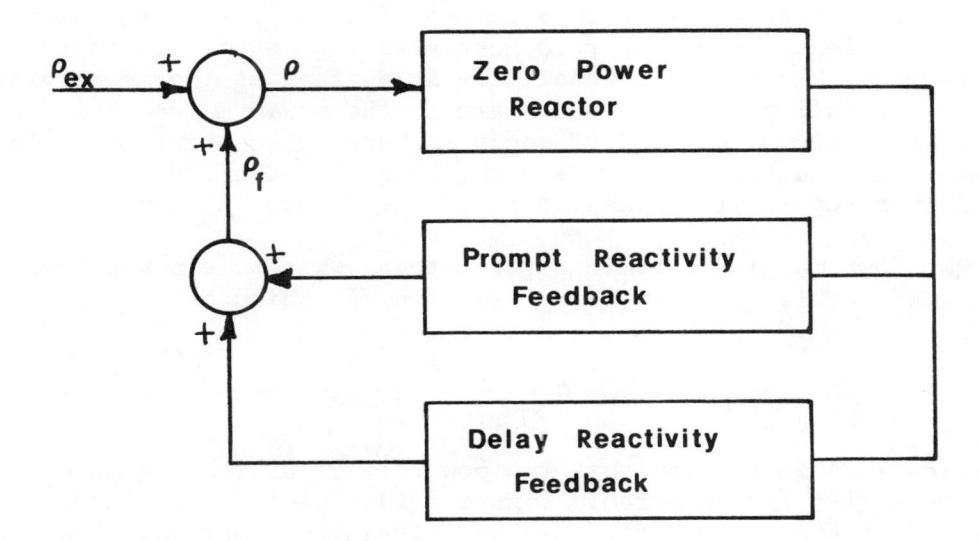

Fig. 2. Reactor with Prompt and Delayed Reactivity Feedback

Applying methods of the control engineer the block diagram reactor model as shown above, and its associated describing equations, is used in stability analysis as a function of various system parameter variations.

FISSION PRODUCT POISONING (4)

Particular attention must be given to the formation of xenon-135 in a power reactor because of its large capture cross section for thermal neutrons (3.4×10^6 barns). Xenon build-up can be viewed as another reactivity feedback effect, which can cause slow power oscillations. Xenon-135 results indirectly from the beta decay of tellurium-135 as follows:

$$Te^{135} \xrightarrow{2m} I^{135} \xrightarrow{6.7 \text{ hr}} Xe^{135} \xrightarrow{9.2 \text{ hr}} Cs^{135} \xrightarrow{2 \times 10^4 \text{ yr}} Ba^{135} \text{(stable)} \quad (11)$$

Tellurium-135 is formed in about 5.6% of thermal neutron fission of U^{235}. In addition, Xe^{135} is formed directly in about 0.3% of fission. In large power reactors it is possible to have oscillation in time and space. The oscillations in time are low frequency on the order of one to two cycles per day which can be handled easily by the reactor control system. A more serious situation can arise due to the uneven concentration of xenon in the reactor caused by a localized perturbation. A large poison build-up in one part of the reactor core will cause some portion of the core to supply more of the load than others and damage can possibly occur. Spatial effects generally become important in large loosely coupled cores in considering transients.

REACTIVITY CONTROL (3)

In the previous sections a discussion of the various ways in which the reactivity of the reactor could change during operation was given. It was seen that these internal changes could be viewed as a feedback effect. An additional change which occurs is due to fuel depletion or burnup. To accommodate such changes the fuel loading must provide sufficient positive reactivity. This excess reactivity must be compensated for by control elements. In addition, sufficient shutdown margin must always be available to reduce the reactor power quickly. Excess reactivity is defined as the reactivity present in the core when all control elements are in the maximum reactivity condition. Shutdown margin refers to the difference in reactivity between criticality and the minimum reactivity state, and the range between the excess reactivity and the minimum reactivity in the worth of the control elements.

The shutdown margin determines the rate at which the power level is reduced on scram. This is shown using equation (7) which can be approximated by

$$n \approx n_o \frac{\beta}{\beta - \rho_o} e^{t\omega_1} \quad (12)$$

From equation (12) it is seen that the power level will be reduced quickly by the fraction $1/(1-\rho_o/\beta)$ on scram. Figure 3 is a plot of the reduction in initial power after scram for various values of shutdown margin. The power does not drop instantaneously as it appears in the figure. This appearance is due to the relatively large divisions used in the time scale. In the design of the scram system a criterion imposed for safety is the so-called stuck-rod criterion. That is, it must be possible to make the reactor subcritical with a specified number of control rods in the fully withdrawn position.

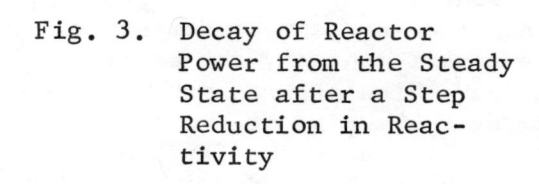

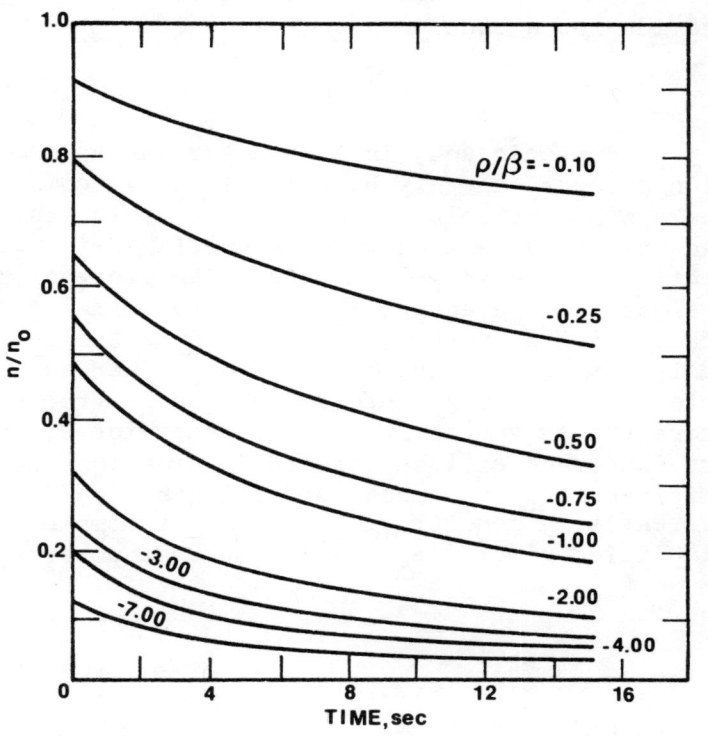

Fig. 3. Decay of Reactor Power from the Steady State after a Step Reduction in Reactivity

Control element functions are usually divided into three categories: 1) power regulation during routine operation to accommodate changes in load demand, 2) shim control which accounts for fuel burnup over time, and 3) safety control which provides negative reactivity quickly under scram conditions. Table 1 gives typical reactivity inventory specifications for the boiling-water (BWR) and pressurized-water (PWR) reactors.

Table I. Reactivity-Inventory Specifications (3)

Control Characteristics	Reactivity (ρ) PWR	BWR
Excess multiplication of clean core (uncontrolled)		
at 68°F	0.293	0.25
operating condition (clean)	0.248	
xenon and samarium equilibrium	0.181	
Worth at control rods	-0.07	-0.17
Worth at borated control curtains	None	-0.12
Total worths of soluble poison	-0.25	None
Total worth of control	-0.32	-0.29

In the PWR control regulation is accomplished through control rod motion, whereas in the BWR the control of coolant flow through the core by the jet pumps provides regulation control up to 25% of the operating power level. Shim control in the BWR is obtained by borated control curtains and control rods, and in the PWR through a soluble absorber, such as boron, placed in the coolant. The loss of the soluble poison due to neutron absorption during operation compensates for fuel depletion. In both PWR and BWR, control rods provide for a shutdown margin at scram.

METHODS OF REACTOR CONTROL

The static and, in particular, the dynamic behavior of the power production process can only be determined by reliable and accurate measurements of the system variables. The application of these measurements by the control and safety systems must be accomplished in a manner which assures proper corrective action as well as providing protection for the equipment and public against extreme accidents. This is normally accomplished by the feedback process where system variables are controlled to a predetermined value, commonly referred to as the setpoint. When measurements deviate from the setpoint, the error is detected by the controller(s) and action is taken to restore the system to its correct setpoint or condition. Some measurements provide trend or feed forward information such that the feedback process will initiate corrective control action before a setpoint limit is violated, thus preventing a reactor trip (scram) or complete loss of load. Figure 4 is a simplified block diagram of a nuclear power system with control.

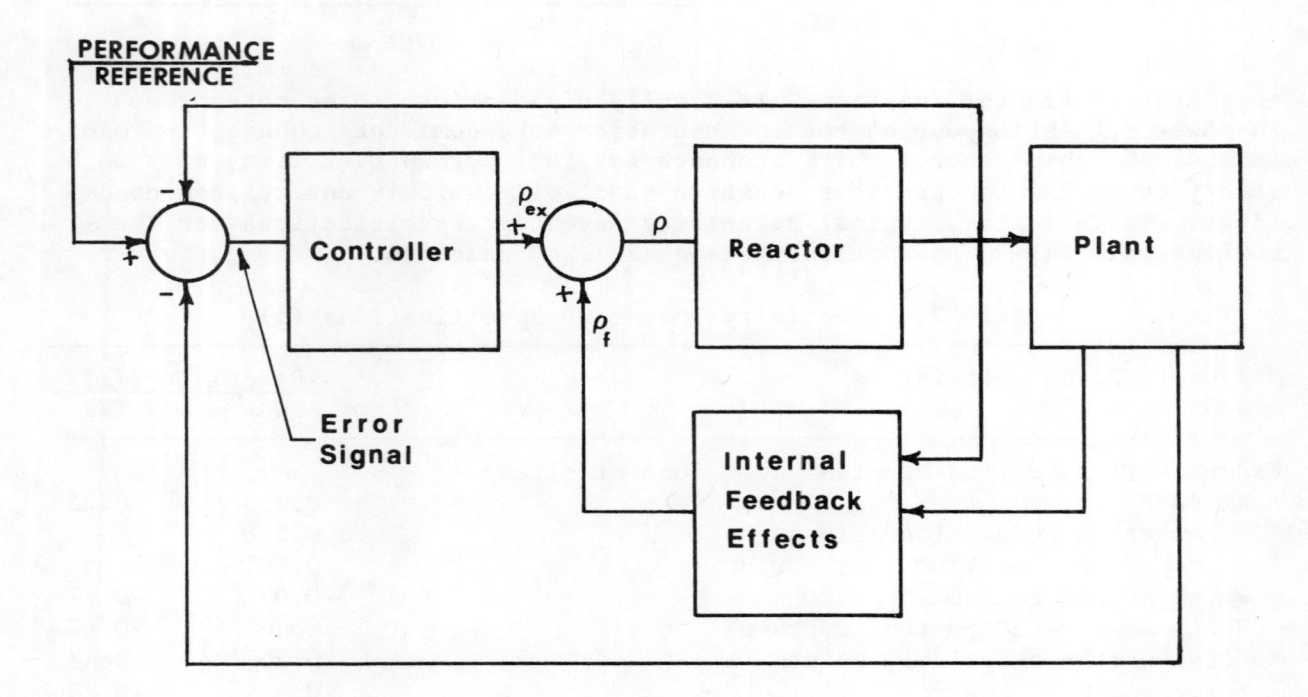

Fig. 4. Nuclear Power Plant with Control

The two main types of nuclear power plants in operation today, PWR's and BWR's, differ somewhat in their methods of control and will be considered separately.

CONTROL DESIGN FOR A PRESSURIZED WATER REACTOR (5)

In discussing the control aspects of a typical Pressurized Water Reactor (PWR), concern will be primarily with the basic criteria used in the functional design of the system for reactor plant control and safety. Emphasis will be placed upon the interrelation between the inherent reactor characteristics and reactor control and safety. Figure 5 shows a simplified diagram of a typical pressurized water reactor plant. The reactor coolant is pumped in a closed system through the reactor vessel where it is heated by the core. From the reactor vessel, the coolant flows through the primary side of the steam generators where it transfers heat to the secondary or steam side. From the steam generators, the coolant flows to the inlet of the reactor coolant pumps and then from the pumps back to the reactor vessel. PWR Steam Supply Systems have one reactor vessel containing one core and two, three, or four reactor coolant loops (depending on the plant rating) with each loop having one pump and one steam generator. Reactor coolant loops and steam generators are, thus, operating in parallel.

The secondary plant consists of the steam systems, turbine, condenser, condensate system and feedwater system. It transfers energy from the steam generators to the turbine.

In the steam generators, steam is produced by boiling feed water pumped into the steam generators from the feedwater system. The reactor coolant flows within tubes in the steam generators. These tubes are surrounded by an inventory of water on the secondary side; the secondary side inventory being maintained by the feed system. Thermal energy from the reactor coolant is transferred through the walls of the steam generator tubes to boil water on the secondary side. The steam generator tubes thus separate the reactor coolant from the secondary side. Steam produced in the steam generators by boiling feedwater passes through steam separators and then to the turbine.

The reactor coolant system is held at a nominal operating pressure of 2250 psia by use of a pressurizer, which is a pressure vessel having a volume approximately 50% filled with water. The remainder of the vessel is filled with steam. Water in the pressurizer is normally heated to the saturation temperature corresponding to 2250 psia. The steam in the pressurizer at steady state conditions is also saturated at 2250 psia and serves to cushion the effects of expansion and contraction of the reactor coolant.

Plant Control Aspects

The reactor control system must satisfy two basic requirements:
1. Deliver steam at acceptable pressures
2. Maintain a stable heat source (reactor).

Steam must be delivered to the turbine at a sufficiently high pressure to maintain turbine plant efficiency. Referring again to Fig. 5, steam is produced in the steam generators at saturated conditions. Thus, the temperature of the water-steam mixture in the steam generators determines the steam pressure. The higher the steam generator secondary side temperature, the higher

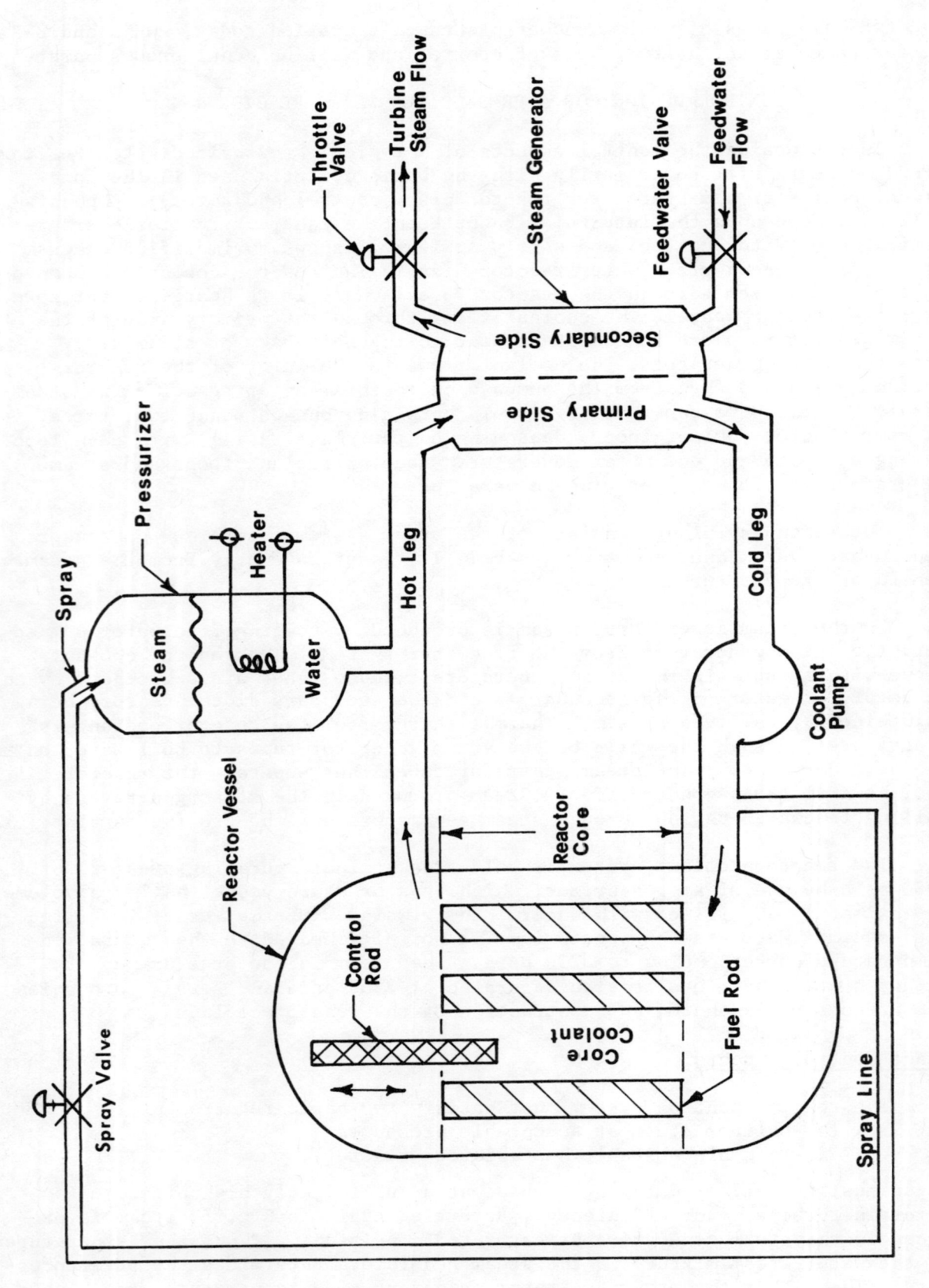

Fig. 5. Pressurized Water Reactor Schematic

the steam pressure. For high secondary side temperatures, a high primary side temperature or a high reactor coolant average temperature is required. Therefore, a high reactor coolant average temperature becomes a desirable goal.

Figure 6 shows that with saturated steam a change in steam temperature results in a sizable change in the steam pressure. Thus, it is desirable to hold the steam temperature constant to avoid a large difference between the no-load steam pressure and the full-load steam pressure. Such a large difference would require the steam piping to be designed for a very high no-load steam pressure and the turbine would have to operate at the low full-load steam pressure.

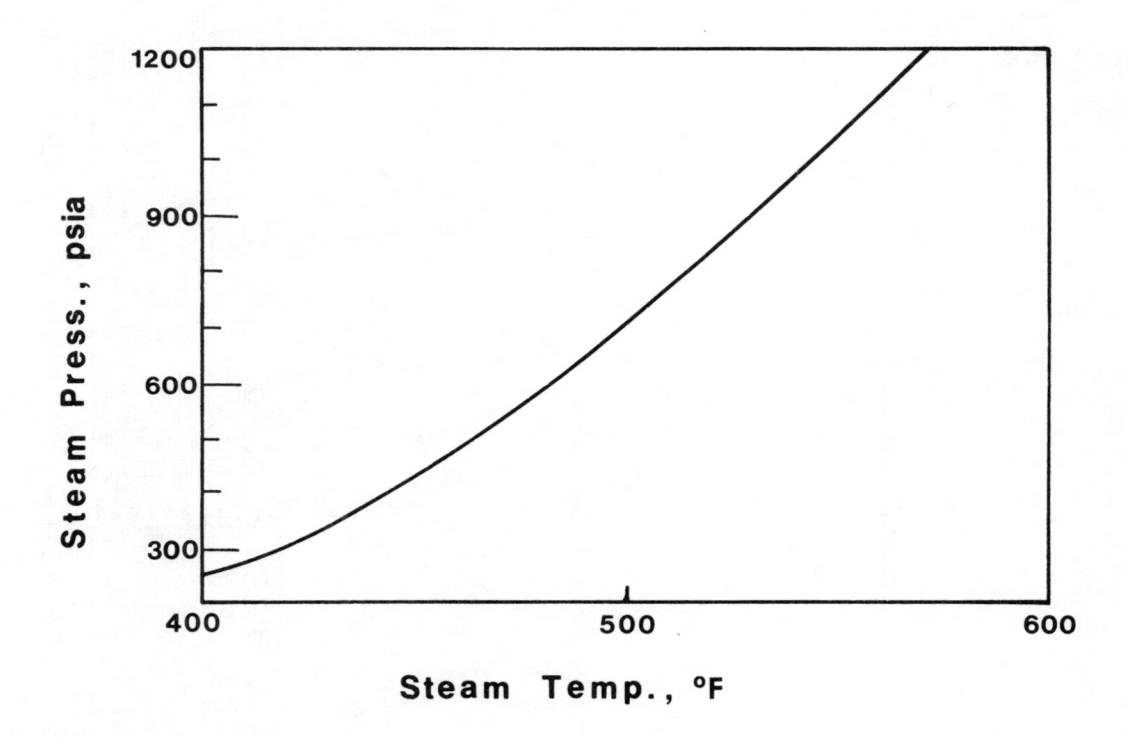

Fig. 6. Steam Pressure as a Function of Steam Temperature

A stable heat source in the reactor is maintained by keeping the reactivity equal to zero. There are three mechanisms for varying reactivity.
1) Control rods. A PWR uses a control-rod-cluster as shown in Figure 7. The control rods distributed within the fuel assembly are stainless steel tubes containing a silver-cadmium alloy absorber material.
2) Moderator coefficient of reactivity. Over most of core life a decrease in moderator density leads to a decrease in core reactivity or as the moderator temperature increases the reactivity decreases.

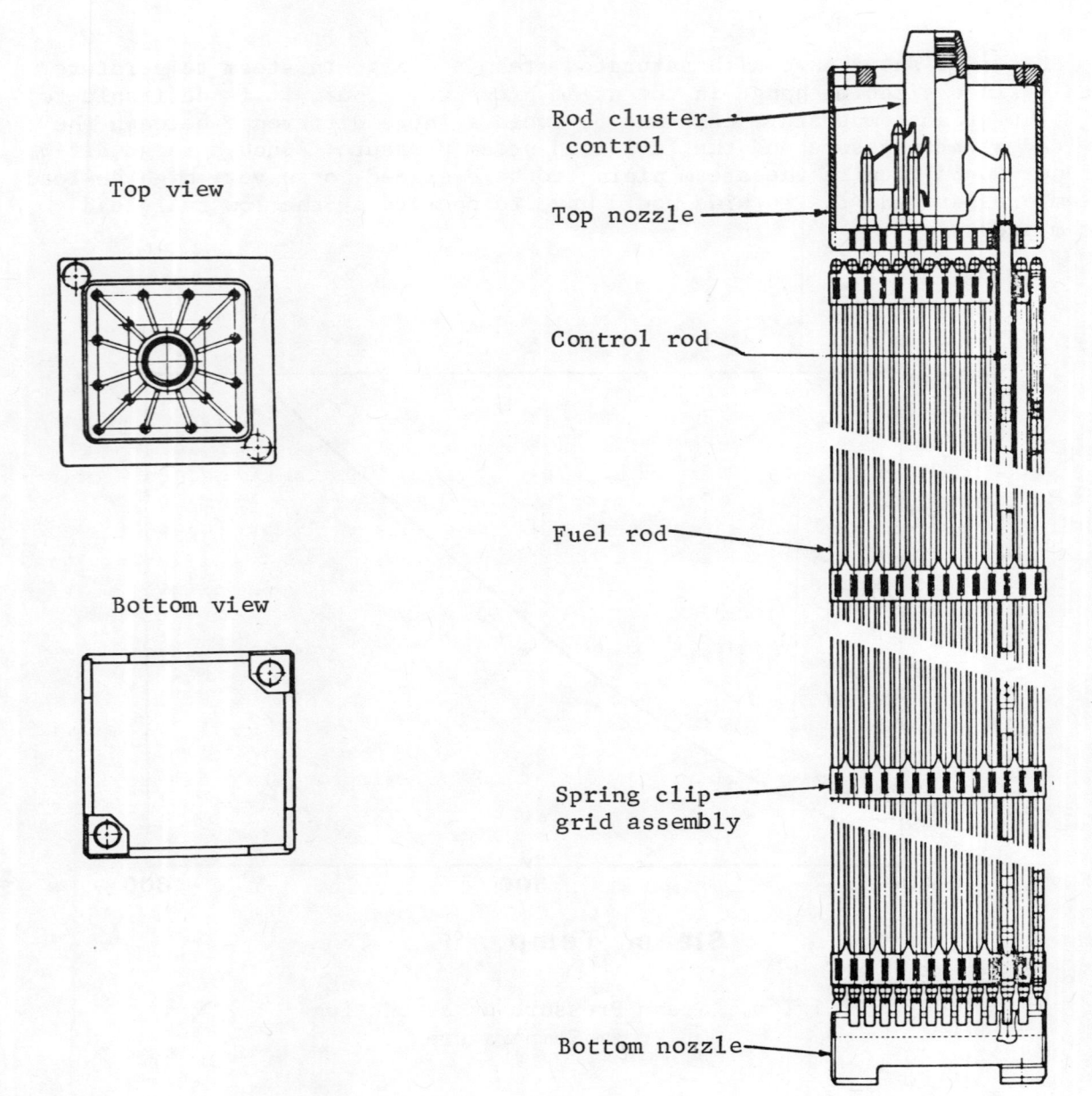

Fig. 7. Pressurized Water Reactor Fuel Assembly

3) Doppler coefficient of reactivity. As discussed previously, the Doppler effect is a reactivity feedback effect dependent upon the fuel temperature. Resonance absorption within the fuel is broadened as the fuel temperature is increased. Thus the Doppler coefficient of reactivity is negative.

Consider now the effect of a power change, a reduction in load. To hold the steam temperature-pressure constant the reactor coolant average temperature must be reduced, i.e.

$$\text{Power} \propto (T_{\text{reactor coolant average}} - T_{\text{steam}}) \qquad (13)$$

From equation (13) it is seen that a constant steam temperature implies a reactor coolant average temperature which falls as power is reduced. The moderator and Doppler reactivity contribution is positive as the power is reduced; therefore, for a constant steam temperature pressure as a function of power, control rods must be inserted as power is reduced to offset the effects of Doppler and moderator reactivity changes. A constant steam temperature pressure as a function of power means considerable investment in control rods and in the pressurizer since the pressurizer must absorb the contractions and expansions of the reactor coolant as the temperature is changed. Consequently a compromise is reached as shown in Fig. 8. The reactor coolant system average temperature is programmed to increase with load. However, the increase is not sufficient to keep the steam temperature and pressure constant.

The programmed reactor coolant average temperature whereby the temperature increases with load requires that the control rod banks be moved out as the power is increased. This is necessary to offset the negative reactivity feedback due to the moderator and Doppler effect. Further, this program results in the thermal energy inventory in the reactor coolant to increase with the load. Thus, at full power there is more thermal energy stored in the coolant than at no-load. When steady state power is increased, energy must be added to the reactor coolant to maintain the desired T_{avg} program. Similarly, when the load is decreased, energy must be removed from the reactor coolant. In particular, following a plant trip stored energy must be removed from the reactor coolant to reach no-load conditions.

The change in reactor coolant average temperature with load results in a change in reactor coolant density with load. As the load and temperature are increased, the density of the reactor coolant water decreases. Density changes in the reactor coolant will cause a change in the pressurizer water level. Therefore, pressurizer water level increases with load and reactor coolant temperature accordingly and decreases as load and reactor coolant temperature decrease.

Heat Sink - Heat Source Mismatch

The core power (heat source) and turbine load (heat sink) are not equal during operational transients. In general if the heat source > heat sink there is an increase in the energy inventory stored in the reactor coolant.

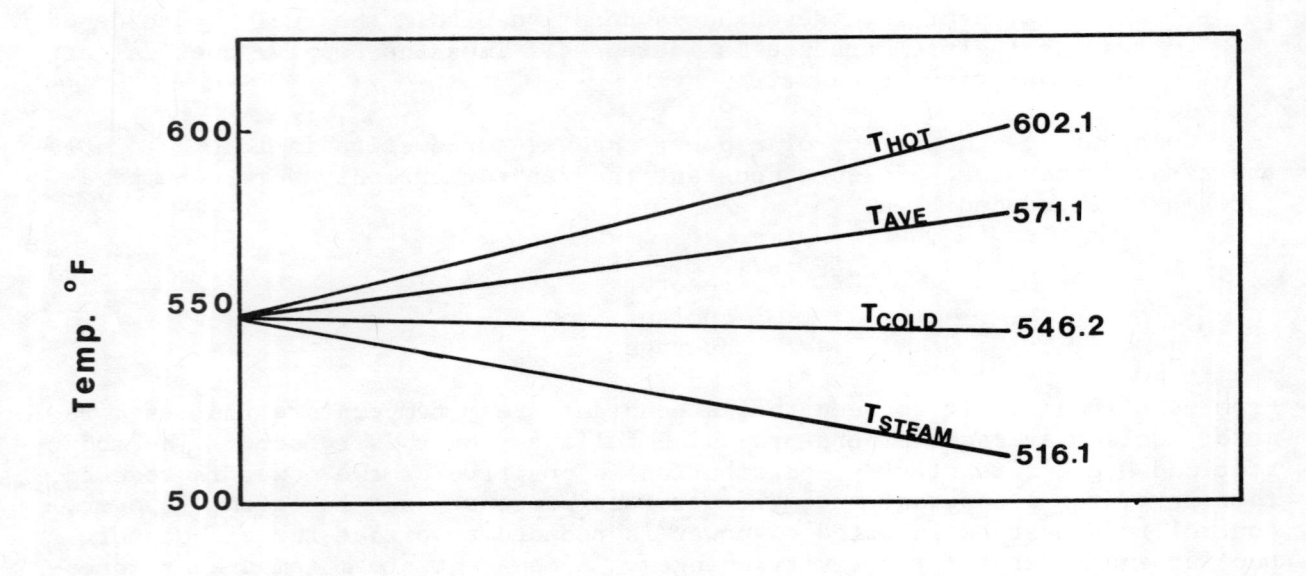

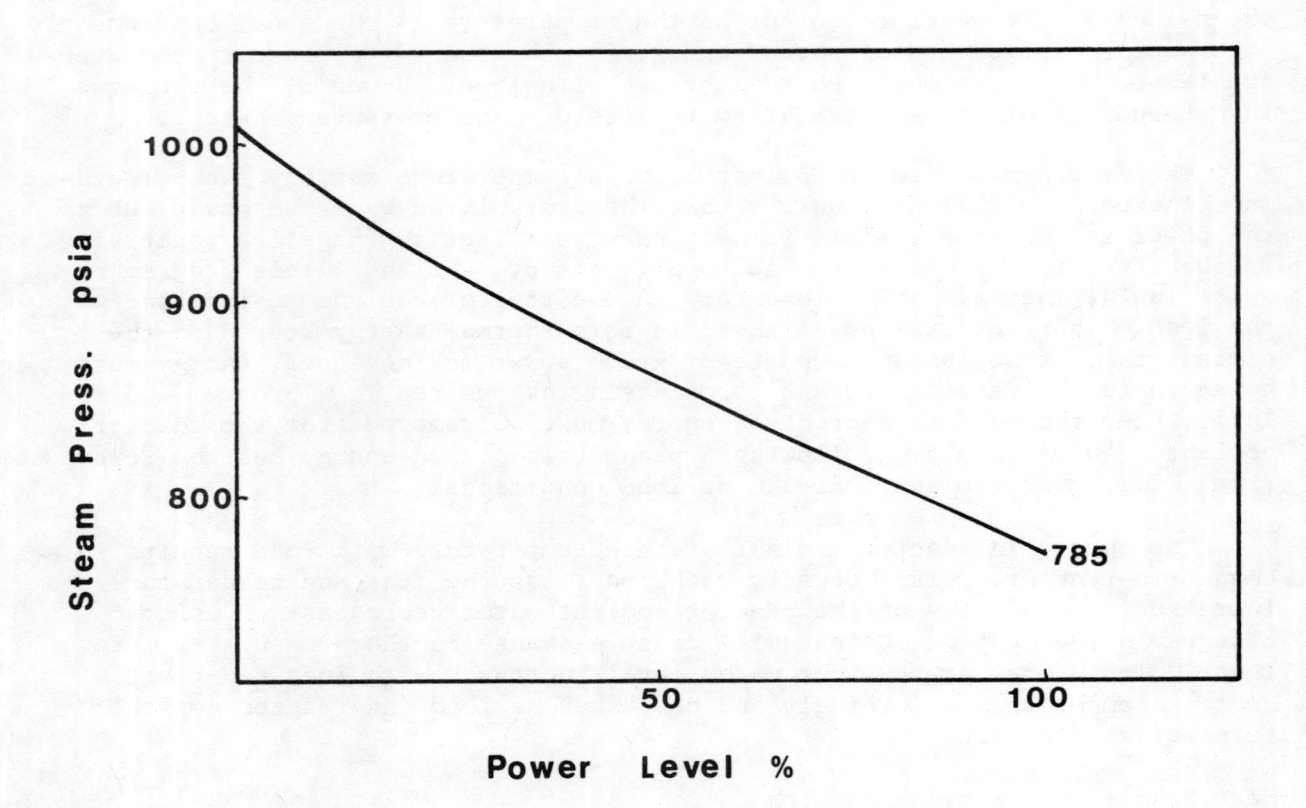

Fig. 8. Temperature and Pressure vs Load

Similarly, if the heat source < heat sink there is a decrease in the energy inventory stored in the reactor coolant.

For an ideal load change, the turbine load would change in a step manner (as shown in Fig. 9) and the core power would change in a similar manner to equal the turbine load. The core power must exceed the turbine load for some period of time in order to heat the reactor coolant water to the new T_{avg} corresponding to the new plant power level. Consequently, there will be some overshoot in the reactor power for a turbine load increase, even in the ideal case.

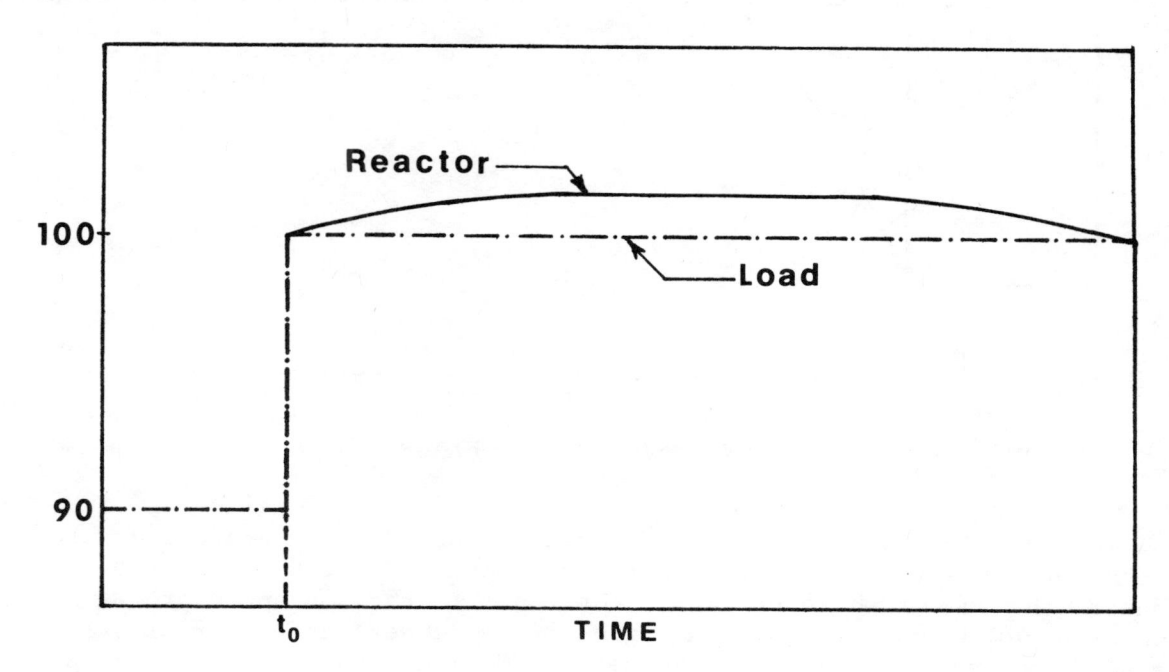

Fig. 9. Ideal Response for a Step Change in Load

Following a step change in turbine load, the reactor power cannot change in a step manner to the new steady state power level. In order to change the reactor power, and the reactor coolant temperature along programmed lines, control rods must be withdrawn to offset the Doppler and moderator reactivity effects. The reactor power is thus less than the turbine load for a period of time. This means that the reactor power must exceed the turbine load later in the transient in order to make up for the energy removed from the reactor coolant water when turbine load exceeds reactor power and in order to heat the reactor coolant to the new programmed T_{avg} corresponding to the new turbine load. The result is that there is an overshoot in the reactor power following a load increase. This is depicted in Fig. 10.

The overshoot in reactor power must be maintained within limits in order to avoid a reactor trip at high nuclear power. This is a design criterion for the reactor control system. The nuclear power overshoot following a design load increase to 100% power must be less than 3% in order to provide margin for the high nuclear power reactor trip normally set at ~ 108%. By moving the control rods at maximum speed early in the transient, the turbine load-reactor power mismatch can be held to a minimum thus aiding in reducing the overshoot when the reactor power nears the final steady state value.

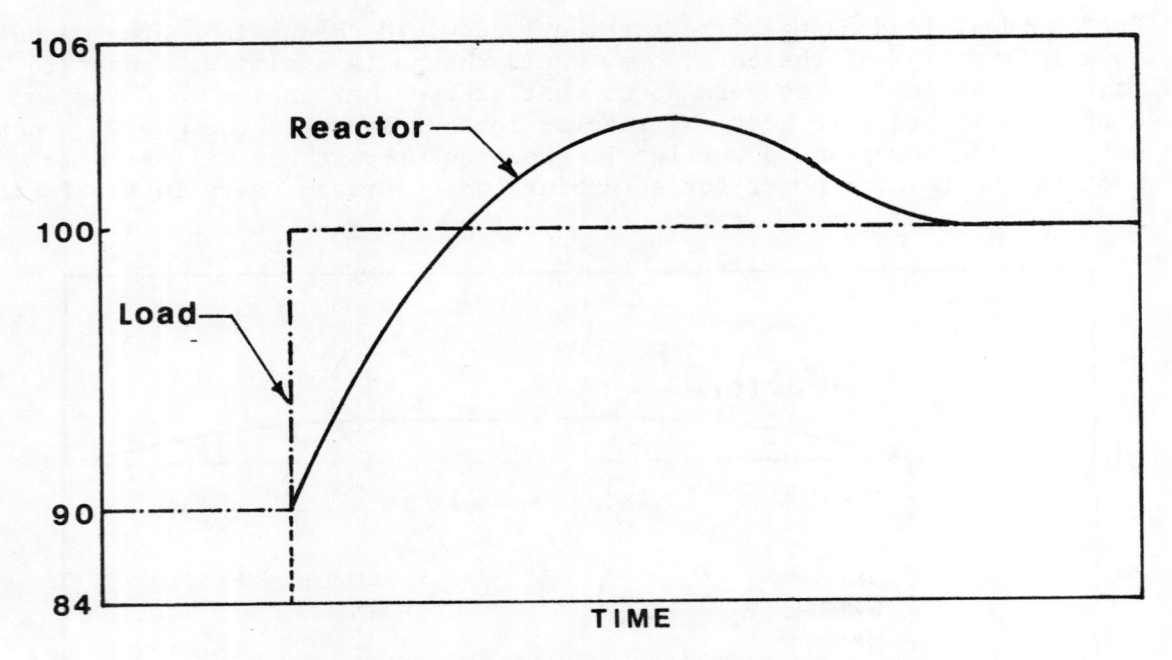

Fig. 10. Reactor Response Following Load Increase

Figure 11 is a block diagram of a reactor coolant temperature controller. The reactor coolant temperature controller consists of two basic error signal channels, the sum of whose output is the input to the rod speed programmer which produces a rod speed demand signal. The temperature error signal is generated by comparing the measured Reactor Coolant System average temperature with the desired average temperature at a particular load. This error signal is used primarily for the purpose of bringing the steady state reactor coolant temperature to the correct programmed value. The power mismatch channel develops an error signal which depends upon the rate of change of the difference between the measured nuclear power and the measured turbine load. This channel provides fast response to a change in load (by means of the turbine load feed-forward signal) as well as control stability (by means of the nuclear power feedback signal) in cases where the moderator coefficient is zero or only slightly negative. The automatic rod control system is designed to maintain a programmed average temperature in the reactor coolant by varying the reactivity within the core. Typically this system is capable of restoring the average temperature to within ± 3.5°F of the programmed temperature including a ± 2°F instrument error and a ± 1.5°F deadband, following design load changes.

The control system is designed to automatically control the reactor in the power range between 15 and 100 percent of rated power for the following design transients:
  1. ± 10% step change in load;
  2. 5%/minute ramp loading and unloading; and
  3. 95% or 50% (depending upon the particular plant) step load decrease

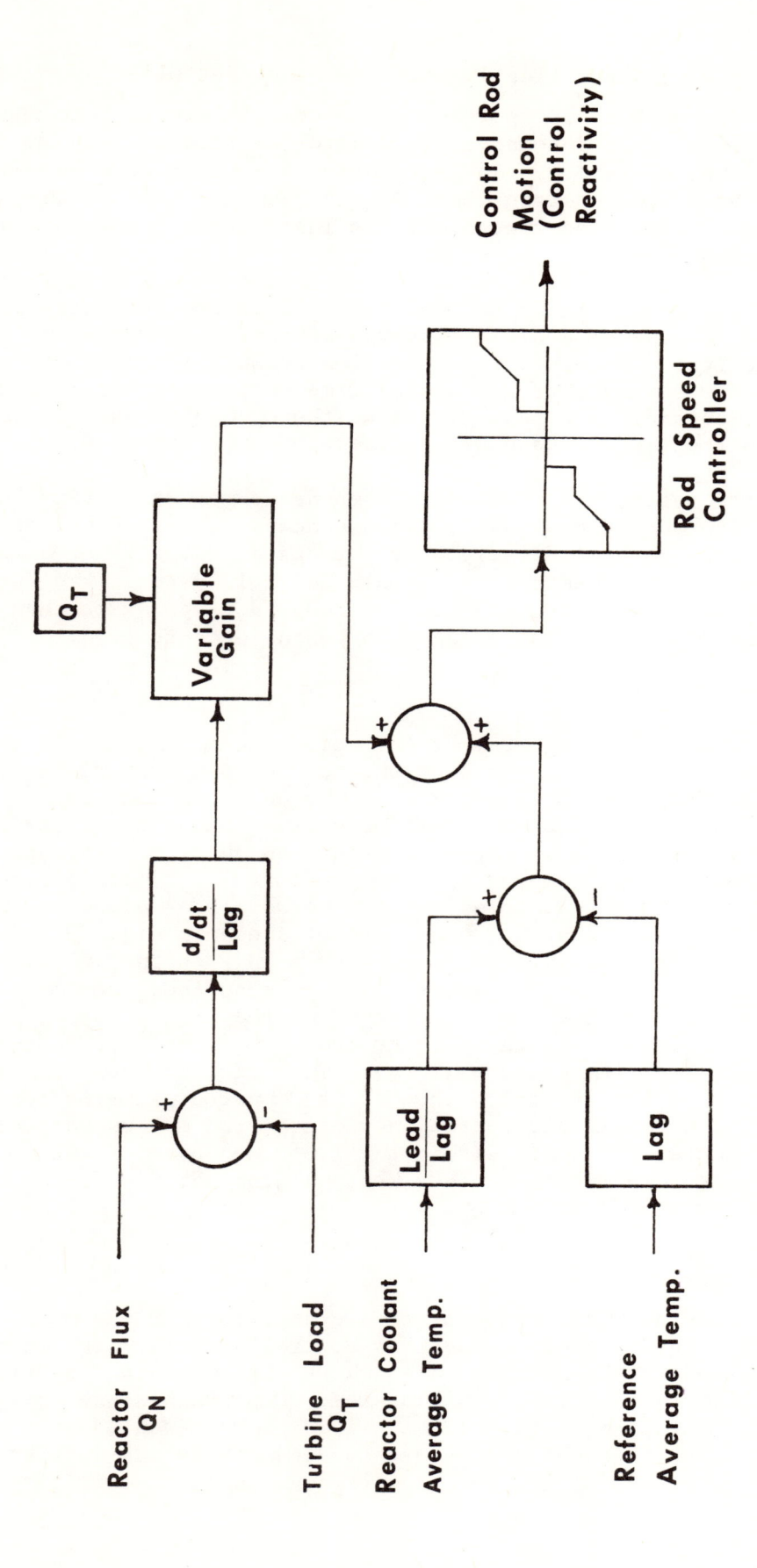

Fig. 11. Reactor Coolant Temperature Controller

with the aid of automatically initiated and controlled steam dump.
A step load change from 90 to 100 percent power or a 5%/minute load increase to 100 percent power must be automatically controlled without tripping the plant on nuclear power overshoot. In particular, a nuclear flux overshoot resulting from the above imposed transients must be limited to less than three percent of full power flux even though all the instrumentation errors are in the adverse direction.

Assurance of stable automatic control below 15 percent of rated power generally requires the use of a different type of controller than the normal power range controller. Since station operation below 15 percent of rated power occurs only for a very short period of time during startup or standby conditions, manual control is acceptable under these conditions. Automatic control below 15 percent of rated power is therefore not provided.

The rod control system will initially compensate for reactivity changes due to fuel depletion and/or changes in xenon concentration. Final compensation for these effects is periodically made through adjustments of boron concentration by means of the Chemical and Volume Control System. The automatic rod control system then readjusts the control rods in response to changes in average temperature resulting from changes in boron concentration.

Average Temperature Channel

One average temperature measurement per reactor coolant loop is provided. This measurement is obtained by averaging the hot leg temperature (T_h) measured at the inlet of the steam generator and the cold leg temperature (T_c) measured at the discharge side of the reactor coolant pump of the associated loop. These average temperatures are passed into an auctioneering unit which generates the highest of the loop average temperature (T_{avg}) signals. The auctioneered T_{avg} signal is sent to a lead/lag network. A second lag network is provided to filter out signal noise and is not shown in Fig. 11. This signal is then compared with a reference temperature (T_{ref}) signal. (The reference temperature is a function of turbine load, as described previously.) Since the steam pressure in the impulse chamber of the high pressure turbine (P_{imp}) is linear with respect to the turbine load, it is used to generate the reference average coolant temperature (T_{ref}). The reference temperature signal is passed through a lag network before it is compared with the compensated T_{avg} signal. During steady-state operation, control rod motion maintains the absolute value of the error signal within the 1.5°F deadband of the rod controller. The average temperature is thus kept within ± 3.5°F of its programmed value, as reported previously.

Power Mismatch Channel

This channel provides fast response to a change in load (by means of the turbine load feed-forward signal) as well as control stability (by means of the nuclear power feedback signal) in cases where the moderator coefficient is zero or is only slightly negative. Turbine load (Q_T) and nuclear power (Q_n) provide the inputs to this channel. Turbine load is represented by the impulse chamber pressure of the high pressure turbine, while the nuclear power signal is obtained from one of the four power range nuclear power signals.

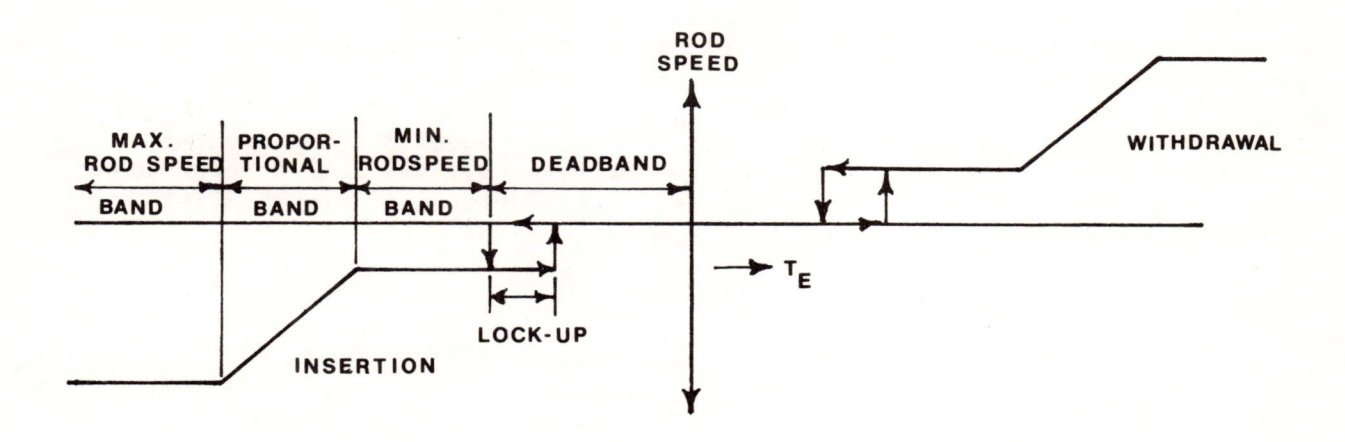

Fig. 12. Rod Speed Program

The deviation between Q_T and Q_n is passed through a rate/lag unit. Since the T_{avg} channel provides fine control during steady-state operation, the power mismatch channel must not produce a steady-state error signal. This is accomplished by derivative action which causes the output of this unit to go to zero during steady-state operation, although the nuclear power and turbine load may not match exactly. A nonlinear gain placed at the output of the impulse unit varies the effect of this channel, with larger load changes having a correspondingly greater effect. Also, since reactivity changes at low power levels have a smaller effect on the rate of change of the nuclear power level than reactivity changes at high power levels, a variable-gain is provided at the output of the power mismatch channel.

This variable gain imposes a high gain on the power mismatch error signal at low power levels, and a low gain at high power levels. The variable gain enables the mismatch channel to provide adequate control at low power levels, as well as stable operation at high power levels.

Rod Speed Program

The total error signal (T_E) sent to the rod speed program is the sum of the outputs of the control channels described above. Figure 12 shows the rod speed program as a function of the total error signal (T_E).

The deadband and lockup are provided to eliminate continuous rod stepping and bistable chattering. Maximum rod speed and the proportional and minimum rod speed bands are identical for rod withdrawal and rod insertion. The rod speed program produces an analog signal which is translated into rod movement by means of the rod stepping mechanism.

Control rods are divided into four banks. Bank A is withdrawn first, followed in order by Banks B, C, and D. Two banks operate simultaneously over certain regions to ensure adequate incremental reactivity.

The control rods are driven by a sequencing variable-speed rod drive control unit. Control rods in each bank are divided into two groups which are moved sequentially: one group at a time. The sequence of motion is reversible; that is, the withdrawal sequence is the reverse of the insertion sequence. Hence, the two groups of a given control bank never deviate from each other by more than one step. Variable speed sequential rod control makes it possible to insert small amounts of reactivity as needed to accomplish fine control of the reactor coolant average temperature within the ± 1.5°F temperature deadband.

Pressurizer Control

Figure 13 shows the fluid systems arrangement used to control the pressurizer level, pressurizer pressure, and coolant pressure. As the pressure decreases below the desired value of 2250 psia the heaters are energized. This heats the water in the pressurizer and boils water to return the pressure to the nominal value. Water in the pressurizer is normally maintained at the saturation temperature corresponding to 2250 psia. Any transient which causes an outsurge of water from the pressurizer leads to flashing of the water in the pressurizer. The flashing retards the fall in pressure. When the pressure increases above 2250 psia spray is used to condense steam in the steam region and return the pressurizer to 2250 psia. For transients where spray is not sufficient, steam is released from the pressurizer by actuation of the pressurizer power operated relief valves. These valves are actuated on high pressurizer pressure and are set to operate prior to reaching the high pressure reactor trip setpoint. The pressurizer power operated relief valves are also actuated prior to reaching the pressure at which the self-actuated safety valves are lifted.

Pressurizer level is controlled by changing the flow from the charging pumps. A continuous small release of coolant occurs via the letdown valve. For decreasing pressurizer levels the charging flow is increased such that charging flow exceeds letdown flow. For increasing pressurizer levels, charging flow is decreased such that letdown flow exceeds charging flow.

As indicated previously, the reactor coolant density will decrease as the average temperature is increased to follow the programmed temperature for increases in load. Assuming a constant water mass in the Reactor Coolant System plus Pressurizer, the pressurizer level will increase (in steady state) as load is increased. In order to minimize the possibility of having to remove water from the Reactor Coolant System when steady state load is increased and of having to add water when steady state load is decreased, the pressurizer level is programmed. The reference pressurizer level (desired value for the level) is programmed as a function of Reactor Coolant System average temperature.

The measured pressurizer level is compared with the programmed level and the error is used to control charging flow with a proportional plus integral controller. On very low pressurizer level, the letdown flow is stopped by

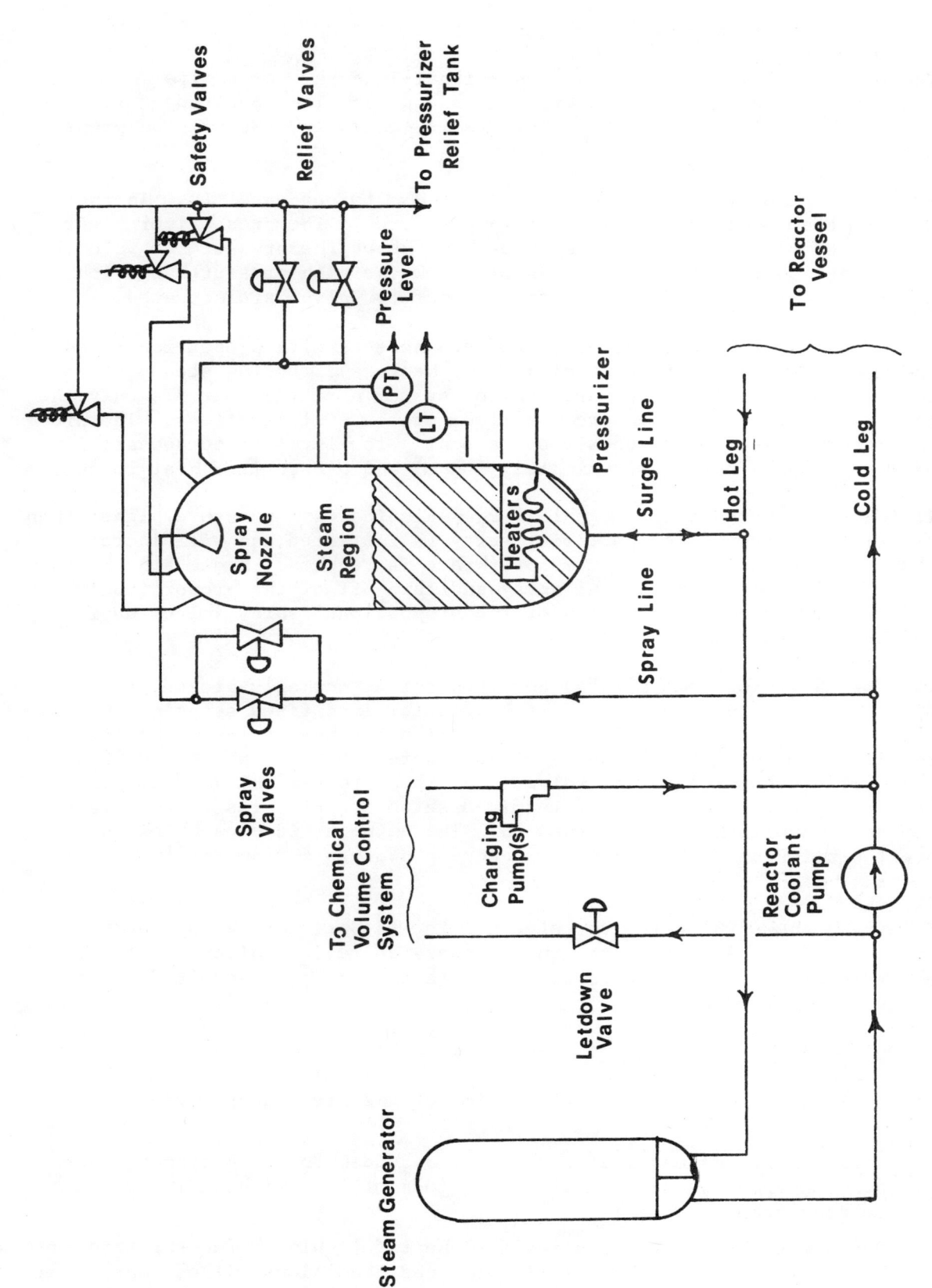

Fig. 13. Pressurizer Pressure and Level Associated Equipment Arrangement

closing an isolation valve in the letdown line.

The error signal is used in a Proportional-Integral-Derivative Controller to control pressurizer spray, heaters, and one pressurizer power operated relief valve. Normally a second relief valve is controlled directly on pressurizer pressure.

Pressurizer spray comes on for high pressures to condense steam in the pressurizer steam space and minimize high pressures. The pressurizer power operated relief valves are opened at pressures above the pressure at which spray is turned on fully but at pressures below the high pressure at which the reactor is tripped or the self actuated safety valves are actuated.

The pressurizer heaters are divided into heaters with proportional control and heaters with on-off control (backup heaters). During steady state operation, the pressurizer pressure control system normally controls only the proportional heaters to compensate for minor pressure fluctuations. The proportional heaters will continuously operate at a low level to compensate for the continuous spray rate (approximately 1 gpm) and pressurizer heat losses.

If the compensated error signal ($P-P_{ref}$) indicates a pressure higher than a predetermined setpoint, proportional spray is initiated and will increase with the pressure until the maximum spray rate is reached. A deadband between the initiation of the proportional spray and turn-off of the proportional heaters prevents frequent operation of the proportional spray valves during minor system pressure variations.

Two (or three for some plants) power-operated, normally-closed, on-off relief valves begin their operation at a predetermined fixed setpoint to limit the system pressure to 2350 psia. The operation of these valves also limits the undesirable opening of the spring-loaded safety valves, which have a higher set point then the relief valves. If the error signal ($P-P_{ref}$) indicates a pressure lower than a predetermined setpoint, all pressurizer heaters (back-up and proportional) are turned on. The setpoint is chosen low enough to prevent continuous switching of the back-up heaters during small pressure variations.

Figure 14 shows the relative setpoints for the pressurizer pressure control system -- i.e. heaters, spray, power operated relief valves. The high pressure reactor trip setpoint is at 2400 psia and the self-actuated safety valves are set at 2500 psia.

Steam Dump and Steam Dump Control

Figure 15 is a schematic representation of the steam dump system. The steam dump:

1. Permits the nuclear plant to accept a sudden 95 to 50 percent (depending upon the particular plant) loss of load without incurring a reactor trip.

2. Removes stored energy and residual heat following a reactor trip and brings the plant to equilibrium no-load conditions without actuation of the steam generator safety valves.

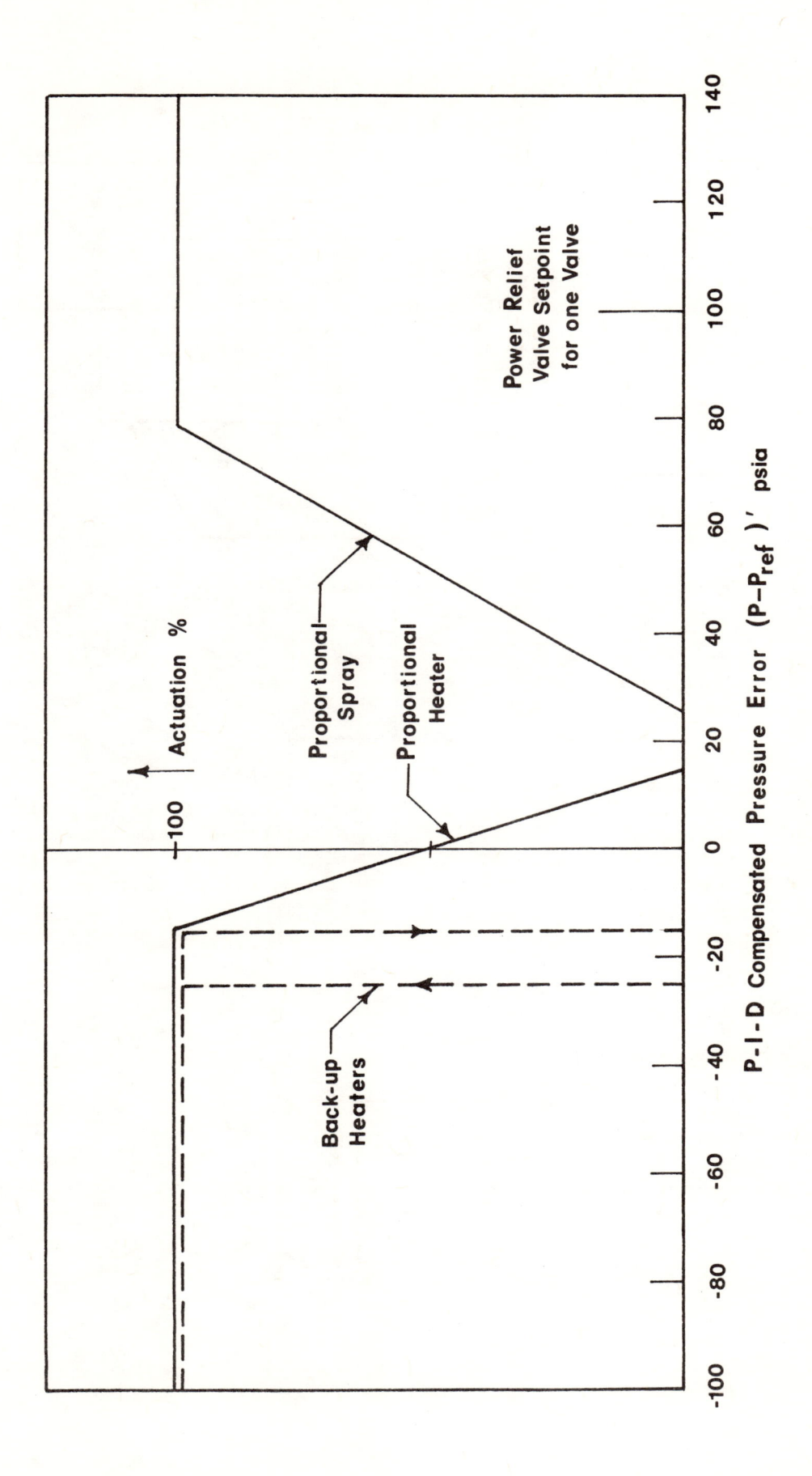

Fig. 14. Pressurizer Pressure Control Scheme

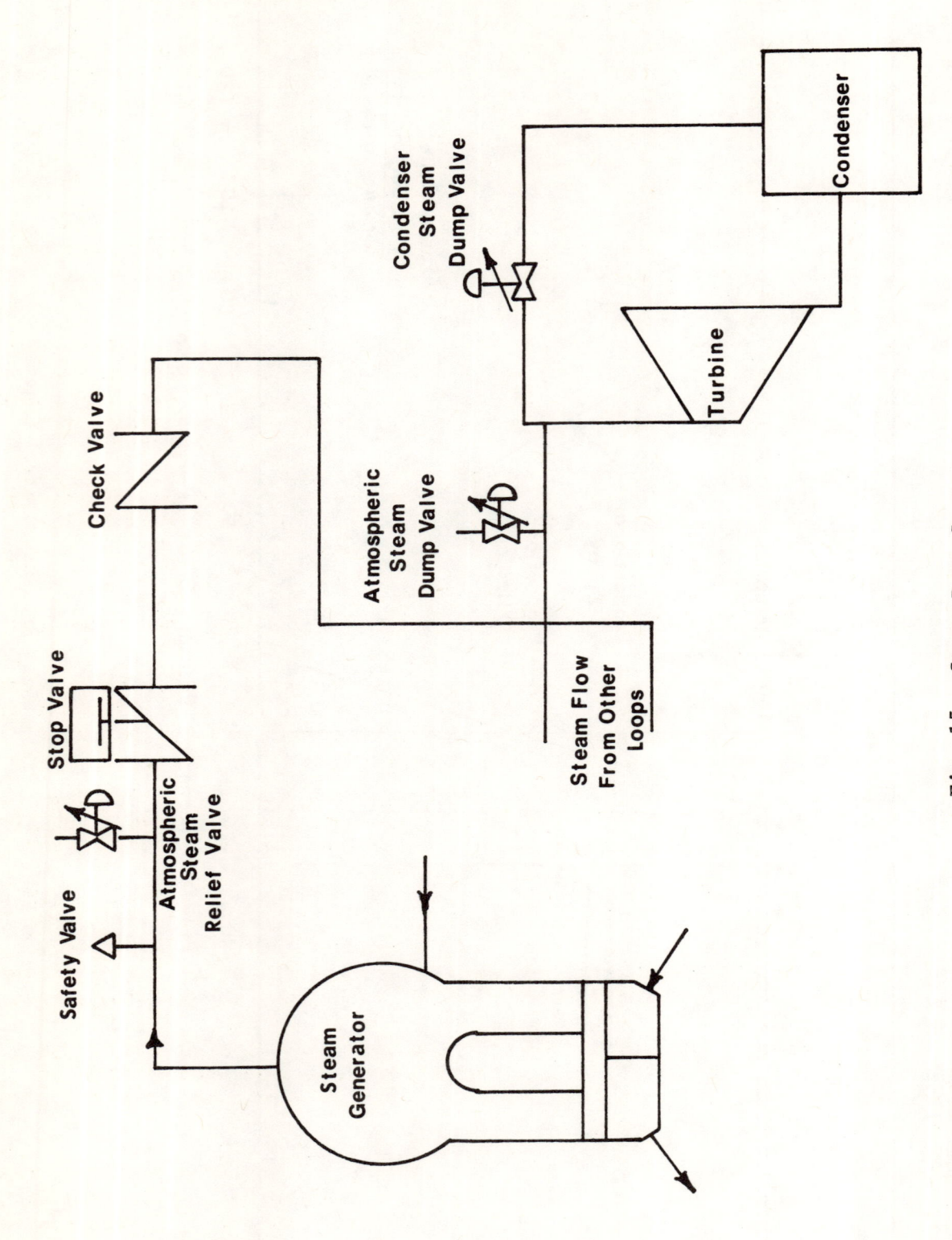

Fig. 15. Steam Dump System

3. Permits control of the steam generator pressure at no-load conditions and permits a manually controlled cooldown of the plant.

Basically, for a large load reduction, the steam dump releases steam either to the condenser or to the atmosphere in order to reduce the magnitude of the decrease in steam flow at the steam generators. The sequence of events is:

A) A load reduction with a sudden large decrease in steam flow to the turbine (up to 95% on some plants, 50% on others).

B) Opening of the steam dump valves within 3 seconds after the load reduction thereby increasing the steam flow from the steam generators to within approximately 10% of the initial value before the load reduction.

The 95% load reduction followed by steam dump thus appears to the steam generators, Reactor Coolant System, and reactor as a step decrease in load of approximately 10%. The steam dump is then modulated closed as the core power is reduced to an equilibrium steady state condition at the new turbine load.

The functional tasks of the steam dump are accomplished by three modes of steam dump control. Requirements 1 and 2 above are met by control of reactor coolant T_{avg}, whereas requirement 3 is met by control of the steam pressure. Interlocks minimize any possibility of an inadvertent actuation of the steam dump system.

A typical steam dump system which would allow a 95% step load reduction without a reactor trip would be comprised of the following:

1. Several valves which can bypass steam to the condenser. These valves would have a total capacity of 45% of the full load steam flow at the full load steam pressure. The valves receive flow from the steam lines downstream of the main steam stop valves.

2. Several valves which can bypass steam to the atmosphere. These valves would have a total capacity of 40% of the full load steam flow at the full load steam pressure. The valves receive flow from the steam lines downstream of the main steam stop valves.

The total capacity of a steam dump system for a plant designed to take a 95% load rejection without a reactor trip is thus 85% of the full load steam flow at the full load steam pressure. It should be noted that all dump (85%) could go to the condenser thereby avoiding any dump to the atmosphere.

For load losses up to about 50% only the condenser dump valves are employed. If the load loss is greater than 50% the condenser dump valves and atmospheric dump valves may be used. The steam dump is not actuated for load losses less than 15%.

Although not a part of the steam dump system, it should be mentioned that there are three steam generator power operated relief valves. These valves have a total capacity of 10% of the full load steam flow. The valves receive flow from the steam lines (one per steam generator) upstream of the main steam stop valves. These valves are set to open before the steam generator self actuated safety valves and are controlled by steam line pressure. The valves can be used to control plant temperature at no load or cool the plant down should the steam dump system not be available for any reason. Normally the steam dump valves and steam generator power operated relief valves are air operated.

Steam Generator Level Control

Figure 16 is a diagram of the steam generator. The steam generator is essentially a boiler where the energy transferred from the reactor coolant flowing on the primary side (within the steam generator tubes) boils water on the secondary side to generate the steam to drive the turbine. The secondary side of the steam generator contains a downcomer region which is an annular region between the steam generator shell and a cylindrical wrapper surrounding the steam generator tubes. These tubes contain the reactor coolant and make up the heat transfer region of the steam generator. Feed water is pumped into the steam generator by the feedwater system. This water flows into the downcomer, down the downcomer, under the tube wrapper, and up the steam generator riser section which contains the steam generator tubes. In the tube or riser region the water is heated and steam is created. At the top of the riser or tube region swirl vanes are used to separate a portion of the water from the steam. The steam continues on up the steam generator to the chevron moisture separators where the remaining moisture content is removed from the steam. Steam leaves the steam generator with a moisture content of less than 0.25%; the water separated from the steam by the swirl vanes and moisture separators flows to the downcomer and is mixed with the feedwater entering from the feed system. Roughly the water in the downcomer is made up of two pounds of water recirculating from the swirl vanes at the top of the tube region per pound of feedwater entering from the feedwater system (at full load).

Each steam generator is equipped with a three element feedwater controller which maintains a programmed water level on the secondary side of the steam generator during normal plant operation. This controller continuously compares measured feedwater flow with measured steam flow, and a compensated steam generator downcomer water level signal with a water level setpoint to regulate the main feedwater valve opening. Manual override of the steam generator level control system is also provided.

A bypass line with a bypass feedwater control valve is provided around each main feedwater control valve. The bypass valve is used to manually control the feedwater flow at loads below approximately 15%, which is outside the load range where automatic control of the main feedwater control valve gives optimum response. The discussion which follows is limited primarily to the steam generator level control system used with the main feedwater control valves.

The three-element valve control functions to maintain the necessary main feedwater valve position so that sufficient feedwater flows into the steam generator to maintain the level at the programmed value. Three signals determine the valve position: the level error signal, the steam flowrate signal, and the feedwater flowrate signal.

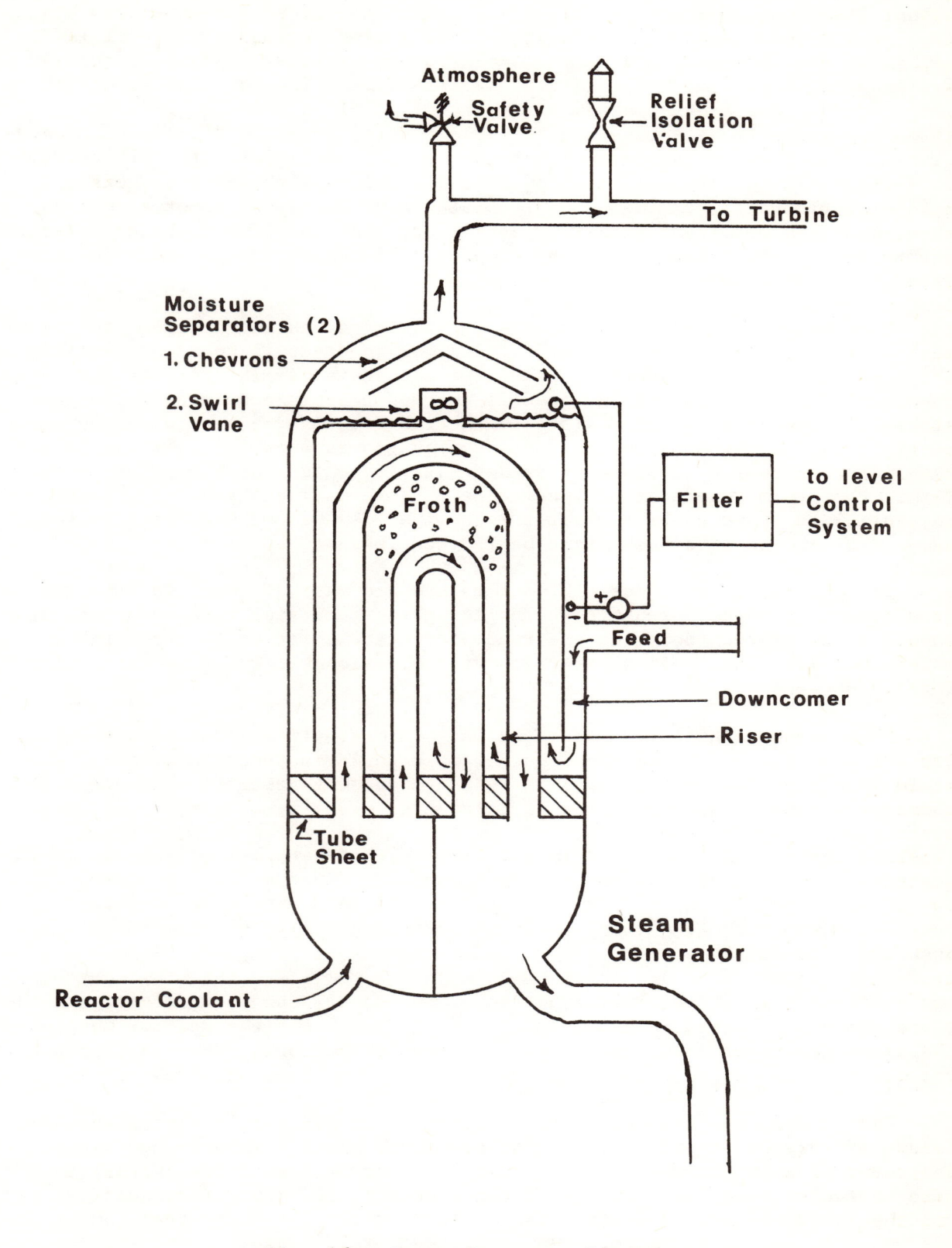

Fig. 16. Steam Generator Diagram

The level error signal represents the deviation of the measured level signal from its programmed level. This level is derived from the firststage turbine impulse pressure. The signal representing the measured level is obtained by sending the signal from the narrow-range level channel through a filter network to dampen the natural oscillations. The two signals are then compared, thereby generating an error signal. This error signal is sent through a proportional-plus-integral (PI) controller which eliminates steady-state level errors. The output from the proportional-plus-integral controller is sent to the three-mode valve controller. This controller subtracts the feedwater flowrate from the steam flowrate, adds the level error signal and then sends the resultant signal to a PI controller to eliminate steady-state errors in feedwater flow. The output of the PI controller is the main feedwater valve position signal.

This system is provided with the capability of having a steam generator level which varies as turbine load varies. When water is boiled rapidly in the steam generator tube region, the region contains a mixture of steam bubbles and water. As turbine load is increased from zero load (where no steam bubbles or steam void exists in the tube region), a greater and greater percentage of the volume in the tube region becomes occupied by steam rather than water. Thus, with constant steam generator water level, the water mass within the steam generator will decrease as load is increased. Typically, the full load steam generator water mass might be 100,000 lbs while at the same level the no load mass might be 150,000 lbs.

At full load, it is desirable to maintain a mass of water in the steam generator to provide a reasonable margin against emptying the steam generator should a malfunction occur in the feedwater system. It is also desirable to maintain a full load mass of water, giving a reasonable total heat capacity for the secondary side.

At no load, high water masses in the steam generator are undesirable, since a rupture of a steam line would lead to a large reactor coolant system cooldown as well as to a large integrated release of steam to the containment should the rupture be within the containment.

For the reasons above, it is not desirable to have a large change in the steam generator water mass as load is changed. Since a constant level at all plant loads would lead to a fairly large reduction in water mass as load is increased from zero load to full load, provisions for level programming have been provided in the system.

Steam generator level is important. Low levels can lead to emptying the steam generators which leads to a loss of heat sink for the Reactor Coolant System. High levels can lead to water carry over through the steam lines to the turbine, which can cause major damage to the turbine and steam lines.

Various reactor trip circuits are provided to trip the reactor should a loss of feedwater lead to a low steam generator level. This is necessary since emptying the steam generators will lead to a loss of the capability to remove reactor core heat. Even after a reactor trip, heat is produced in the core by the decay heat from the fission products. This decay heat cannot be "turned off" and is of the order of magnitude of 1% of full power even hours after a reactor trip.

As a result of the production of decay heat, an auxiliary means of supplying feedwater to the steam generators is provided in the event that a major malfunction of the main feedwater system occurs.

Four means are provided to override the control signals from the steam generator control system:

1. Manual control;
2. High level override, which closes the feedwater valve;
3. Low T_{avg} and reactor trip, which closes the feedwater valve; and
4. Safety Injection Actuation, which closes the feedwater valve.

The main feedwater control valve can be controlled manually from the control room. A high level override exists to prevent excessive moisture carryover to the turbine. Actuation of this signal closes the main feedwater valve. When this signal ceases, control reverts to the previous means of control.

Upon actuation, the low T_{avg} signal causes the main feedwater valves to close following a reactor trip. This override prevents excessive reactor coolant system cooldown.

Actuation of the Safety Injection System (SIS) will cause all main feedwater control valves to close. This is a portion of the action required to ensure that main feedwater is blocked to all steam generators following SIS actuation. It is included because the severity of certain accidents which cause SIS actuation can increase with additional feedwater flow.

BOILING WATER REACTOR CONTROL (3,6)

Many of the control aspects of the PWR, as discussed previously, apply to the BWR. This section will be concerned with those control features which are unique to the BWR. The BWR steam supply system differs from the PWR in that steam is produced by boiling in the reactor core instead of in the secondary side of a heat exchanger as in a PWR. The BWR is referred to as a Direct-Cycle Reactor System and the PWR as an Indirect-Cycle Reactor System. Figure 17 is a schematic of a typical BWR system. Because of boiling in the core, control rods are inserted from the bottom since the liquid-vapor mixture in the upper portion of the core reduces local reactivity. Poisoning from below produces a more even power distribution along with flexibility in shaping the neutron flux as the zone of voids and position of the control rods are independently adjusted. Also, the steam-separator and dryer is placed directly above the core which would make control rod insertion from the top difficult.

The control-rod-drive for a BWR is hydraulically operated using water under pressure, compared with that of a PWR which uses a rack and pinion driven by a motor through a gear box or by a magnetic-jack type system. In an emergency the hydraulic system drives all control rods into the core with a velocity on the order of 5 ft/sec. In comparison, PWR control-rod-drive provides for a gravity scram.

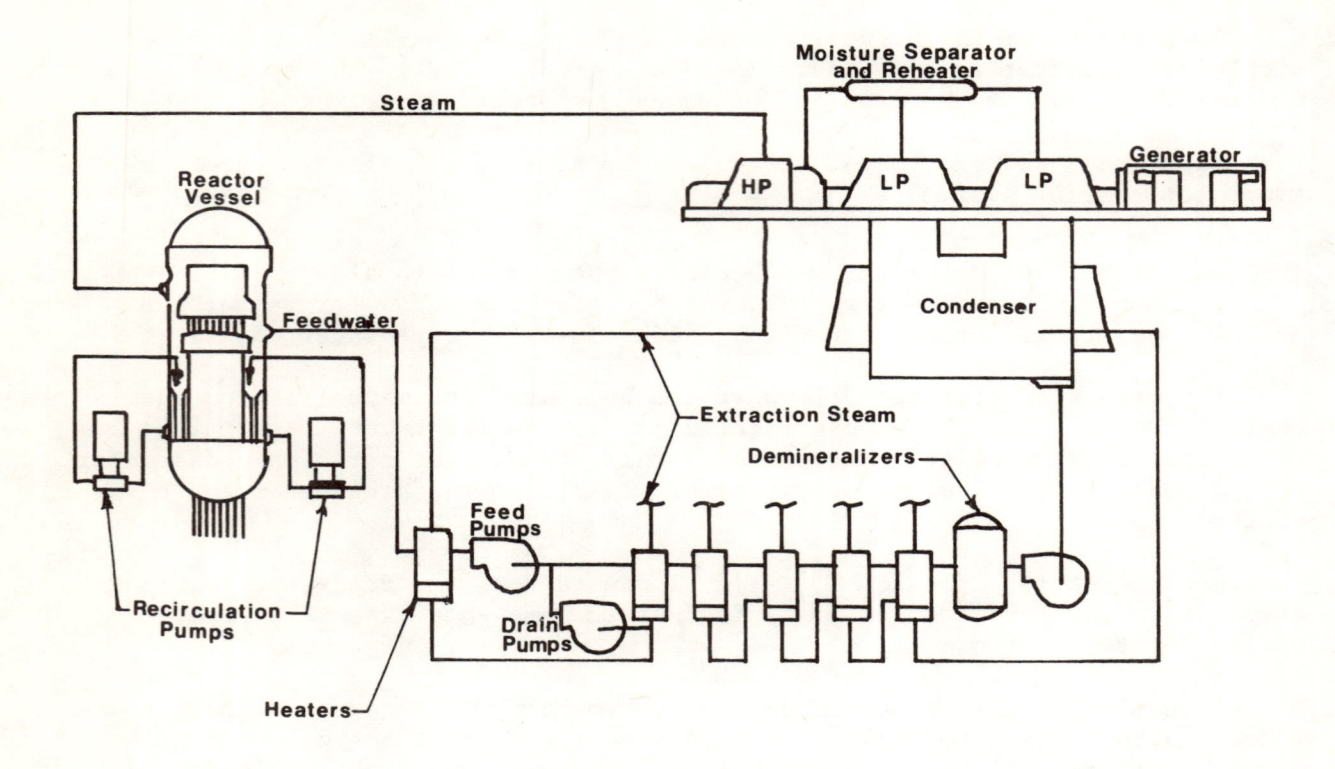

Fig. 17. Direct Cycle Reactor System

A BWR control rod is a cruciform shaped blade as shown in Fig. 18. The blade consists of an array of sheathed stainless steel tubes filled with B_4C powder. Control blades are inserted between fuel bundles as depicted in Fig. 19.

Power Level Adjustment

As the control rods are withdrawn, the reactor power increases until the increased steam formation just balances the change in reactivity caused by withdrawal of the control rods. Control rods can be operated in groups or one at a time. Load rate changes up to about 60 MW_t per minute can be obtained using control rods. This is the normal method of changing power level for increases and decreases of more than 25% of rated power. For power level changes less than 25% of operating power, adjustments are made by varying the recirculation flow without any movement of control rods. This is the normal method used in load following and allows for rates of up to 1% of rated power per second. Figure 20 shows the Jet Pump Recirculation System.

Power changes are caused by the void coefficient of reactivity. An increase in recirculation flow temporarily reduces the number of steam voids and consequently increases the reactivity of the core which causes the power level to increase. This increase in power generates more steam which reduces reactivity and a new steady state power level is reached. When recirculation flow is reduced the power level is reduced in a similar manner. Operation at

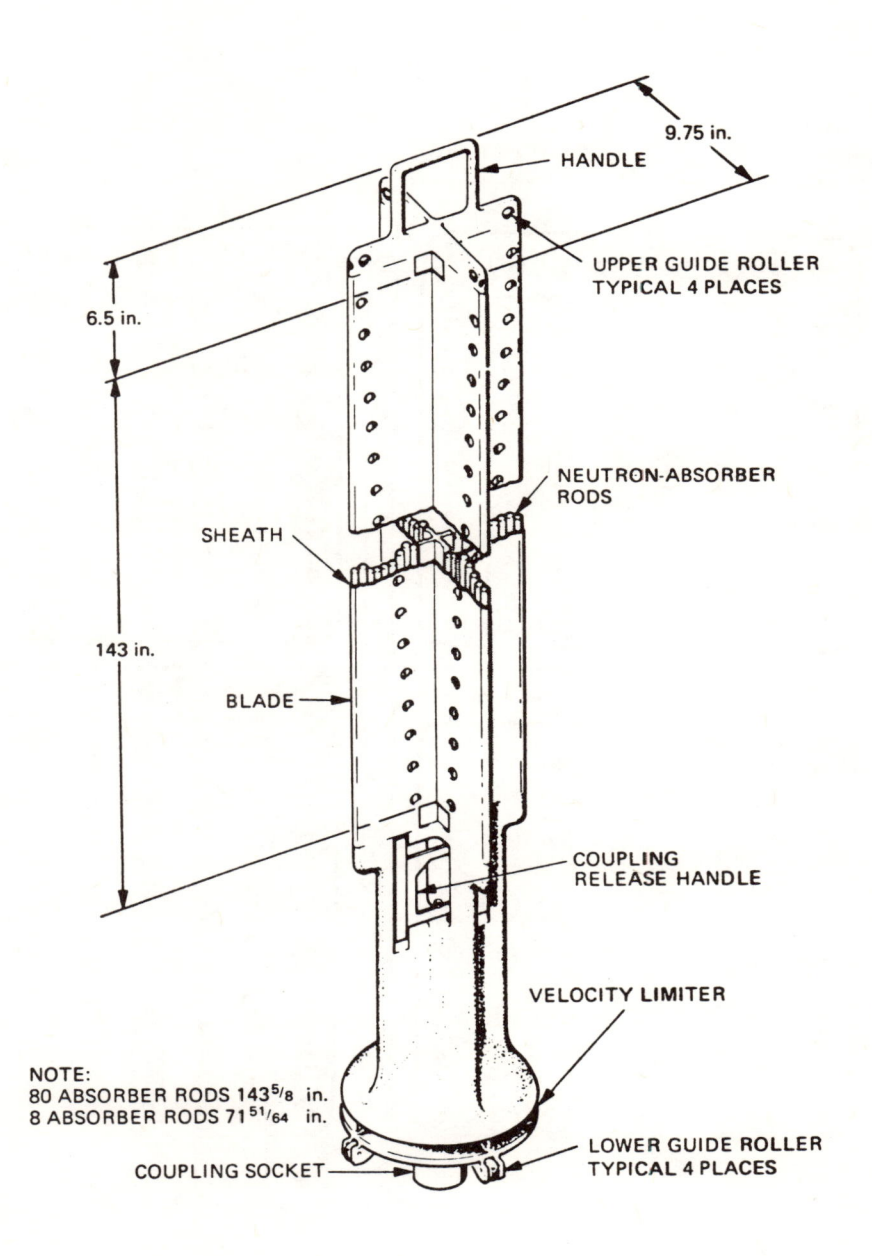

Fig. 18. Boiling-Water-Reactor Control Rod with Cruciform-Shaped Blade

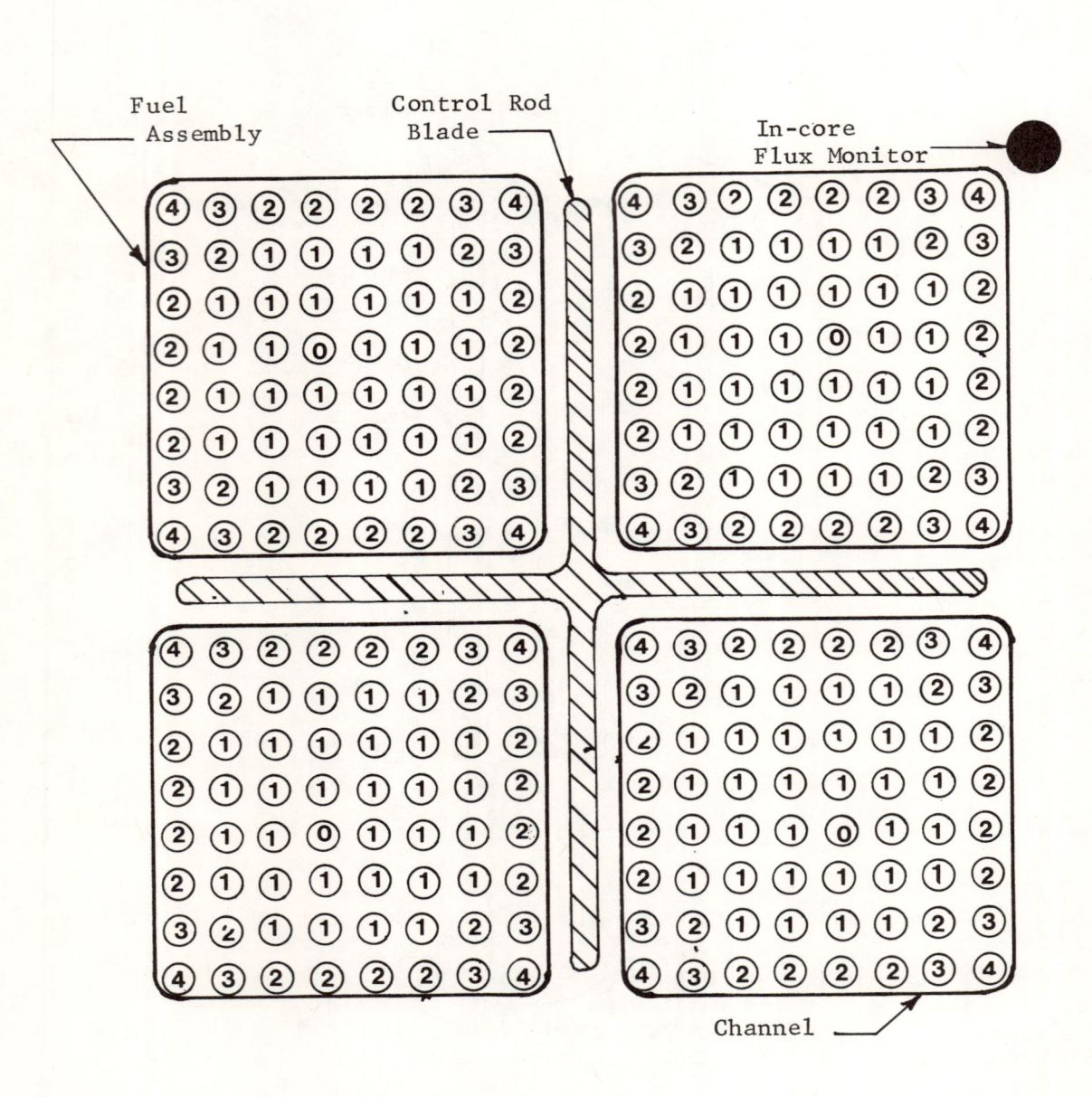

Fig. 19. Core Lattice

lower recirculation flow gives a better neutron economy and more favorable fuel cost. Operating curves are established for various recirculation flow rates and rod patterns.

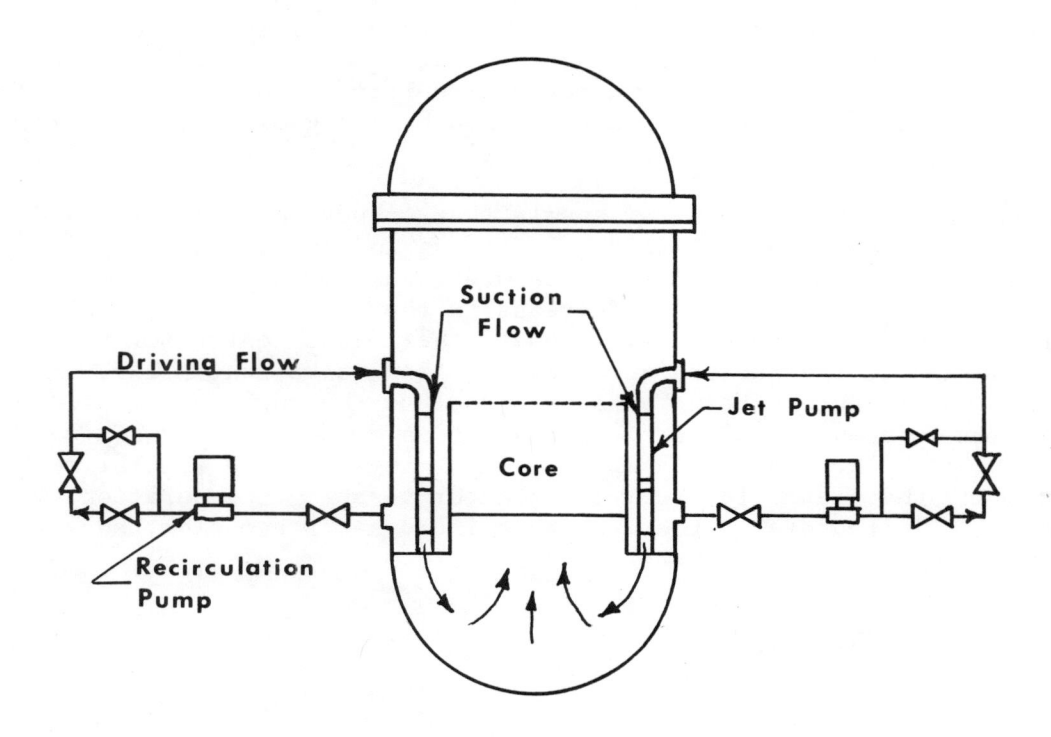

Fig. 20. Jet Pump Recirculation System

During operation the reactor power level may be changed by varying the recirculation flow, control rod position or a combination of the two using the operating curves. These curves are periodically evaluated during startups.

A schematic diagram of the Reactor-Turbine Control is shown in Fig. 21. To change reactor power, a signal from the master controller adjusts the settings of the controller for each valve until the error signal between the actual valve position and the master controller signal is zero. The master controller receives its command signal from the operator or a load/speed error. The change in reactor power as a result of the change in recirculation flow causes the initial pressure regulator to reposition the turbine control valves. Automatic load control is obtained by supplying a speed-load error signal from the turbine governor to the master controller. To increase the speed of response of the system to a load change, an automatic temporary change in the setpoint of the pressure regulator is produced by the demand for a change in the turbine output. An increase in load demand lowers the pressure setpoint and water in the reactor flashes to produce extra steam for the turbine. A decrease in load raises the pressure setpoint which causes the turbine control valve to move toward the closed position.

System Control

The two main functions which must be controlled for normal operation are the system pressure and reactor coolant flow rate. System pressure is maintained nearly constant by the pressure regulator which provides a signal to the turbine admission valves. Reactor recirculation flow required to meet the system power requirements is established by a signal from the speed-load indicating mechanism to the master controller.

Consider now the sequence of events that occurs when there is a decrease in turbine speed under normal load. Referring to Fig. 21, a positive speed-load signal is sent to the initial pressure regulator and the master controller which causes a momentary decrease in the pressure setting of the initial pressure regulator and also causes the master controller to increase the flow demand to the recirculation system flow valve controller. A decrease in the pressure setting sends a signal to the turbine admission valves causing them to open rapidly proportional to an amount and for a time which is a function of the speed-load error. This gives a rapid initial response by increasing the steam flow from the reactor vessel. The duration of this transient is limited due to the fact that increased steam flow tends to reduce reactor pressure and power level.

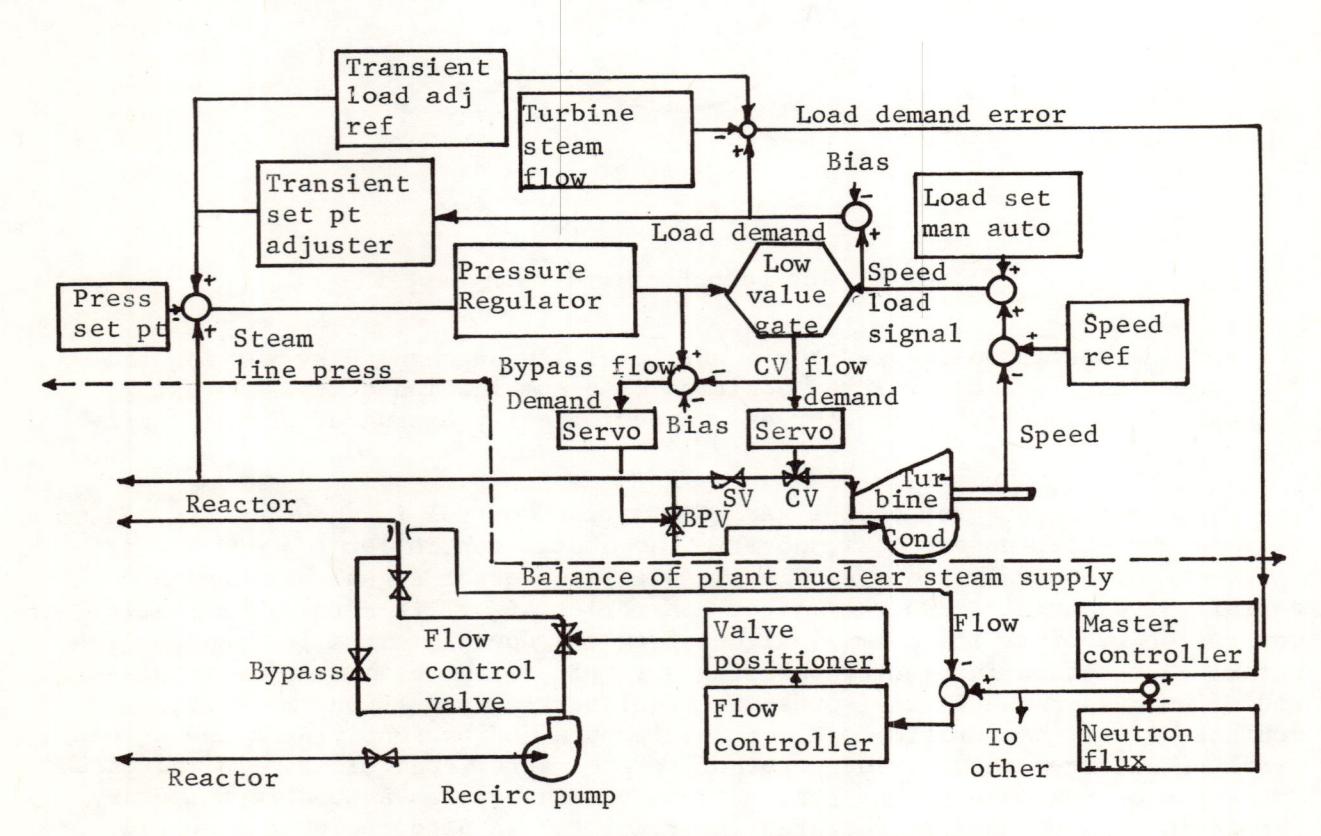

Fig. 21. Reactor - Turbine Control

The increased flow demand causes the flow control valve to open wider increasing the reactor recirculation flow and thus reactor power. The increase in reactor power will cause a slight increase in pressure which will result in the turbine admission valves opening sufficiently to increase the turbine output until the speed error is canceled.

If the turbine speed increases, the control system operated in the opposite direction to that described above.

Turbine Bypass Valve

The turbine bypass valve performs three basic functions.

1) Reduces the rate of reactor pressure increase when the turbine admission valves are moved rapidly in the closing direction.
2) Controls reactor pressure during turbine startup, which allows the reactor power level to be held constant while the turbine steam flow is varied as the turbine is brought up to speed.
3) Helps control reactor pressure after a turbine trip. It is used to discharge the decay heat to the condenser and to control the rate of cooling of the reactor system.

Pressure Relief System

To control large pressure transient a pressure relied system is provided. Safety/relief valves are operated under the following conditions 1) closure of the main steam isolation valves, 2) sudden closure of the turbine admission or step valve, and 3) failure of the turbine bypass system to relieve excess pressure. The safety/relief valves discharge steam from the steam lines inside the drywell to the suppression chamber. Each valve is operated from its own overpressure signal.

Feedwater Control System

The flow of feedwater into the reactor vessel is automatically controlled by the feedwater control system which maintains the level of water in the reactor within specified levels during all phases of plant operation. The system uses signals from the reactor vessel water level, steam flow and feedwater flow.

The feedwater flow control system accommodates feedwater pump failure by reducing the reactor water recirculation flow.

PLANT PROTECTION (5,6)

There are three main barriers which protect the environment from radioactive contamination: 1) fuel cladding, 2) reactor vessel and primary system, and 3) containment building. These three lines of defense so to speak are depicted in Fig. 22.

Fission products which build up in the fuel are contained within the clad. Therefore, maintaining clad integrity is important. To help insure against a breach of fuel cladding, the limits listed below are set.

1. Minimum DNB ratio equal to or greater than 1.3.
2. Fuel center temperature below the melting point of UO_2.
3. Internal gas pressure less than the nominal external pressure even at end of core life.
4. Clad stresses less than the yield strength.
5. Clad strain less than 1%.
6. Cumulative strain fatigue cycles less than 80% of design strain fatigue life.

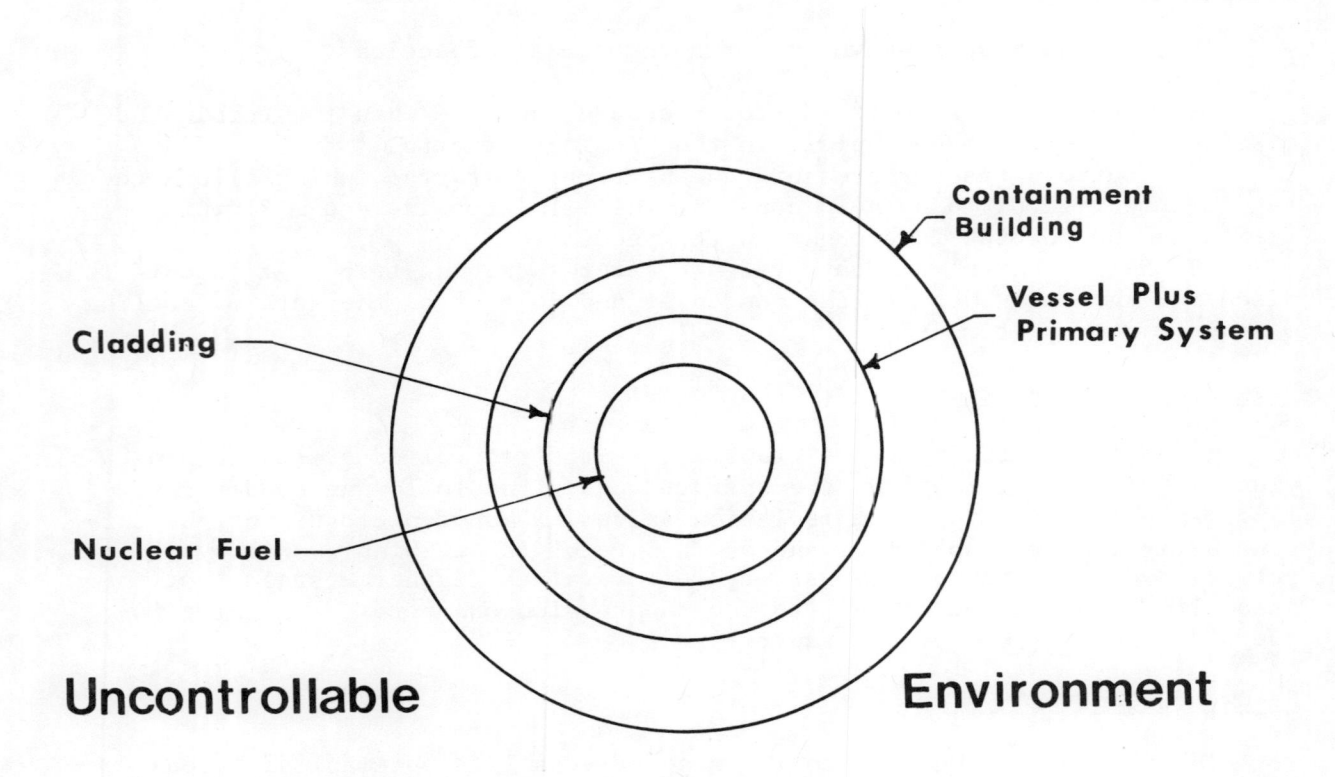

Fig. 22. Protection Barriers Against Contamination of the Environment

The DNB (Departure from Nucleate Boiling) describes a situation in which a sharp reduction of the heat transfer coefficient from clad surface to coolant occurs with the possibility of clad failure due to high clad temperature. To avoid this situation the hot regions of the core are operated within the nucleate boiling regime of heat transfer wherein the heat transfer coefficient from clad surface to water is very large and the clad surface temperature is only a few degrees above the coolant saturation temperature.

The DNB is not observable. Therefore, the observable parameters, thermal power, reactor coolant temperature, and pressure, have been related to DNB. The local DNB heat flux ratio, defined as the ratio of the heat flux that would cause DNB at a particular core location to the local heat flux, is indicative of the margin to DNB. The minimum value of DNB ratio, DNBR, during normal operational transients and anticipated transients is limited to 1.30. A DNB ratio of 1.30 corresponds to a 95% probability at a 95% confidence level that DNB will not occur and is chosen as an appropriate margin to DNB for all operating conditions.

In general the heat flux at which DNB occurs becomes lower for:

1) High reactor coolant enthalpies
2) Low reactor coolant flows
3) Low reactor coolant pressures

The maximum heat flux limit as well as the minimum DNB ratio limit must be met at the most restrictive location within the core. It is a fundamental characteristic of a nuclear reactor that the nuclear power is not produced uniformly at all locations within the core. The power distribution within the core depends upon the geometry of the reactor, the spatial distribution of enriched fuel, the location of neutron absorbers such as control rods or burnable poison, the relative proportion of fuel to moderator in the different regions of the core. Density variations of the coolant moderator within the core as well as fuel temperature distributions also affect power distributions. The ratio of maximum heat flux to average heat flux ranges between 2 and 3.

Reactor Scram

This section will discuss a number of situations which result in a reactor scram (rapid insertion of control rods).

To keep the maximum heat flux within a specified limit and to maintain the DNB ratio above 1.3, the reactor is scrammed if the difference between the reactor vessel outlet and inlet temperatures exceeds allowable limits. These two functions are referred to as the overpower scram and the overtemperature scram, respectively.

The reactor is scrammed if the pressure in the reactor drops below a lower limit. This scram limits the pressure on the low end over which the overtemperature protection has to be effective in preventing the DNB ratio from going below 1.3.

The neutron flux level in the core is determined by monitoring neutron leakage from the core. This is accomplished by detectors in the primary shield. The reactor is scrammed whenever the nuclear power reaches approximately 108% of the value at steady state full power conditions. It should be noted that the design limit on maximum heat flux is not the same as a limit on the nuclear power. The heat flux is the power being transferred from the clad surface to the reactor coolant at a given time. It changes during any transient and lags behind the nuclear power due to the heat capacity of the fuel and clad. In some cases the nuclear power may for a short period of time during a transient exceed the maximum design limit on heat flux by a considerable amount. The heat flux, however, will still remain within the allowed limit because of the time lag required to heat the fuel and clad. The design limit on power for the core applies to heat flux and not to the nuclear power.

Several types of neutron detectors are used to monitor the leakage neutron flux from a completely shutdown condition to 120% of full power. The power range channels are also capable of recording overpower excursions up to 200% of full power. The neutron flux covers a wide range between these extremes. Therefore, monitoring with several ranges of instrumentation is necessary. The lowest range or "source range" usually covers six decades of leakage neutron flux. The lowest observed count rate depends on the strength of the neutron sources in the core and the core multiplication associated with the shutdown reactivity.

The next range or "intermediate range" normally covers eight decades. Detectors and instrumentation are chosen to provide overlap between the higher portion of the source range and the lower portion of the intermediate range. The highest range of instrumentation or "power range" covers slightly more than two decades of the total instrumentation range. The power range is a linear range that overlaps with the higher portion of the intermediate range.

There is a source range nuclear overpower scram which can be bypassed manually when above a specified level in the intermediate range. Also, there is an intermediate range nuclear overpower scram which can be bypassed manually when above a specified level in the power range (10%). In the low power range there is a scram at about 25% power which can be bypassed when the power level is above 10%.

As previously indicated, low reactor coolant core flows reduce the margin to DNB. For this reason the reactor is scrammed on

a) low measured coolant flow
b) reactor coolant pump bus underfrequency
c) reactor coolant pump bus undervoltage

In the PWR water must be maintained in the secondary side of the steam generators in order to remove the heat produced in the core. Reactor scrams are provided on signals indicating a possible loss of the steam generators as heat sinks. A complete loss of all feedwater without a reactor scram would lead to emptying the steam generators, possible over-pressurization of the Reactor Coolant System, and possible clad damage. A reactor scram occurs when measured feedwater flow is less than measured steam flow by more than 20%, if the measured steam generator level is below a specified value. Also, a reactor scram occurs when the measured steam generator level falls below a specified low setpoint value.

This trip signal is used to start auxiliary feedwater pumps which are powered by diesel generators and steam from the steam generators via a turbine. These auxiliary feedwater pumps can be operated independently of outside electrical power. The pumps are used to insure that water can be supplied to the steam generators for the purpose of removing core decay heat.

High pressure in the reactor will cause a scram because this indicates abnormal conditions in the primary system. Normal transient pressures will not cause a scram.

If the water level in the reactor falls below a point where improper cooling of the fuel might result, a scram will occur.

The reactor is scrammed following a turbine trip occurring above 10% load. This scram anticipates the high pressure transient which follows a turbine trip.

An excessive release of fission products into the reactor water will cause a reactor scram.

The reactor is scrammed on signals indicating a possible break of the reactor coolant system. Scrams also occur if the pressure in the containment is abnormal. Signals are provided to automatically isolate all auxiliary fluid systems to and from the containment should there be indications that reactor coolant has been released to the containment.

Finally, a scram will occur on loss of auxiliary power.

CONCLUDING REMARKS

In general, the instrumentation associated with the reactor plant is divided into that for control and that for protection. The control instrumentation is used in maintaining steady state operating conditions, in reactor startup and normal shutdown, and in permitting the plant to automatically sustain the design load change transients without parameters exceeding values requiring a shutdown of the reactor. The protection instrumentation is that used to insure the reactor is rapidly shutdown when any parameter approaches a safety limit (i.e. the limit on the DNB ratio or the limit on the maximum heat flux). The protection instrumentation is also used to automatically actuate systems required for safety -- emergency core cooling, auxiliary feed, or containment spray.

The control systems are designed to allow flexible operation of the plant with reasonable allowances made for equipment malfunctions and failures.

The protection systems are designed to very rigid criteria concerning the effects of equipment malfunctions and failures. The system used scrams as coincidence of two out of four measurements for each variable. It is redundant in design so that failure of any one element will not interfere with a required scram.

Assumed protection system actions are an important part of the Safety Analysis Report required by the NRC. The reader is referenced to Appendix A Accident Analysis for further discussion on the subject.

REFERENCES

1. L. E. WEAVER, <u>Reactor Dynamics and Control</u>, American Elsevier Publishing Co., Inc., New York (1968).

2. D. L. HETRICK, <u>Dynamics of Nuclear Reactors</u>, University of Chicago Press, Chicago (1971).

3. A. SESONSKE, <u>Nuclear Power Plant Design Analysis</u>, USAEC Technical Information Center, Oak Ridge, Tennessee (1973).

4. L. E. WEAVER, <u>System Analysis of Nuclear Reactor Dynamics</u>, Rowman and Littlefield, New York (1963).

5. R. J. JOHNSON, Unpublished course notes, Georgia Institute of Technology.

6. GENERAL ELECTRIC COMPANY, "General Description of a Boiling Water Reactor," 12th Printing, Nuclear Energy Division, San Jose, California (May 1974).

FLUID FLOW AND HEAT TRANSFER IN WATER COOLED REACTORS

Roger W. Carlson
School of Nuclear Engineering
Georgia Institute of Technology
Atlanta, Georgia 30332

I. INTRODUCTION

The public health and safety as well as safe operation of a power plant are the concern of all nuclear engineers. This entails special responsibilities when considering a nuclear power plant for the large inventory of fission products that is contained in the core of the reactor would have catastrophic consequences if completely released to the environment (1). The probability of complete release is very low (2) but the public awareness of the consequences makes even small releases unacceptable (3). Such awareness coupled with the requirements of the Nuclear Regulatory Commission have resulted in the thorough analysis of the consequences of the limiting accident in all categories of accidents and transients that can occur in a nuclear power plant (4,5). The possible abnormal operating conditions have been divided into four groups that are identified in Table I where the frequency of occurrence and the difficulty of recovery have been employed to define the categories.

Each operational condition implies differing design constraints to reflect the necessary protection of the plant, the environment, and the public after the transient has proceeded to its conclusion. For example, frequent transients are limited to no departure from nucleate boiling to prevent cladding overheating and deterioration, while in the loss of coolant accident departure from nucleate boiling is expected and the analysis is directed toward assuring that the chemical reaction between the cladding and steam does not produce an explosive concentration of hydrogen.

The analysis of the consequences of operational transients and accidents takes two forms. The first is experimental where a power plant or a simulated power plant is forced to endure the transient of interest and the consequences of the transient are observed. This approach has obvious limitations when the transient being considered is expected to preclude recovery of the plant and where a number of possible initial conditions and system designs must be considered. The second form of analysis relies upon the development of predictive capability to permit the analysis of the consequences of transients with the aid of the large digital computers or analog computers. The predictive ability must be predicated upon the ability of the analytical techniques to reproduce the observed responses to the available experiments from actual plant experience, isolated flow loops, and semiscale mockups of reactor power plants.

Table I. Transient Classification for Analysis Purposes (3,4)

Condition	Description	PWR Example	BWR Example
I	Normal Operation at Steady State and Full Power	Normal Operation	Normal Operation
II	Frequently Occurring Faults	Loss of Load	Loss of Load
III	Infrequently Occurring Faults	Incorrectly Loaded Fuel Assembly	Uncontrolled Control Rod Withdrawal of Power
IV	Improbable Faults that Represent Worst Possible Cases	Loss of Coolant Accident	Loss of Coolant Accident

Accommodation or Recovery

Condition I is accommodated by margins in the design.

Condition II is accommodated by margins in the design, or at most a shutdown with a return to power after correcting the fault.

Condition III is accommodated by the cooling system, and only a small fraction of the fuel shall be permitted to fail. The fault shall not be capable of initiating a condition IV incident. Recovery is anticipated after a shutdown for minor repairs.

Condition IV is accommodated by the emergency core cooling system, and all consequences shall be proven to represent no undue risk to public health and safety. Return to normal operation will be possible only after major repairs in the reactor core and primary coolant system.

It is the intent of this paper to review the basic principles that are necessary for prediction of the response of a reactor to the transients which are of interest for an assessment of reactor safety. To this end, a discussion of the sophisticated portions of the thermal and hydraulic analysis of condition I and IV occurrences will be presented. Where possible, developments that are available in basic textbooks on heat transfer and fluid flow will merely be mentioned; no elaboration will be presented.

II. CONDITION I -- STEADY STATE OPERATION

The thermal and hydraulic analysis of the steady state operation of a nuclear power plant is directed toward the identification of the maximum power that can be removed from the reactor without violating any of the design limitations that prevent failure of any of the components or design limitations which prevent the development of intolerable initial conditions for postulated accidents. Margin is provided in the design process to allow for any uncertainties in the calculations and to accommodate any unforeseen deterioration in the performance of the reactor.

The analysis of steady state operation may not be directly relevant to the discussion of reactor safety analysis but many of the analytical techniques and assumptions are relevant to the prediction of the consequences of accidents.

The design limitations that are of most significance from a thermal and hydraulic point of view are restrictions upon the maximum fuel and cladding temperatures. The fuel temperature is restricted to less than melting to prevent any unpredictable motion of the fuel which could possibly result in an increase in the multiplication factor of the reactor (3). The cladding temperature is restricted to temperatures slightly above the boiling temperature of the coolant by requiring that there be no departure from nucleate boiling resulting in excessive cladding temperatures which reduce the cladding strength and initiate a chemical reaction of the zirconium and water, further weakening the cladding.

The factors that complicate the determination of the maximum fuel temperature, heat transfer resistance of the gap between the fuel and cladding, departure from nucleate boiling, and the flow distribution within the reactor will be discussed in the following sections.

A. Maximum Fuel Temperature

The limitation on the maximum fuel temperature restricts operation to conditions that do not induce centerline melting of the fuel. The melting point of uranium dioxide is 5080°F or 2805°C (5). The development of the central void and the inclusion of fission products in the fuel matrix causes the melting temperature to reduce during burnup at a rate of approximately 58°F per 10,000 MWD/MTU (5).

The determination of the maximum fuel temperature is complicated by several effects. The classical textbook solution of the heat conduction equation for steady state conditions in a fuel rod is given by (6)

$$T_{\mathcal{L}} = T_s + q''' R^2/4k, \qquad (1)$$

where

$T_{\mathcal{L}}$ = the temperature at the center of the fuel rod
T_s = the temperature at the surface of the fuel rod
k = the thermal conductivity
R = the radius of the fuel
q''' = the volumetric heat generation.

The development of this solution required the following assumptions (6):

1. Thermal conductivity is not a function of position or temperature within the fuel rod.
2. The volumetric heat generation rate is uniform across the fuel rod.
3. The properties of the fuel are uniform across the fuel rod and there are no cracks nor other nonuniformities in the structure of the fuel.
4. There is no void at the center of the fuel rod.
5. All heat transfer was in the radial direction.

In essence each of these assumptions is invalid for a fuel rod of a nuclear reactor that is composed of uranium dioxide and clad with an alloy of zirconium.

The thermal conductivity of uranium dioxide is definitely a function of temperature as shown in Fig. 1. The incorporation of the temperature dependent thermal conductivity (7) into the solution of the heat conduction equation requires the introduction of a new variable θ which is the integral of the thermal conductivity over temperature given by

$$\theta(T) = \int_0^T k(\tau) d\tau. \qquad (2)$$

Substituting θ for T in the steady state heat conduction equation results in a differential equation that can be solved analytically with the solution given by

$$\theta_{\mathcal{L}} = \theta_s + q''' \pi R^2/4\pi \qquad (3)$$

where the symbols are as before. This solution is coupled with the integral of the thermal conductivity data, shown in Fig. 2, where the integral at the surface temperature is determined, the difference in the integral is evaluated from the above equation and added to the integral at the surface to determine the integral at the center which is then used to determine the temperature at the center.

It has been shown that the inclusion of the temperature dependence of the thermal conductivity results in a lower center temperature if the surface temperature is below about 2000°F (8). This is a consequence of the increased thermal conductivity at the surface and the resulting reduction in the resistance to heat flow at the surface.

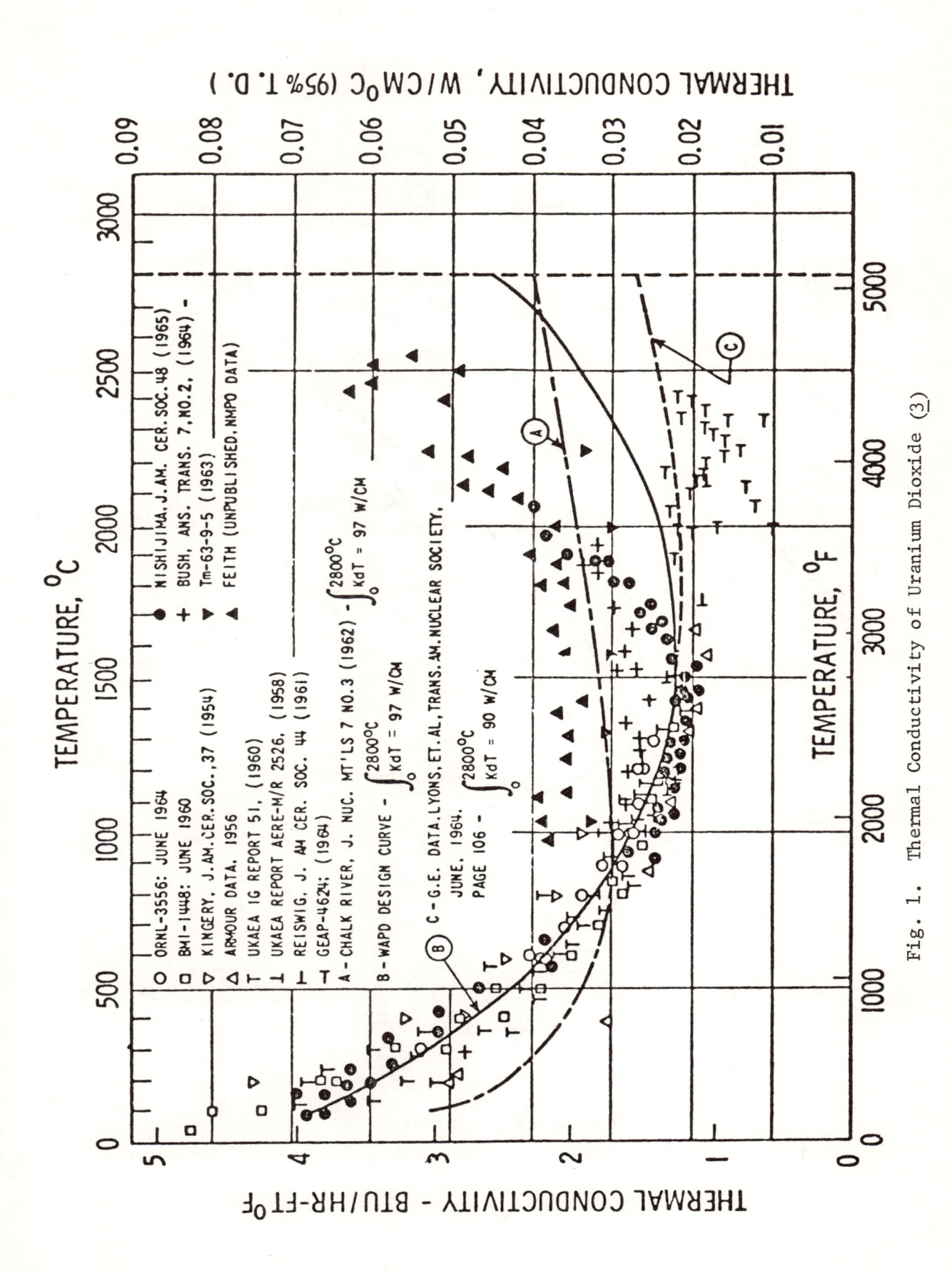

Fig. 1. Thermal Conductivity of Uranium Dioxide (3)

213

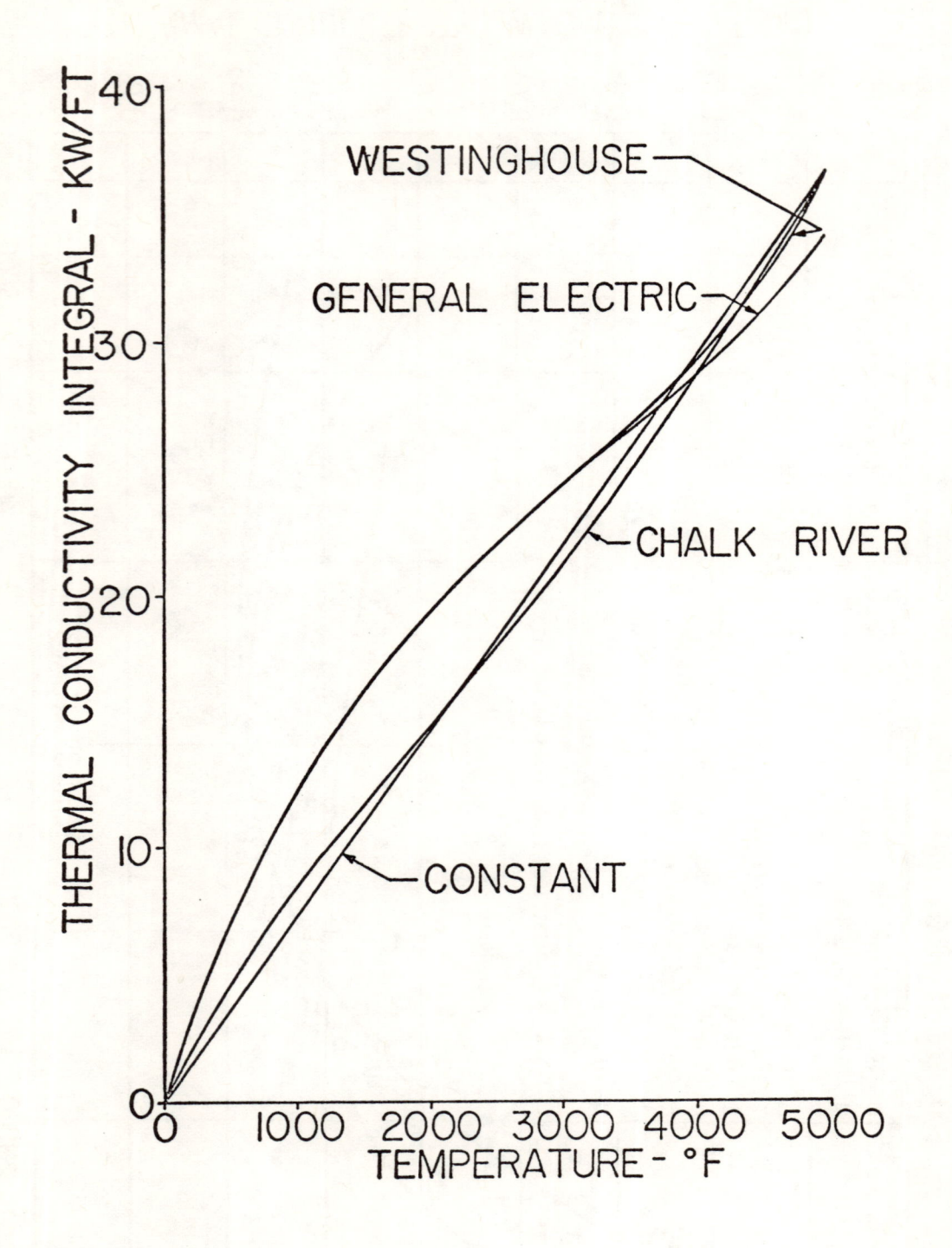

Fig. 2. Thermal Conductivity Integral for Uranium Dioxide (8)

The fact that the volumetric heat generation rate is not uniform across the radius of the fuel has been shown by nuclear analysis of the flux distribution in an absorbing medium (9). The magnitude of the depression is about 10% as shown in Fig. 3, where the results of a LASER (10) calculation for the first core of the Yankee reactor (Rowe, Mass.) are presented (11). The data presented are the fractional depletion which is essentially proportional to the radial power distribution within the fuel during burnup. The empirical data presented confirm the accuracy of the trends predicted by the calculations. The incorporation of the depression in the heat generation rate is accomplished by either resorting to a numerical solution of the heat conduction equation (12) or by introducing a factor that modifies the analytic calculation of the fuel temperature to simulate the effect of including the depression in the heat generation rate (3).

Studies of the behavior of uranium dioxide during irradiation have shown that the structure of the fuel undergoes considerable change during burnup. The fuel initially is a uniform composition with porosity evenly distributed within the fuel. In the region above 1400°C the fuel is restructured and the porosity migrates toward the center of the fuel rod forming a central void and leaving a region of columnar grains (13). In the portion of the fuel that is above 1300°C the restructuring of the fuel is not so radical but the grains are transformed into equiaxed grains with a redistribution of the porosity (13). The remainder of the fuel remains unrestructured and retains its original characteristics. A photomicrograph of a cross section of a fuel rod after irradiation (14) is shown in Fig. 4 which displays each of the regions described above.

The dimension of the central void is computed from a balance of the volume between the void and columnar grain region. Since the central void represents the porosity that has migrated out of the columnar grain region the radius of the central void is given by (15)

$$r_v = r_{cg} \sqrt{\epsilon} \qquad (4)$$

where
r_v = radius of the central void
r_{cg} = radius of the columnar grain region
ϵ = original fractional porosity of the fuel pellet.

The boundaries of the columnar grain region and the equiaxed regions are determined by the location of the isothermal contours that represent 1600 and 1300°C, respectively (13).

The phenomenon known as densification has resulted in the development of gaps in the stack of fuel pellets and possibly the coolant pressure can cause the clad to creep into the gap in the fuel stack. Under these conditions the assumption of one-dimensional heat transfer in the radial direction is invalid. However, for conservatism, the one-dimensional heat transfer model is retained to provide margin to compensate for the uncertainties in the geometry of the fuel and clad in the vicinity of the gap in the fuel stack (16).

The variation of the thermal conductivity of uranium dioxide with porosity is given by the Maxwell-Euken relationship (17)

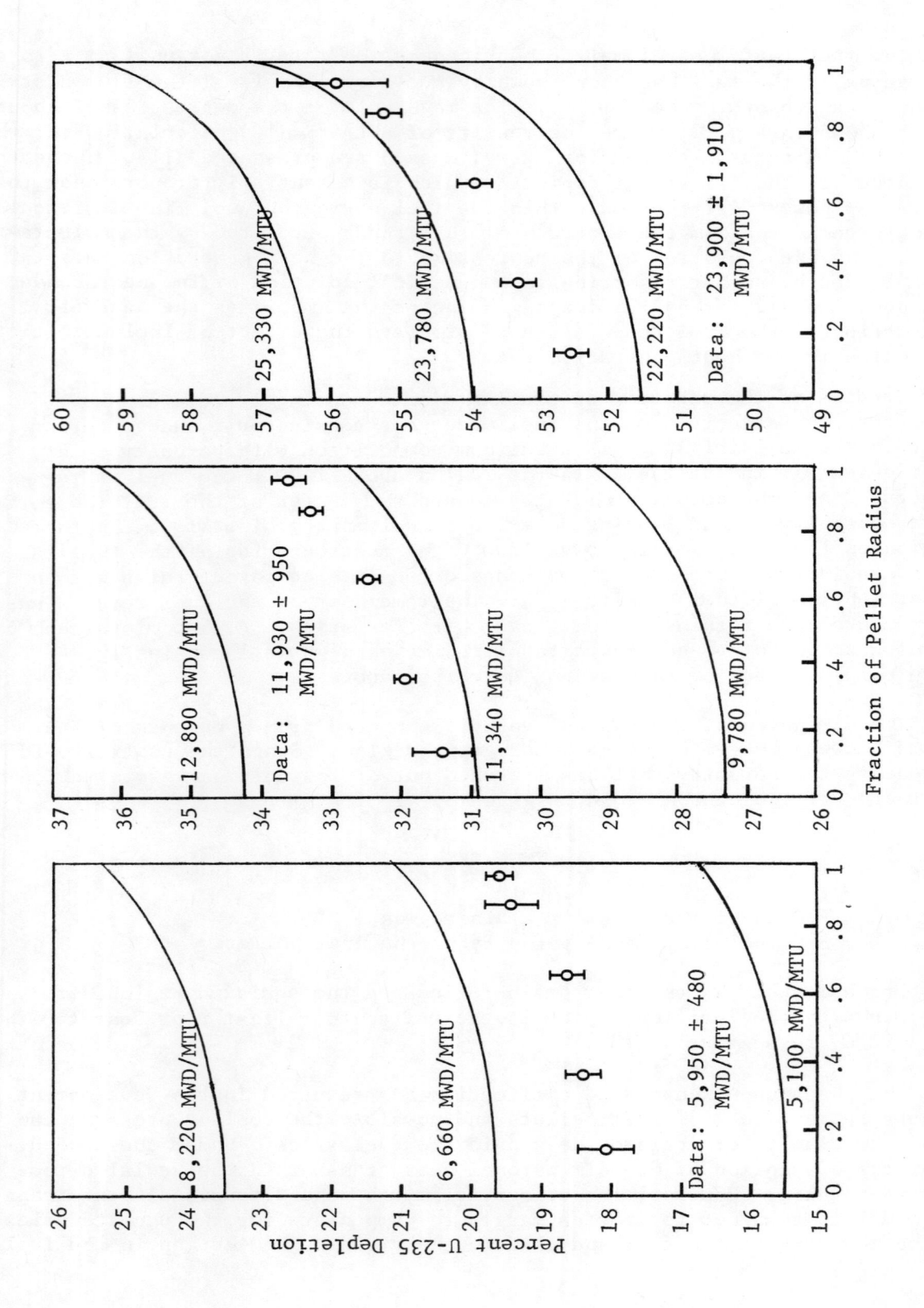

Fig. 3. Radial Distribution of U-235 Depletion (11)

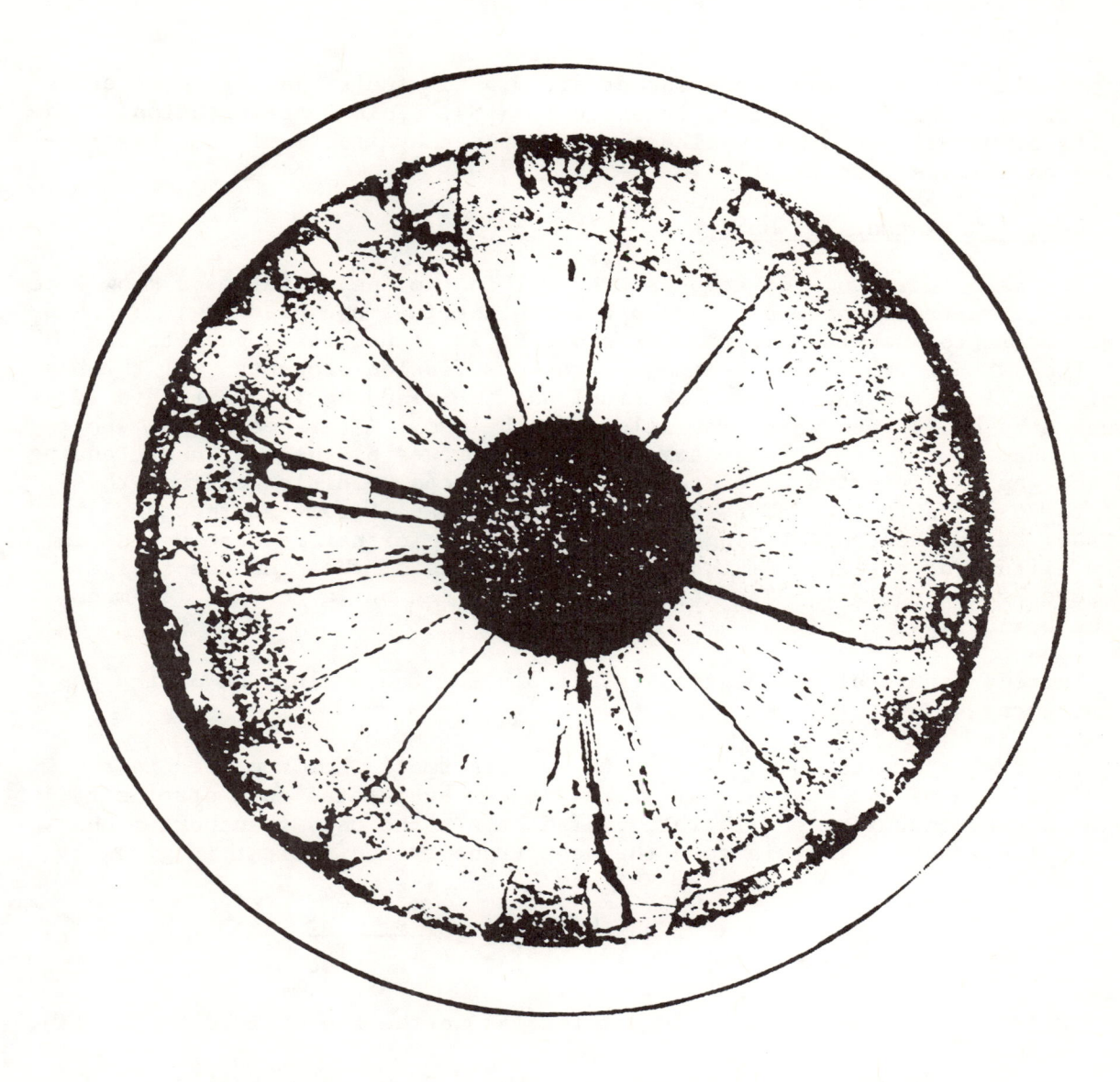

Fig. 4. Cross Section of Highly Irradiated Mixed-Oxide Fuel Pin Showing the Central Void, Columnar Grain Zone, Equiaxed Grain Growth Zone, and Undisturbed Zone Next to the Cladding

$$k_p = \frac{1-\epsilon}{1-\beta\epsilon} k_o \qquad (5)$$

where
- k_p = thermal conductivity of the porous fuel
- k_o = thermal conductivity of the solid fuel
- ϵ = volume fraction of the porosity
- β = shape factor.

Recommended values of the shape factor are 0.5 for fuels that are greater than 90% dense and 0.7 for less dense fuels (18). This representation of the effects of porosity can be applied to determine the effects of fuel restructuring as well as densification.

B. Thermal Resistance of Gap

The calculation of the temperatures within the fuel requires a knowledge of the thermal resistance of the gap between the fuel and cladding. The gap is difficult to analyze analytically since it is of uncertain geometry and is filled with a mixture of helium and gaseous fission products. At the beginning of life the gap between the fuel and the cladding is about 8 to 10 thousandths of an inch which is filled with helium. After heatup the gap is reduced due to the differing thermal expansion rates of the fuel and cladding and the gap is distorted by pieces of fuel that can be dislodged from the pellets during shipping and by the cracks that result from the thermal stresses during heatup. During burnup the gap is further reduced by the combined effects of swelling of the fuel and creeping of the clad due to the pressure of the coolant. The cladding and fuel can be in contact for much of the burnup (3).

There are several correlations available for the evaluation of the gap conductance.

In cases where the dimension of the gap is small when compared to the mean free path of the gas molecules, the gas can be treated as a stationary medium where conduction and radiation of heat are the only transport mechanisms of importance. In this case the heat transfer coefficient is given by (19)

$$h_c = (2k/\delta)_g + \sigma \left(\frac{1}{\epsilon_s} + \frac{1}{\epsilon_{sc}} - 1 \right) \frac{T_s^4 - T_{sc}^4}{T_s - T_{sc}} \qquad (6)$$

where
- $(2k/\delta)_g$ = thermal conductivity of the gas in the gap divided by the gap width, Btu/hr-ft^2-°F
- σ = Stephan-Boltzmann constant, 0.173×10^{-8} Btu/hr-ft^2-°R^4
- ϵ_s = emissivity of the fuel surface, dimensionless
- ϵ_{sc} = emissivity of the cladding inner surface, dimensionless
- T_s = absolute temperature of the fuel surface, °R
- T_{sc} = absolute temperature of the cladding inner surface, °R.

Two additional formulations of the gap conductance are (3)

$$h = \frac{k_g}{\delta/2 + 14.4 \times 10^{-6}} \qquad (7)$$

and
$$h = 1500 \, k_g + \frac{4.0}{0.006 + 12\delta} \tag{8}$$

where the symbols have their previous meanings. In evaluating the temperature drop across the gap, the Westinghouse procedure is to use whichever of these two expressions yields the larger gap conductance (3).

These three models for the same quantity represent the evaluation of the different data or theories that are available. The first is purely theoretical and is only valid if the gas that occupies the gap is stationary, which is not likely. The second model is also based upon a theoretical approach but has been modified by the incorporation of empirical constants and the deletion of the relatively insignificant radiation transport term. The third model is based entirely on empirical data that have been derived from fuel irradiations with thermocouples introduced into the fuel and also from examination of fuel samples which identified the location of the equiaxed and columnar grain regions with their characteristic temperatures.

It should be noted that the thermal conductivity of the gas in all of these models is based upon the mixture of the helium and gaseous fission products and must include an allowance for the accommodation of the helium atoms between the larger fission product atoms (20,21).

During irradiation the cladding and fuel can come into contact, at which time the modeling of the gap must change to reflect the improved heat transport at the points of contact. The contact will not be perfect in that the fuel surface will be very irregular due to fuel cracking and swelling.

The modeling of the heat transfer coefficient with contact follows the work of Ross and Stoute where the heat transfer coefficient has two components, namely, the heat transfer through the gas and heat transfer through the areas in contact. Mathematically, this is expressed as

$$h = h_s + h_g \tag{9}$$

where
- h_s = heat-transfer coefficient representing the heat transport through the points in contact
- h_g = heat-transfer coefficient representing the heat transport through the gas.

Both heat-transfer coefficients are based upon the entire surface area. The heat-transfer coefficient for the points of contact is given by (22)

$$h_s = \frac{2 \, k_f k_c \, p}{(k_f + k_c) a_o (R_f^2 + R_c^2)^{\frac{1}{4}} H} \tag{10}$$

where
- p = contact pressure
- k = thermal conductivity of respective material
- a_o = empirical constant 0.08 ft$^{1/2}$
- R = surface roughness of the respective material
- H = Meyer hardness of the softer material.

The subscripts f and c refer to fuel and cladding. The heat-transfer coefficient for the areas separated by gas is given by (22)

$$h_g = \frac{k_g}{(2.75 - 1.7 \times 10^{-4} p)(R_f + R_c) + (g_f + g_c)} \tag{11}$$

where

k_g = thermal conductivity of the gas
g = temperature jump distance for the respective material.

The temperature jump distance is the distance that the gap would have to be increased to duplicate the effect of the resistance to heat transfer at the interface between the gas and solid.

Another representation of the heat transfer coefficient when the fuel and cladding are in contact is given by (23)

$$h = 0.6 p + \frac{k_g}{14.4 \times 10^{-6}} \tag{12}$$

where the symbols are as above. This representation follows the form of the relationship of Ross and Stoute and has the surface properties replaced by empirical constants. Equation 12 is in agreement with measured temperatures within fuel rods during irradiation.

The combination of the calculation of the heat transfer in the fuel with the heat transfer across the gap and the introduction of the textbook representations of the temperature drop across the cladding and interface between the cladding and coolant results in the calculated temperature distributions presented in Fig. 5 (24). Figure 6 (3) presents the variation of the important temperatures with the linear heat rate of the fuel. The data in Fig. 6 represent the beginning of life conditions which produce the highest fuel temperatures. Burnup effects such as contact between the cladding and the fuel and the formation of a central void within the fuel serve to reduce the central temperature.

C. Departure from Nucleate Boiling

The term departure from nucleate boiling and critical heat flux are synonymous and represent the deterioration of the heat removal capability of boiling that results in a large increase in the surface temperature of the cladding (6). During nucleate boiling, the bubbles are created in the crevices and microscopic imperfections of the heated surface. As the bubbles grow they detach from the surface and move into the main stream of the flowing coolant. When the heat flux is sufficiently large, the number of bubbles that develop can become so large that they coalesce prior to detaching from the surface of the cladding. The cladding temperature rises rapidly since the heat must be conducted across a blanket of steam. This phenomenon is shown in Fig. 7 for both PWR and BWR conditions.

Heat transfer after the departure from nucleate boiling is of no importance in steady state analysis since the high cladding temperatures cannot be tolerated; however, during the analysis of the loss of coolant accident it is important since the highest cladding temperatures occur after departure

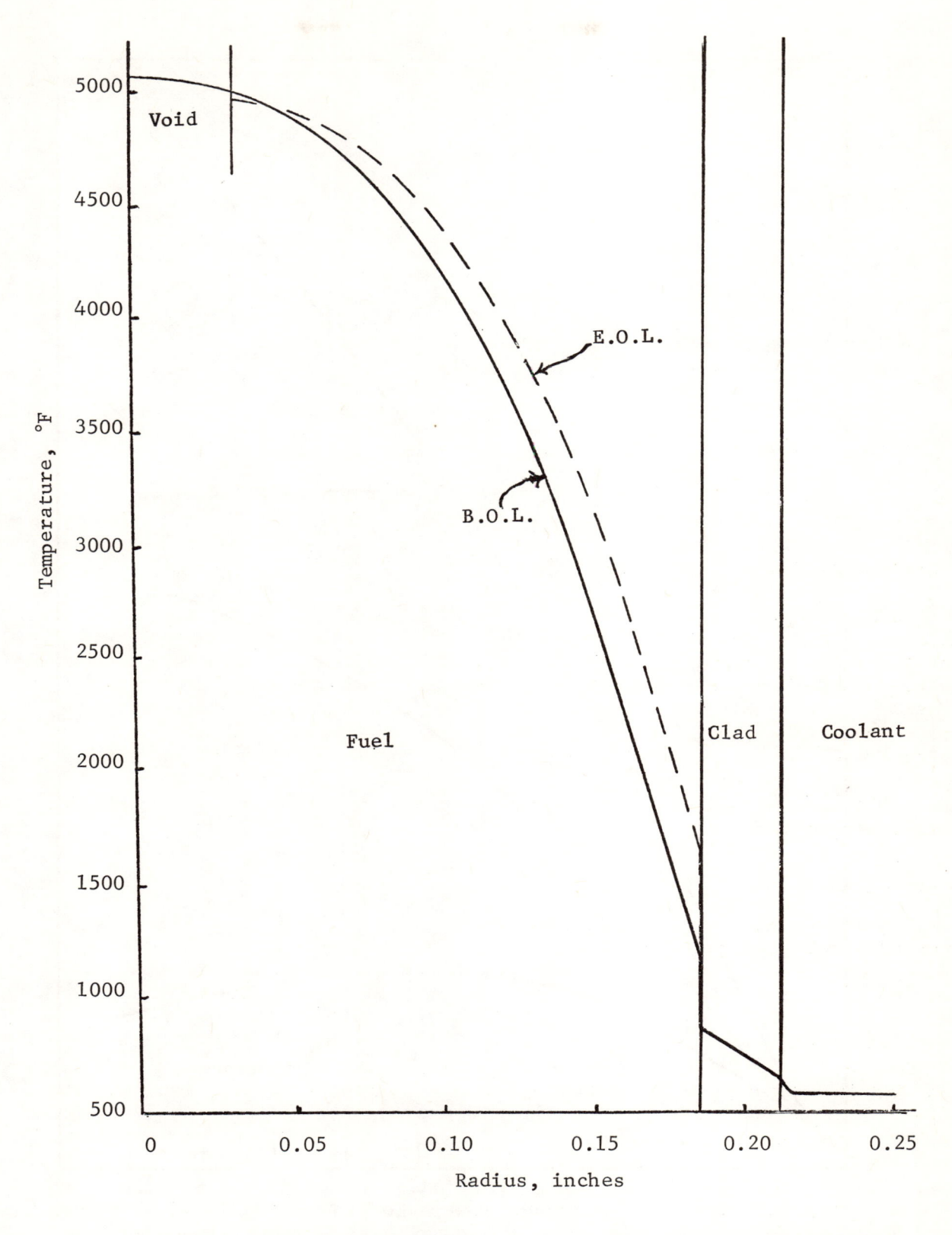

Figure 5. Typical Fuel Pin Temperature Profile at 26 kw/ft (24)

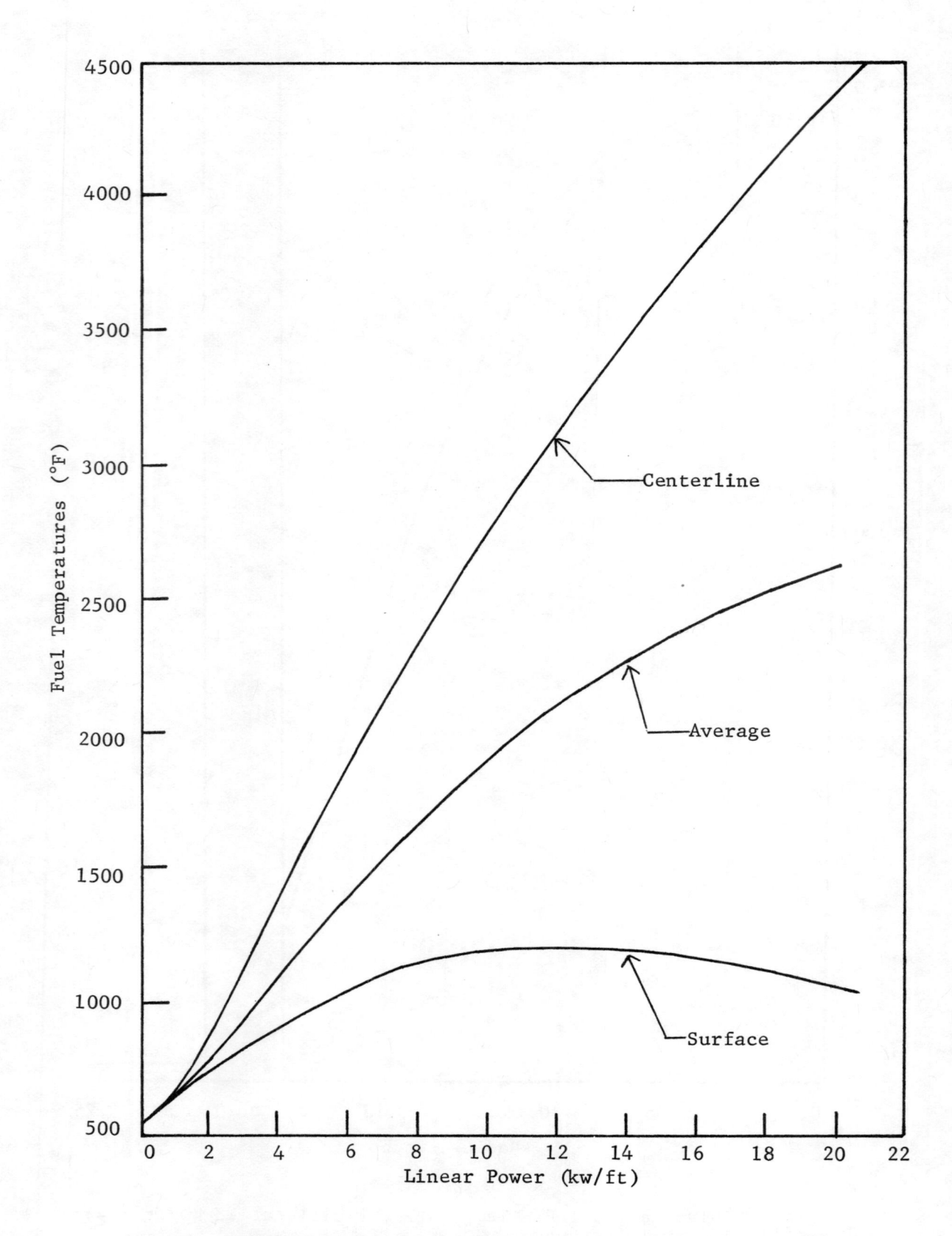

Fig. 6. Fuel Temperature Dependence on Linear Heat Rate (3)

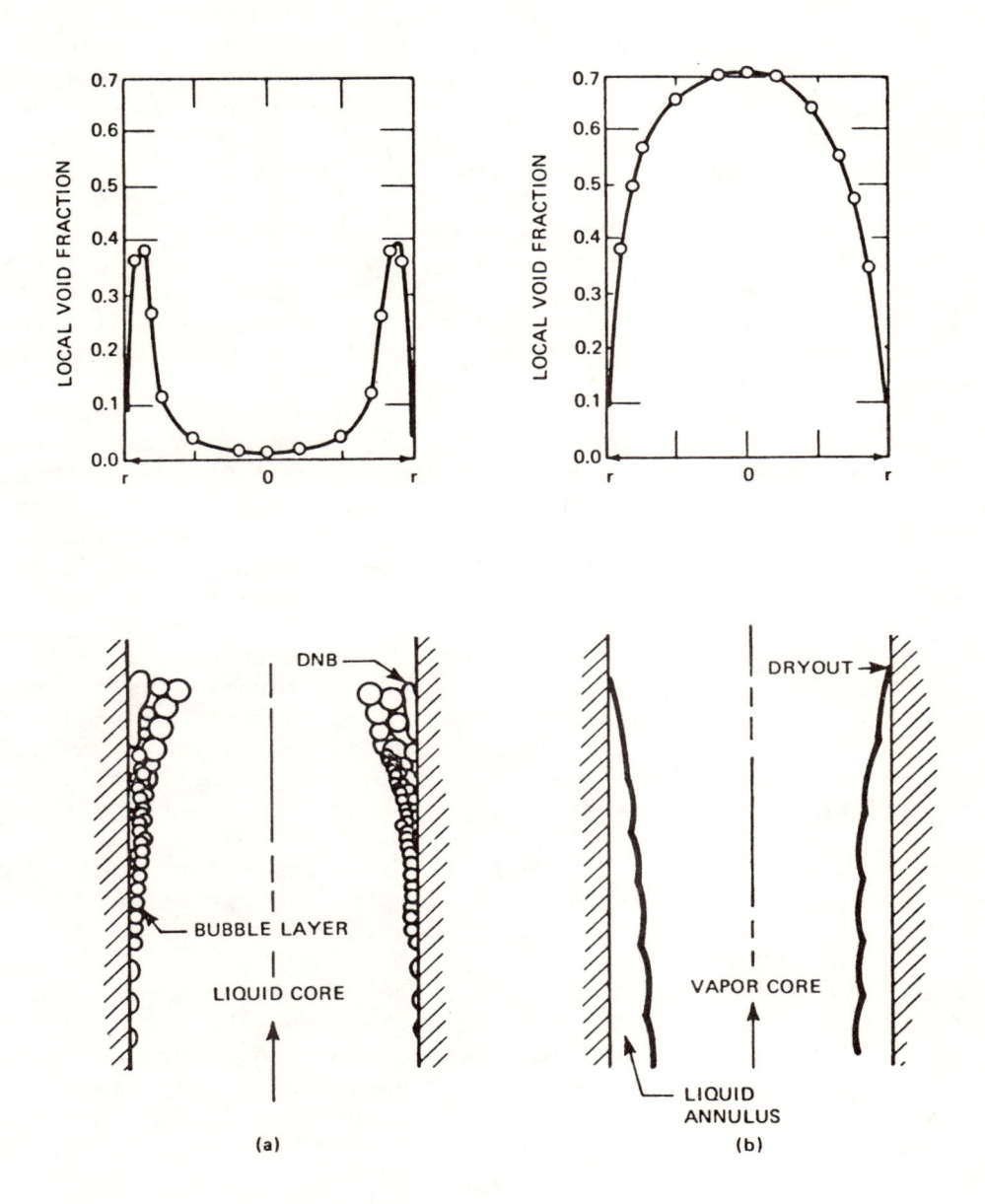

Figure 7. Comparison of Boiling-Crisis Mechanisms in Various Flow Patterns--(a) Subcooled bubbly flow, (b) Annular flow (25)

from nucleate boiling.

The prediction of the onset of the departure from nucleate boiling has resisted all attempts at theoretical analysis from first principles. Rather, empirical data have been gathered and correlations developed to represent the data. There are two basic approaches to the correlation of the departure from nucleate boiling data that are employed. The approach employed by Westinghouse is to develop a least-squares type fit to all of the data and then to establish sufficient margin to include all of the uncertainty in correlating the measured data. The philosophy adopted by General Electric was to establish an envelope type of correlation so that the correlation would predict the smallest conceivable value for the heat flux that leads to departure from nucleate boiling. The design margin that accompanies this philosophy should be very small with allowances needed only for the measurement uncertainties.

An even more significant difference in the approach to the correlation of the data is embodied in the parametric dependence. General Electric assumed that the local parameters of flow rate and quality and void fraction were controlling and coupled the correlation of the departure from nucleate boiling heat flux with its model for calculating the local flow distribution. This resulted in a rather simple correlation but required the combination of the correlation and the flow distribution model for the application of the correlation. The approach adopted by Westinghouse attempted to correlate the departure from nucleate boiling heat flux as a function of system parameters since the flow distribution and axial power shape in a pressurized water reactor are nearly always similar. This resulted in a far more complex correlation but eliminated the need for any sophisticated calculation of the local flow conditions at the point of departure from nucleate boiling (1).

However, the evolution of the Westinghouse correlation resulted in the adoption of local conditions as the basis for their correlation (25) but the philosophy of a least-squares type of fit was retained.

The evolution of the departure from nucleate boiling correlations has paralleled the development of the test facilities to make the necessary measurements. The first facilities were capable of measurements on a single heated rod in a tube and the departure from nucleate boiling caused melting of the heated rod. The development of higher power facilities and a nondestructive method of detecting the onset of departure from nucleate boiling permitted the extension of the tests to multiple rod configurations. This simulated the conditions in an array of fuel rods as in a fuel assembly. Still further development permitted the consideration of axial power shapes and then the incorporation of spacers to permit the study of the effect of engineering decisions on the departure from nucleate boiling heat flux. Also the inclusion of a nonheated surface was introduced into the studies and still further development of the test facilities resulted in the studies of transients in both flow and heat flux.

This chronology applied to both pressurized water reactor tests as well as boiling water reactor tests. Each new development in the pressurized water reactor technology resulted in a new correlation or in the correlation of a correction factor to incorporate the effects of the new phenomena.

Since the boiling water reactor correlation was an envelope type correlation closely tied to the calculation of the flow distribution there were few modifications to the correlation required to extend the ability of the correlation to reproduce the data.

The Westinghouse correlation, known as the W-3 correlation, is given by (3)

$$\frac{q''_{DNB,EQ}}{10^6} = [(2.022 - 0.0004302p) + (0.1722 - 0.0000984p) e^{(18.177 - 0.004129p)x}]$$
$$\times [(0.1484 - 1.596x + 0.1729|x|)G \times 10^{-6} + 1.037] \times [1.157 - 0.869x]$$
$$\times [0.2664 + 0.8357 e^{-3.151 D_e}] \times [0.8258 + 0.000794(h_f - h_i)]. \qquad (13)$$

This correlation covers the ranges: $G = 0.5 \times 10^6$ to 5.0×10^6 $lb_m/hr\ ft^2$, p = 800 to 2,000 psia, channel length L = 10 to 79 in., D_e = 0.2 to 0.7 in., heated-to-wetted perimeter ratio = 0.88 to 1.0, inlet enthalpy $h_i \geqq 400$ Btu/lb_m, local quality x_{loc} = -0.15 to +0.15 (negative quality means subcooled liquid).

For nonuniform axial heat flux the above correlation is divided by a factor F

$$q''_{DNB,N} = \frac{q''_{DNB,EU}}{F} \qquad (14)$$

where F is given by (3)

$$F = \frac{C}{q''_{l_c}(1 - e^{-Cl_c})} \int_o^{l_c} q''(z) e^{-C(l_c - z)} dz \qquad (15)$$

where
$C = 0.44(1 - x_c)^{7.9}/(G \times 10^{-6})^{1.72}$, in.$^{-1}$
l_c = channel length at which DNB takes place, in.
q''_{l_c} = heat flux at l_c, Btu/hr ft^2
z = variable distance from channel entrance, in.
x_c = quality at DNB.

When a spacer is present the above correlations are multiplied by a modified spacer factor F'

$$q''_{PRED} = q''_{DNB,N} \times F'_S \qquad (16)$$

where F'_S is given by (3)

$$F'_S = \left(\frac{p}{225.896}\right)^{0.5} (1.445 - 0.0371\ L)[e^{(x+0.2)^2} - 0.73] + K_S \frac{G}{10^6} \left(\frac{TDC}{0.019}\right)^{0.35} \qquad (17)$$

where
  p = the primary system pressure, psia
  L = the total heated core length, feet

x = the local quality expressed in fractional form
G = the local mass velocity, $lb_m/hr\text{-}ft^2$
TDC = the thermal diffusion coefficient
K_S = the axial grid spacing coefficient which has the following values:

Grid Spacing, inches	K_S
32	0.027
26	0.046
20	0.066

The range of experimental conditions was:

Axial grid spacing	20 inches, 26 inches, and 32 inches
Local DNB quality	−15 to +15 percent
Local mass velocity	1.6×10^6 to 3.7×10^6 $lb_m/hr\text{-}ft^2$
Local inlet temperature	440°F to 620°F
Pressure	1490 to 2440 psia
Local heat flux	0.3×10^6 to 1.1×10^6 Btu/hr-ft^2
Axial heat flux distribution	Nonuniform (cos u and u sin u)
Heated length	8 and 14 feet
Heater rod O.D.	0.422 inch

The presence of a cold wall in the flow channel experiencing departure from nucleate boiling is incorporated by multiplying the above correlations by a factor CWF

$$q''_{PRED,CW} = q''_{PRED} \times CWF \tag{18}$$

where CWF is given by (3)

$$CWF = 1.0 - Ru\left[13.76 - 1.372e^{1.78x} - 4.732\left(\frac{G}{10^6}\right)^{-0.0535} - 0.0619\left(\frac{p}{1000}\right)^{0.14}\right.$$

$$\left. - 8.509 D_h^{0.107}\right] \tag{19}$$

and
Ru = 1 − D_e/D_h
D_e = hydraulic diameter
D_h = heated equivalent diameter.

A plot of the departure from nucleate boiling heat flux that demonstrates the ability of the correlation to reproduce the empirical data is presented in Fig. 8 (3). The General Electric correlation, known as the Hench-Levy correlation, for boiling water reactors is given by (6)

For 1,000 psia,

$$q''_c = 0.705 \times 10^6 + 0.237G \qquad \text{for } x < x_1 \tag{20}$$
$$q''_c = 1.634 \times 10^6 - 0.270G - 4.71 \times 10^6 x \qquad \text{for } x_1 < x < x_2$$
$$q''_c = 0.605 \times 10^6 - 0.164G - 0.653 \times 10^6 x \qquad \text{for } x > x_2$$

where
$x_1 = 0.197 - 0.108 \times 10^{-6} G$
$x_2 = 0.254 - 0.026 \times 10^{-6} G$

At other pressures p, psia,

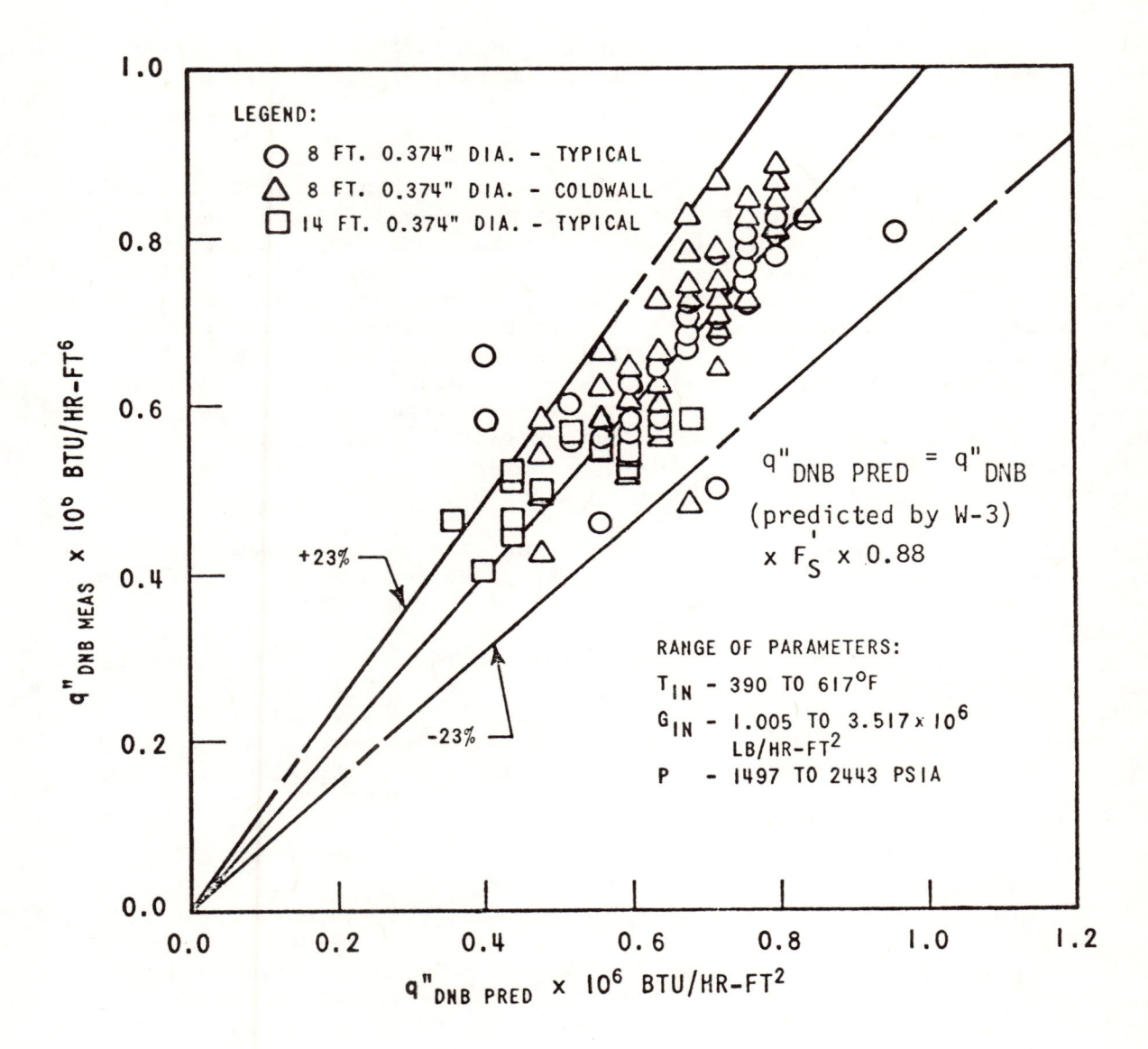

Fig. 8. Comparison of All 17 x 17 DNB Data (3)

$$q_c'' = q_c''(\text{at } 1{,}000 \text{ psia}) + 440(1000 - p) \tag{21}$$

The above correlations apply over a pressure range of 600 to 1450 psia, $G = 0.4 \times 10^6$ to 6.0×10^6 $lb_m/hr\ ft^2$, quality x = negative (subcooled water) to 0.45, D_e = 0.245 to 1.25 in., and channel length L = 29 to 108 in.

The parametric effects of the variables that influence the departure from nucleate boiling heat flux are summarized in Table II (25).

D. Flow Distribution Calculations

The calculation of the flow distribution in an open array of fuel rods is very complicated. The complete approach relies upon the solution of the differential equations for the transport of mass, momentum, and energy throughout the coolant. This results in a coupled set of nine partial differential equations that are nonlinear (26). The geometry does not lend itself to analytical representation and the flow is turbulent throughout the primary coolant system. Each of these complications by itself is sufficient to prevent the analytic solution of these differential equations. A nodal type of approach is employed where the transport of mass, momentum, and energy across the boundaries of each node are computed and balanced (27). By employing nodes that have dimensions that are large compared to the characteristic lengths of turbulence, the turbulence is reduced to a boundary condition.

A graphical representation of the balance of mass, momentum, and energy transport across the boundaries of a node is presented in Fig. 9 (27). The term ω represents the diversion crossflow which is the transfer of flow from one flow channel to another to satisfy the momentum equations. The term ω' is the turbulent crossflow which is the flow interchange between channels because of the turbulence of the flow. By the nature of turbulence there will be no net transfer of mass by turbulent crossflow but there will be a net transfer of both momentum and energy since a nodel model assumes the properties are constant over the entire node and a sudden discontinuity occurs at the interface between nodes (27).

Initial attempts to calculate the diversion crossflow dramatically showed that the balance equations for the pressure drop between adjacent flow channels are unstable for numerical approximation. To compensate for this it was necessary to implement a method of solution that requires the pressure in all flow channels to be equal at the exit of each axial node (28). This approach eliminates all of the diversion crossflow equations and simultaneously reduces the determination of the diversion crossflow to the determination of the total change in the mass flow rate within each flow channel. Thus, the knowledge of the origin or destination of the diversion crossflow is destroyed. Consequently, it must be assumed that those channels that are gaining flow have only inflow. Additionally, the properties of the incoming flow are taken as the average of the properties of the flow in the surrounding channels that are losing flow. For cases where there are large differences in the properties of adjacent channels, this can represent a significant uncertainty in the calculation since only the coarsest knowledge of the diversion crossflow is available from such calculations.

The model described above is based upon the development of COBRA and is

Table II. Summary of Parametric Effects (25)

Parameter	Effects	Reference
	On DNB in Bubbly Flow	
Pressure	$[T_{wall} - T_{sat} + (V/4)]$ is a function T_r	Bernath, 1960
	High pressure reduces bubble diameter (Fig. 2.7, equilibrium bubble size)	Cumo and Palmieri, 1967
Local enthalpy	$CHF/CHF_{sat} = 1 + 0.01 \Delta T_{sub}$ at 200 to 1500 psia	Povarnin and Semenov, 1960
Velocity	CHF increases lightly in vortex flow	Matzner et al., 1966
	$CHF/CHF_{G=0} = 1 + 1.45 (G/10^6)$ at 500 to 1500 psia (Fig. 2.1)	Griffel and Bonilla, 1965
	Very-high velocity induces upstream DNB	Kutateladze, 1952
	Direction of flow has no effect on DNB	Jacket, Roarty, and Zerbe, 1958
Heat flux and transient	CHF depends on local heat flux and immediate upstream flux	Tong, 1965; Styrikovich, Miropol'skii, and Eva, 1963
	CHF is independent of heater wall thickness	Aladyev et al., 1961
	CHF increases in fast power transient and in flow depressurization	Moxon and Edwards, 1967; Schrock et al., 1966; Tong, 1969; Tong et al., 1965
Geometry:		
Length	CHF is independent of length for L/D > 10	Ornatskii, 1969
Diameter	CHF is independent of diameter for D > 2mm	Ornatskii, 1969
	CHF remains constant for square- and triangular-lattice rod bundles	Tong et al., 1967a
Cold wall	CHF is less affected by cold wall in subcooled region	Tong, 1965
Grid	$CHF_{grid}/CHF_{no\ grid} = 1 + 0.03 (G/10^6)(TDC/0.019)^{0.35}$	Tong, 1969
Surface condition	CHF increases slightly with surface deposit	Tong, 1965
	CHF increases with 0.013-in. roughness but remains constant with 0.0025-in. roughness	Durant and Mirshak, 1959; Tong et al., 1967b
	On Dryout in Annular Flow	

Table II. Continued

Parameter	Effects	Reference
On Dryout in Annular Flow		
Pressure	CHF decreases with pressure increase at a given exit quality (Fig. 2.6)	Aladyev et al., 1961
	High pressure reduces bubble diameter (Fig. 2.7)	Cumo and Palmieri, 1967
Local enthalpy	$CHF/CHF_{X=0} = 1 - 1.12\chi$, where χ is in fractions (Fig. 2.1)	Griffel and Bonilla, 1965
Velocity	CHF increases strongly in vortex flow (Fig. 2.15)	Matzner et al., 1966
	$CHF/CHF_{G/10^6=1} = 1 - 0.25\,(\Delta G/G)$, where $\Delta G = G - 10^6$ (Fig. 2.1)	Griffel and Bonilla, 1965
	High velocity ($G > 5 \times 10^6$) induces upstream	Waters et al., 1965
	Critical stream velocity = 52 ft/sec at 1000 psia	Bennett, Collier, and Lacey, 1963
	Direction of flow has no effect on dryout	Jacket, Roarty, and Zerbe, 1958
Heat flux and transient	CHF depends chiefly on average heat flux and little on immediate upstream flux	Tong, 1965; Styrikovich, Miropol'skii, and Eva, 1963
	CHF increases slightly with wall thickness in a long test section	Lee, 1965
	Critical exit quality increases with boiling length	Bertoletti et al., 1965
Geometry:		
Length	CHF decreases strongly with length at high qualities	Shitsman, 1966; Proskuryakov, 1965
Diameter	CHF is independent of gap between heater rods	Towell, 1965; Lee and Little, 1962
Cold wall	CHF reduces near a cold wall at a given local quality	Tong, 1965
Grid	Grid reduces liquid-film thickness but increases deposition rate	Tong, 1965
Surface condition	CHF decreases with 0.0025-in. roughness	EURAEC, 1966
	CHF decreases with 0.0003-in. roughness	Janssen, Levy, and Kervinen, 1963
On Pool Boiling Crisis		
Pressure	$CHF/CHF_{pr=0.1} = 1.70 - 3.9T_r - 0.04T_r^2 + 2.41T_r^3$	Cobb and Park, 1969

Table II. Continued

Parameter	Effects	Reference
	On Pool Boiling Crisis	
	$+ 7.58T_r^4 + 5.20T_r^5 - 12.9T_r^6$ (Fig. 2.9)	Semeria, 1962
	Bubble departure diameter (mm) = 1/atm (Fig. 2.7)	Farber, 1951
Local enthalpy	$CHF/CHF_{sat} = 1 + 0.033 \Delta T_{sub}$ at 14.7 psia	Lienhard and Keeling, 1969
Velocity	CHF depends on the induced flow pattern near heating surface	Tachibana, Akiyama, and Kawamura, 1968
Heat flux and transient	CHF depends on local heat flux	Carne and Charlesworth, 1966
	$CHF/CHF_{kt=10^{-3} W/°C} = 1 + 0.4 \; 3 + \ln(kt)$	
	CHF increases in fast transient owing to simultaneous activation of nucleation sites	Tachibana, Akiyama, and Kawamura, 1968
Geometry: Diameter	CHF varies inversely with the diameter or width of the heater	Lienhard and Keeling, 1969
Surface condition	CHF increases with deposit on heater wire but remains constant on upward horizontal flat surface	Ivey and Morris, 1965; Farber and Scorah, 1948; Berenson, 1962

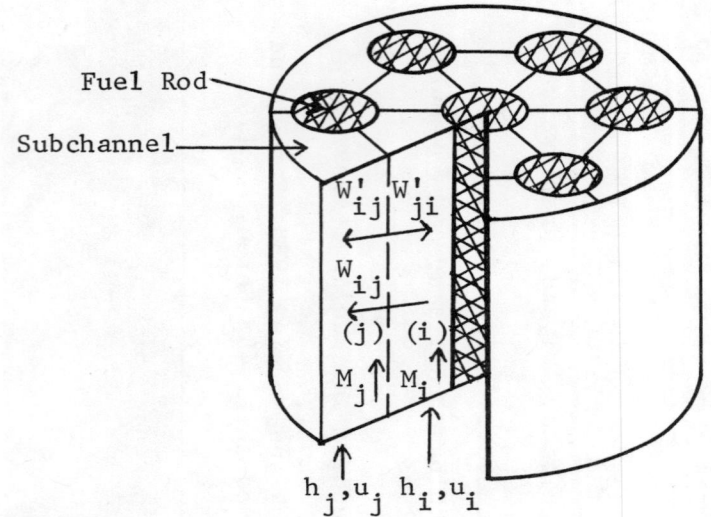

Continuity Equation

$$A_i \frac{\partial \rho_i}{\partial t} + \frac{\partial m_i}{\partial x} = - \sum_{j=1}^{N} w_{ij} \; ; \quad i = 1,2,3,\ldots N.$$

Energy Equation

$$\frac{1}{u_i''} \frac{\partial h_i}{\partial t} + \frac{\partial h_i}{\partial x} = \frac{q_i'}{m_i} - \sum_{j=1}^{N} (t_i - t_j) \frac{c_{ij}}{m_i}$$

$$- \sum_{j=1}^{N} (h_i - h_j) \frac{w_{ij}'}{m_i} + \sum_{j=1}^{N} (h_i - h^*) \frac{w_{ij}}{m_i}$$

where

$$u_i'' = \frac{m}{A\left(\rho - h_{fg} \frac{\partial \psi}{\partial h}\right)}$$

Axial Momentum Equation

$$\frac{1}{A_i} \frac{\partial}{\partial t} m_i - 2u_i \frac{\partial \rho_i}{\partial t} + \frac{\partial p_i}{\partial x} = -\left(\frac{m_i}{A_i}\right)^2 \left[\frac{v_i f_i \varphi_i}{2D_i} + \frac{K_i v_i'}{2 \Delta x} + A_i \frac{\partial}{\partial x}\left(\frac{v_i'}{A_i}\right)\right]$$

$$- g\rho_i \cos\theta - f_T \sum_{j=1}^{N} (u_i - u_j) \frac{w_{ij}'}{A_i} + \sum_{j=1}^{N} (2u_i - u^*) \frac{w_{ij}}{A_i} .$$

Transverse Momentum Equation

$$\frac{\partial w_{ij}}{\partial t} + \frac{\partial (u^* w_{ij})}{\partial x} = \frac{s}{\ell} (p_i - p_j) - F_{ij}$$

Figure 9. Subchannel Definitions and Balance Equations in COBRA (27)

very similar to the models employed in other flow redistribution codes. A summary of the flow redistribution codes that have been developed is presented in Table III (29).

The magnitude of the diversion crossflow and turbulent crossflow have to be established empirically since the modeling relies upon the introduction of constants for both the magnitude and effect of both crossflow terms. A summary of the experiments that are directed toward the determination of crossflow data is presented in Ref. 29.

The time dependence of the flow distribution is included in the solution of the flow distribution so that the representation of transients is possible with the currently available codes. Application of this type of code to the solution of the detailed flow distribution during the reflooding phase of the loss of coolant accident has not been attempted due to the lack of knowledge of the precise transitions in flow pattern as the heat transfer changes from convection to steam to film boiling to nucleate boiling and to convective heat transfer (1).

The computed mass velocity distribution for a flow channel in a boiling water reactor fuel assembly during a loss of flow transient is presented in Fig. 10 (8). This calculation is typical of the type of analysis that is possible with a flow redistribution model.

III. CONDITION IV -- LOSS OF COOLANT ACCIDENT

The loss of coolant accident is the design basis accident for the fourth class of operational transients. This class of accidents is considered unlikely and recovery after the accident is not anticipated without major repairs. Hence, the analysis of the accident is predicated upon the demonstration of the total containment of the fission products and, consequently, no hazard to the population or environment (42).

The loss of coolant accident results from the escape of coolant from the primary system at a rate that exceeds the capacity of the coolant makeup system. Since failure of the pressure vessel is considered extremely unlikely, the spectrum of accidents is restricted to failures of the piping system which range up to the double ended guillotine rupture of the largest pipe in the primary coolant system (43). Once a rupture has occurred, the primary coolant undergoes decompression and blowdown from a subcooled state. The rate of decompression is controlled by the sonic velocity of the coolant, the area changes in the piping, the length of piping, and the resistances to flow in the primary coolant system. The subcooled decompression is followed by a saturated blowdown period that lasts considerably longer. The depressurization phase continues until the pressure in the primary coolant system reaches equilibrium with the pressure in the containment and the coolant stops escaping through the ruptured pipe (42).

The differential pressure across the core is determined by the rates of decompression of the upper and lower plenums and is very important because it determines the magnitude and direction of the flow of coolant through the core. The decompression rates for the two plenums will differ because the coolant must pass through the steam generator before it can escape through

Table III. Codes for Subchannel Analysis (29)

Author	Code	Reference	Natural Mixing - Turbulent Interchange	Natural Mixing - Diversion Cross-flow	Forced Mixing - Flow Scattering	Forced Mixing - Flow Sweeping	Molecular Conduction
Kattchee & Reynolds	HECTIC-II	(30)	B	NC	NC	NC	NC
Armour and Smith	MANTA	(31)	B	NC	NC	NC	A
Lowe	(None)	(32)	B	NC	NC	NC	NC
Schramm and Berland	Al	(33)	NC	A	NC	NC	NC
Oldaker	G-20	(34)	NC	NC	NC	B	NC
Sanders	U-3 Modified	(35)	NC	NC	NC	B	NC
Eifler, Nijsing, Airola	(Unnamed)	(36)	NC	NC	NC	B	NC
Schraub	MIXER	(37)	B	NC	B	NC	NC
Aernick et al.	THINC	(38)	A	A	NC	NC	NC
	THINC-II		A	NC	B	*	*
Rowe	COBRA	(27,39)	A	B	B	B	A
St. Pierre	SASS	(40)	A	A	B	B	NC
Bowring	HAMBO	(41)	B	A	B	B	NC

NOTE: A—Model predicts or code calculates internally
B—Same as A but empirical data provided as program input
NC—Not considered
*—Not published

234

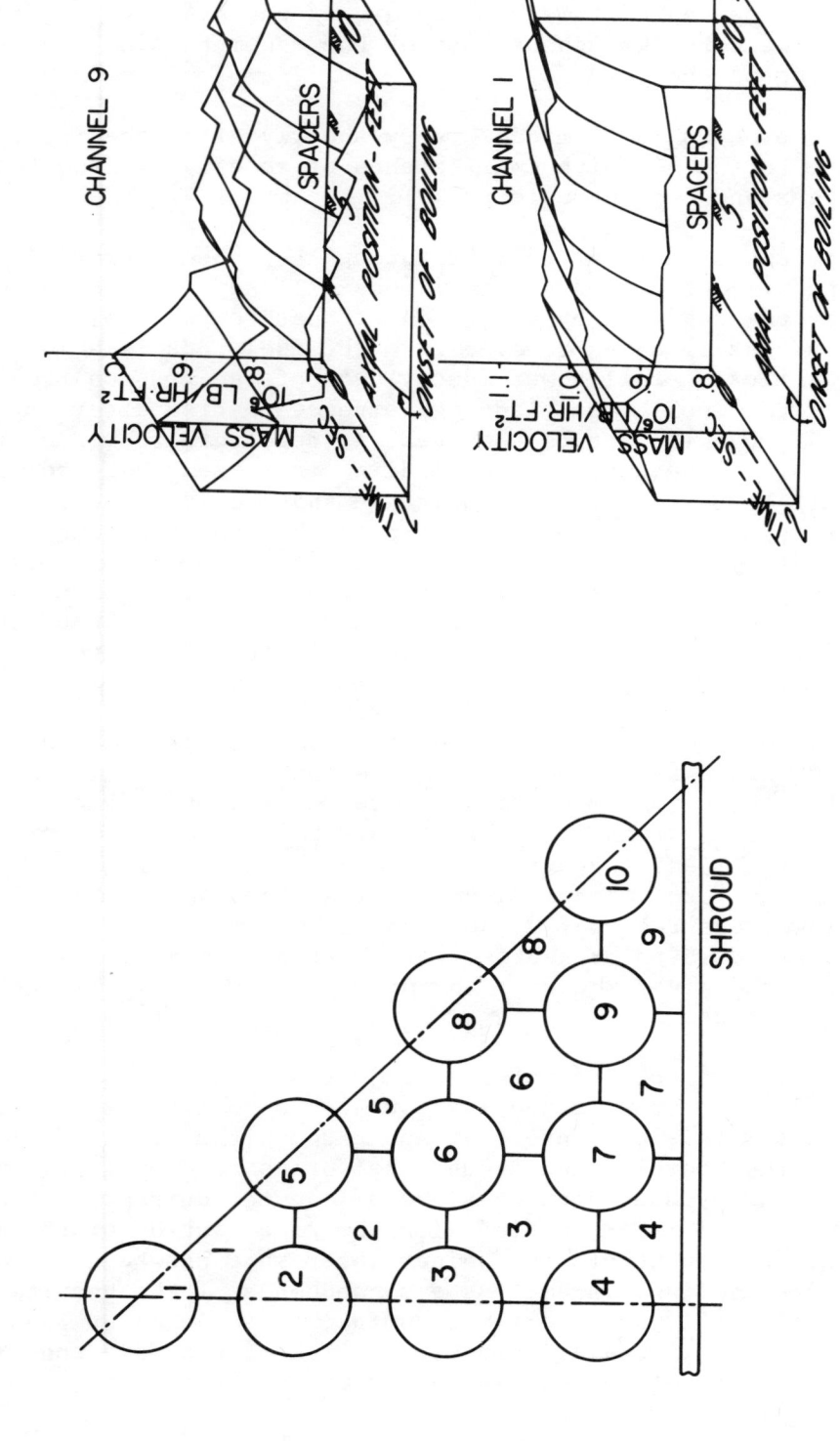

Fig. 10. Coolant Mass Velocity During Overpower Transient (8)

the ruptured pipe for one of the plenums while the other plenum will be essentially in direct communication with the rupture in the pipe (43). The rupture of the cold leg of the primary coolant system is the worst accident because the lower plenum will suffer the more rapid decompression and the flow of coolant will be reversed causing complete loss of cooling of the core after a very short period of time. A rupture of the hot leg would draw coolant through the core throughout most of the decompression and provide at least some cooling of the core (43).

The fission process is terminated very shortly after the initiation of the accident due to the negative coefficients of reactivity from fuel temperature increases and coolant voiding (3).

During the loss of flow of coolant through the core, departure from nucleate boiling will occur and the fuel rods will have to be cooled by transfer of heat to steam and by radiation of heat to the cooler fuel rods. Because the heat that is being removed from the fuel rods is a modest fraction of the heat flux at full power (decay heat of the fission products and stored energy of the fuel), the cladding temperature rise after departure from nucleate boiling will not be as severe as departure from nucleate boiling at full power. Consequently, the cladding is not expected to melt and calculations must be performed to demonstrate that no large scale loss of cladding integrity occurs due to overheating. When the cladding temperature exceeds approximately 2000°F, the zirconium reacts exothermically with the steam to produce zirconium oxide. This reaction releases approximately 140 calories per gram mole and the energy production must be included with the energy released by the decay of the fission products and the release of stored energy of the fuel (42).

During blowdown, the reactor instrumentation detects the low coolant pressure coupled with a reduction in the level of the water in the system and initiates the injection of emergency core cooling water into the primary system. Two types of systems are present for coolant injection with one type being an active system and the other being a passive system. The active systems employ pumps to force the emergency coolant into the primary system at both high and low pressures, while the passive systems use an accumulator that contains borated water pressurized with nitrogen and the coolant is injected through check valves whenever the primary system pressure is below the accumulator pressure.

The emergency core cooling water is injected into the cold leg of each primary coolant loop to provide the fastest access to the core and the most restrictive path for this coolant to escape through the ruptured pipe. However, this flow might be diverted around the core shroud by a phenomenon known as accumulator bypass which consists of counter current flow of the two phases of water. This is coupled with the condensation of the steam in the lower plenum as it contacts the colder water that has been injected by the emergency core cooling system. As a consequence of the uncertainty of these two processes, it is customarily assumed that the coolant injected during the blowdown phase is unavailable for cooling of the core and is not included in the inventory of coolant in the primary system at the end of the blowdown phase (44).

At the conclusion of the blowdown phase the emergency core cooling systems continue to inject water into the primary system and the lower plenum begins to fill with water. When the level of the water reaches the bottom of the fuel the reflood stage begins. This stage is characterized by the transition of the core coolant from steam to a steam and water mixture to complete water. The heat transfer mechanism changes from convection to steam to unstable film boiling to stable film boiling to nucleate boiling to convection to water. The quench front that accompanies the rewetting of the fuel rods is a very dynamic phenomenon that is characterized by large oscillations which include the initiating of cooling by droplets that are in advance of the quench front and that are essentially random in their contact with the superheated walls of the fuel rods. The analysis techniques employ uniform quench rates and ignore the oscillations in the quenching processes in an attempt to introduce conservatism into the analysis at a point where considerable predictive uncertainty exists (42).

The motion of the quench front is hindered by the process known as steam binding. This is the back pressure on the liquid which is caused by the resistance that the steam encounters as it is forced out of the upper plenum. Steam binding is more severe in a cold leg rupture since the steam in the upper plenum must flow through the steam generator before it can escape through the rupture and make room for the cooling water (43).

The limits that have been established for acceptable response to a loss of coolant accident are identified in Table IV (45,46). These limits are established to prevent loss of integrity of the cladding, to limit the production of explosive hydrogen due to the zirconium-water reaction, and to assure that the reactor remains coolable so that the heat from the decay of the radioactive fission products can be safely removed from the core for a long period of time.

The thermal and hydraulic phenomena which are important to the analysis of a loss of coolant accident are the heat transfer after departure from nucleate boiling and the heat transfer by radiation (47). Both of these phenomena along with a description of the principles employed in the computer codes that are used to evaluate the consequences of a loss of coolant accident will be discussed in the following sections.

The other thermal and hydraulic phenomena that are part of the analysis of the loss of coolant accident are not included in the discussion here because they are primarily empirical and are largely dependent upon the simultaneous occurrence of several phenomena which are each not predictable in an analytical sense. Examples of such phenomena are the determination of the velocity of the quench front during the rewetting phase and the determination of the accumulator bypass during the blowdown phase. The experimental study of the loss of coolant accident is the subject of the next paper and these empirical phenomena will be dealt with then.

A. Heat Transfer after Departure from Nucleate Boiling

The maximum cladding temperature occurs after the fuel rods have experienced departure from nucleate boiling and are surrounded by a film of steam. Heat transfer through steam is so poor that the cladding temperature rises

Table IV. Acceptance Criteria for Emergency Core Cooling Systems (45)

Limit Parameter	Limit	Purpose
I Cladding Temperature	2200°F	Prevent excessive loss of cladding integrity
II Cladding Oxidation	17% of cladding thickness	Prevents excessive loss of local cladding ductility and strength
III Hydrogen Generation	1% of amount if all Zr reacted	Prevents the generation of an explosive concentration of hydrogen, and prevents combustion of hydrogen that escapes to the containment
IV Geometry	Remains coolable	Prohibits the failure of cladding so as to block the coolant passages during the accident
V Geometry	Coolable for a long time	Assures that the decay heat can be removed from the core to prevent any further damage from overheating due to buildup of the decay heat

Note: These limitations must be satisfied when the calculations employ the assumptions set forth in the NRC emergency core cooling system evaluation model (46).

rapidly to provide a sufficient thermal gradient to conduct the heat away from the fuel rod. Only the low power that is present during the loss of coolant accident prevents the melting of the cladding.

It is possible to conservatively estimate the maximum cladding temperature by postulating no heat transfer after the departure from nucleate boiling (1) but this is overly conservative. Rather, the evaluation of the heat transfer during the film boiling period provides a more realistic maximum cladding temperature.

An important parameter in the analysis of heat transfer after departure from nucleate boiling is the Leidenfrost temperature. The Leidenfrost temperature is the lowest wall temperature where the surface of an entrained droplet of liquid evaporates so rapidly that the droplet can never reach the wall and appears to "dance" on the heated surface. When a fuel rod has departed from nucleate boiling the entrained droplets in the steam blanket will randomly approach the heated surface. If the surface temperature is below the Leidenfrost temperature the droplets will provide additional cooling of the surface while surface temperatures above the Leidenfrost will prevent any enhancement of the heat transfer (19). For steam-water mixtures at 30 psia and wall superheats less than 50°F, heat-transfer coefficients have been observed to be three to six times greater than for the flow of dry steam. For wall superheats exceeding 50°F the heat-transfer coefficients were almost identical to those for dry steam. Additional data indicate that wall superheats necessary for reaching the Leidenfrost temperature increase with increasing pressure. For steam-water mixtures at 250 psia the wall superheat required is about 160°F and 250°F at 1000 psia (18).

For film boiling at high mass velocities when wall temperatures are below the Leidenfrost temperature, the following correlation is recommended (18):

$$\left(\frac{hD_e}{k}\right)_f = 0.0193 \left(\frac{D_e G}{\mu}\right)_f^{0.8} \left(\frac{c_p \mu}{k}\right)_f^{1.23} \left(\frac{\rho_g}{\rho_{bulk}}\right)^{0.68} \left(\frac{\rho_g}{\rho_f}\right) \qquad (22)$$

where the subscript f outside the parentheses indicates the properties are evaluated at the film temperature $(T_w + T_b)/2$. This correlation represents data covering the following ranges:

$q_w'' = 0.11 \times 10^6$ to 0.61×10^6 Btu/hr-ft^2
$G = 0.88 \times 10^6$ to 2.5×10^6 lb$_m$/hr-ft^2
$p = 580$ to 3190 psia
$D_e = 0.10$ to 0.32 inch
$T_b = 483$ to $705°F$
$T_w = 658$ to $1100°F$

For lower mass velocities and pressures, Ref. 18 presents additional correlations for predicting film boiling heat-transfer coefficients.

When wall temperatures exceed the Leidenfrost temperature, heat-transfer coefficients can be evaluated using correlations for steam. McEligot, et al. (48) recommend

$$\frac{hD_e}{k} = 0.020 (N_{Re})_b^{0.8} (N_{Pr})_b^{0.4} (T_b/T_w)^{0.5} \tag{23}$$

where subscript b indicates bulk temperature and all fluid properties are for steam only.

Appendix K of 10 CFR 50 (44) specifically states that three correlations are acceptable for use in the post departure from nucleate boiling. These three correlations are identified in references 49, 50, and 51. The form of these correlations is very similar to the correlation presented above for heat transfer when the surface temperature is above the Leidenfrost temperature. Additionally, Appendix K prohibits the use of these correlations when the temperature difference between the cladding and steam is less than 300°F.

B. Radiation Heat Transport

Heat is being transported by radiation at all times during the operation of a nuclear reactor. However, during steady state operation the heat transferred from the fuel rod to the coolant is essentially balanced by the heat that is transferred from the coolant to the fuel rod. Additionally, the magnitude of the heat transfer is very small compared to the amount of heat that is transferred by convection and conduction. During a loss of coolant accident the coolant flashes to steam and the cladding temperatures become large giving rise to conditions where radiation becomes a significant means of heat removal from the fuel rods.

The consideration of heat transport by radiation is important since it is most effective when the emergency core coolant is beginning to reflood the core. The first portion of the core to be cooled is the shroud or control rod guide tubes (42) so that they can serve as a heat sink for radiant heat transfer. This is also the time when the hottest cladding temperatures occur so the most correct calculation of the heat transfer is justified. Heat is transferred from fuel rod to fuel rod by radiation and eventually to the shroud or guide tubes where the emergency core cooling water acts as a heat sink.

The transport of heat by radiation is controlled by the Stephan-Boltzmann relationship:

$$q_b'' = \sigma T^4 \tag{24}$$

where
  q_b'' is the energy emitted from a black body
  T is the absolute temperature
  σ is the Stephan-Boltzmann constant 0.1713×10^{-8} Btu/ft^2-hr-°R^4

When heat is being transferred from one body to another body the net heat flux is the difference between the energy that is emitted by each body which is expressed as:

$$q_{net}'' = \sigma(T_1^4 - T_2^4) \tag{25}$$

where
  1 and 2 as subscripts identify the two bodies transferring heat.

This expression is appropriate for only very idealistic cases and must be modified by factors to account for realistic conditions that serve to reduce the amount of heat transfer. There are four factors that reduce the heat transfer which are: (1) the bodies are gray bodies, (2) the medium between the bodies absorbs some of the radiation, (3) the radiation leaving one body does not all reach the other body, and (4) the surface conditions of the bodies cause the radiation to be somewhat temperature dependent beyond the Stephan-Boltzmann relationship (52).

The emissivity of a surface is denoted by the symbol ϵ and represents the energy being emitted by a body as a fraction of the energy that would be emitted by a black body. Similarly, a factor F can be introduced which represents the fraction of the energy leaving one body which reaches the second body. Introducing these factors and rearranging the relationship to solve for the net heat transferred from one body to another gives (53):

$$q/A = \frac{\sigma(T_1^4 - T^4)}{\frac{1-\epsilon_1}{\epsilon_1 A_1} + \frac{1-\epsilon_2}{\epsilon_2 A_2} + \frac{1}{A_1[F_{12}+(1/\tau_1+A_1/A_2\tau_2)^{-1}]}}, \quad (26)$$

where

F_{12} is the view factor which represents the fraction of heat that leaves body 1 and reaches body 2

τ_1, τ_2 are the fractions of energy that leave body 1 and body 2 which are absorbed by the transmitting medium.

The analysis is further complicated when one realizes that the heat transfer between each rod and all of its neighbors must be calculated simultaneously since the heat transport from any fuel rod depends upon the temperatures of all of the surrounding fuel rods.

The most complicated of the correction factors is the view factor or the factor that accounts for the failure of all of the energy radiated by one body to reach the second body. This factor can be computed by the appropriate integrals of the geometry since this factor is simply the fraction of the solid angle that is subtended by the second body. In practice, the determination of this factor is not simple for in an array of closely spaced fuel rods each fuel rod will "see" its nearest neighbors and a few of the fuel rods that are one row away. However, this relationship will be altered for the rods that are near a shroud or a guide tube since these fuel rods will "see" the shroud in addition to its neighboring fuel rods.

The presence of water vapor as the transmitting medium also complicates the problem of determining the heat transfer by radiation since it emits and absorbs radiation in limited spectral ranges (54). The transmittance of radiation through the water vapor and the emissivities of the fuel rods and the shroud and the guide tubes are empirically determined constants which are introduced into the analysis.

The difficulty of solution of such a system of equations is sufficient to frequently cause the adoption of a Monte Carlo type of solution. The Monte Carlo technique follows packets of thermal energy in much the same manner that neutrons or gamma rays are tracked in shielding problems. At each point along the track of a packet a random number is selected to determine whether the packet is to be absorbed by the medium or continue along the path until a solid surface is encountered. Repetition of this process for enough packets of energy will yield a statistically significant calculation of the energy transport throughout the fuel assembly and identify the cladding temperatures (55).

When all of the components of radiation heat transfer are combined the net effect of radiation is to flatten the temperature distribution within a fuel assembly and reduce the effect of the power distribution within the assembly at the time of the initiation of the loss of coolant accident.

C. Coolant Escape from Ruptured Pipes

The rate of escape of coolant from a ruptured pipe is very important to the analysis of the consequences of the loss of coolant accident. The reduction in the coolant inventory determines the time when the flow in the core reverses and also determines the flow rate prior to the reversal which is necessary for the heat removal from the fuel rods. Thus, the escape rate of the primary coolant ultimately controls the maximum temperature of the cladding (42).

The flow through the ruptured pipe will be critical in that the flow rate will be independent of the downstream or containment pressure. The coolant velocity will fall below the sonic velocity after much of the coolant has escaped from the primary coolant system and the flow rate will then be determined by the difference in pressure across the break.

The critical flow in the loss of coolant accident will be either single phase or two phase depending upon the pressure of the coolant in the pipe upstream of the break. In a PWR the initial escape of the coolant will be single phase but will change to two phase when the pressure falls below the saturation pressure that corresponds to the temperature in the pipe. The conditions in a boiling water reactor will result in two phase critical flow at all times during the accident (42).

Because of its simpler nature single phase critical flow will be discussed first and then the complexities of two phase critical flow will be considered. The concept of critical flow is most easily introduced by examining the discharge from a pipe that is connected to an infinite reservoir of high pressure and high temperature fluid. The fluid escaping through the pipe will have a velocity that is dependent upon the difference between the reservoir pressure and the environment pressure until the environment pressure is lowered below a certain point. For all pressures below this pressure the flow rate will be a constant and is identified as the critical flow rate (6). The fluid escapes from the pipe at a velocity that just equals the velocity of the pressure wave which is traveling into the reservoir. The environmental pressure is not felt within the reservoir since the fluid that is transmitting this information is also the fluid that is escaping. The velocity of the fluid as it escapes from the pipe and the velocity of any pressure waves in the fluid are the sonic velocity (28).

The relationships that characterize the critical flow rate and the sonic velocity are determined from the one-dimensional momentum conservation equation for a single phase fluid. The momentum equation is (6):

$$d(pA) + \frac{\dot{m}}{g_c} du = dF \qquad (27)$$

where
- $\dot{m}$ = mass flow rate
- F = frictional force
- u = velocity of the fluid
- p = pressure
- A = cross section area
- g_c = conversion constant.

The conservation of mass equation is also necessary and is given by (6):

$$\dot{m} = \rho u A \qquad (28)$$

where ρ is the density of the fluid and the other symbols are defined as above.

When assumptions of isentropic flow (entropy is constant and hence the friction term is zero) and noting that the derivative of the mass flow rate is zero for critical flow, these equations reduce to (6):

$$G_c^2 = -\frac{g_c}{\frac{dv}{dp}} \qquad (29)$$

where
G_c is the critical mass velocity
v is the specific volume and is equal to the reciprocal of the density and all other symbols are as defined above.

This simple expression gives the critical mass velocity in terms of the properties of the fluid and the difference in pressure across the exit does not appear which is consistent with the physical interpretation of this phenomenon.

When the flow at the exit from the pipe is two-phase, the simple derivation above will not hold because the momentum equation for each phase must be developed and a term for the transfer of fluid from one phase to the other must be included. Such a solution is very complex and it is customary to first consider two extremes. The first case represents the conditions where there is no mass transfer between the two phases and the other case where the slip ratio is determined to give the maximum possible mass velocity at the exit from the pipe. The first model is known as the frozen state model and the second model is referred to as the equilibrium slip model.

The frozen state model is developed by assuming that there is no transfer between the two phases during the transit through the rupture and hence each phase behaves independently. These assumptions are appropriate for bubbly flow with low steam volume fractions.

The development of the frozen state model begins with the one dimensional momentum conservation equations for each phase which are (6,18)

$$d(pA_f) + \frac{1}{g_c} d(\dot{m}_f u_f) = 0 \qquad (30)$$

$$d(pA_g) + \frac{1}{g_c} d(\dot{m}_g u_g) = 0$$

where the symbols are as defined previously and the subscripts f and g refer to the liquid and gaseous phases, respectively.

Combining these equations and introducing the conservation of mass equation and noting that the average specific volume is given by (6,18)

$$v = v_f \frac{1-x}{1-\alpha} = v_g \frac{x}{\alpha} \qquad (31)$$

where
α = steam volume fraction
x = quality or steam weight fraction

gives as the expression for the critical mass velocity

$$G^2 = \frac{g_c}{\frac{dv}{dp}} = -\frac{g_c}{\frac{dv}{dp}} \qquad (32)$$

where the symbols are as defined above. Equation 32 is identical to Eq. 29 for the critical mass velocity in a one-phase fluid which reflects the assumption that the phases are independent.

The development of the equilibrium slip model by Fauske (56) also begins with the momentum conservation equations and employs a revised expression for the average specific volume of the mixture which incorporates the effects of the slip or difference in velocity between the two phases. The expression for the average specific volume is (56,18)

$$v = \frac{1}{S} [v_f(1-x)S + v_g x][1 + x(S-1)] \qquad (33)$$

where
S is the slip ratio or the ratio of the velocity of the vapor to the velocity of the liquid
and all other symbols are as defined above.

Taking the derivative of the specific volume with respect to the slip ratio gives (18)

$$\frac{dv}{dS} = (x-x^2)\left(v_f - \frac{v_g}{S^2}\right) \qquad (34)$$

where the symbols are as defined above.

The slip ratio that gives the maximum flow is obtained by equating the derivative to zero and solving for the slip ratio which gives:

$$S^* = \sqrt{\frac{v_g}{v_f}} \qquad (35)$$

Combining this result with the expression for the average specific volume and introducing both into the expression for the critical mass velocity gives (18):

$$G_{max}^2 = -g_c S^* \left\{ [(1-x + S^*x)x] \frac{dv_g}{dp} + [v_g(1+2S^*x-2x) \right.$$
$$\left. + v_f(2xS^*-2S^*-2xS^{*2}+S^{*2})] \frac{dx}{dp} \right\}^{-1} \qquad (36)$$

where the symbols are as defined above.

In this expression the derivatives of the specific volume of the liquid phase with respect to the pressure have been set equal to zero since they are very small when compared to the derivative of the specific volume of the vapor phase with respect to pressure.

A plot of the critical mass velocity as a function of the stagnation enthalpy (enthalpy when the fluid velocity is zero) is presented in Fig. 11 (57). The data in this figure represent the results of calculations based upon the slip-equilibrium model described above.

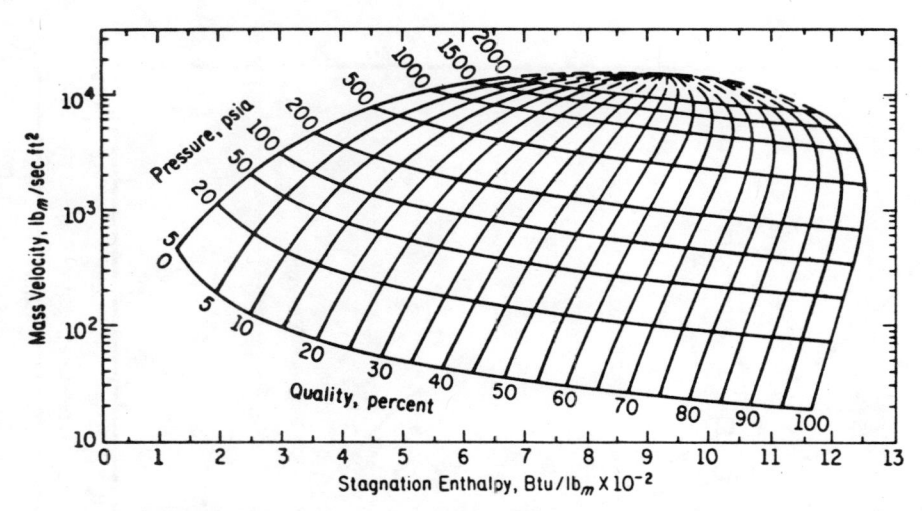

Fig. 11. Predictions of Critical Steam-Water Flow Rates with Slip Equilibrium Model (57)

This development of the critical mass velocity gave excellent agreement with measured data for pressures below 400 psia. However, at higher pressures a new phenomenon becomes important that this model does not include. The experimental evidence suggests that choking of the flow occurs at the area contraction where the pipe is attached to the reservoir. The upstream choking was associated with the flashing of the liquid to steam in the pipe and resulted in considerable alteration of the properties of the fluid that determined the critical mass velocity at the exit of the pipe. The critical mass velocity is presented in Fig. 12 (58) as a function of the critical pressure. These data show excellent agreement with the expressions developed by Fauske

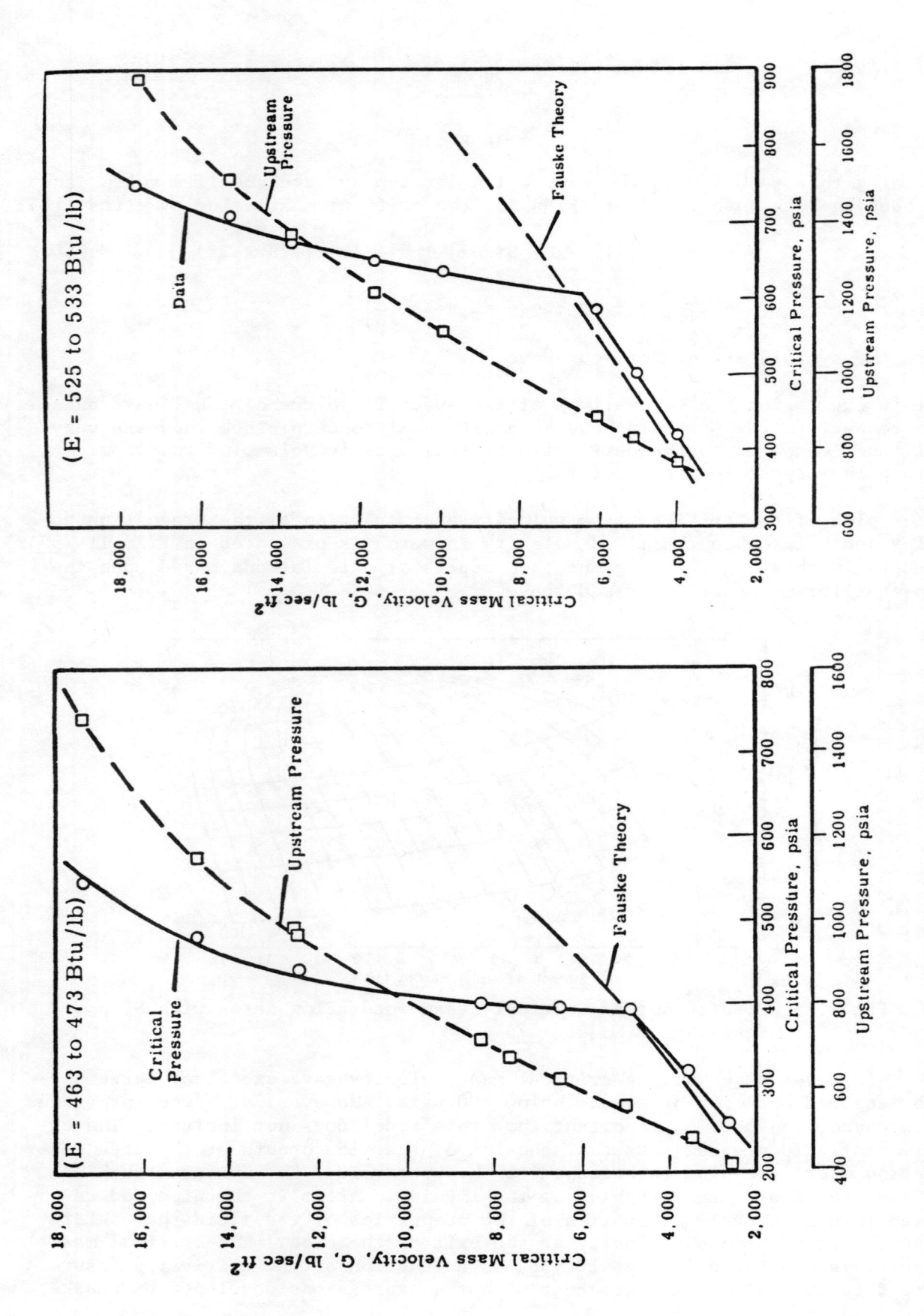

Fig. 12. Correlation of Data Assuming Choking at Exit of Test Section (58)

up to definite critical pressures and then marked deviations due to the upstream choking (58).

Several authors have presented models for the critical flow rate and the NRC has identified the results of Moody (59) as the preferred approach to calculate the critical mass velocity for the analysis of the loss of coolant accident. In this model the critical mass velocity is similar to that of the equilibrium slip model of Fauske with the exception that the optimum slip ratio is determined to maximize the energy which results in a slip ratio of (59)

$$S^* = (v_g/v_f)^{1/3} \tag{37}$$

where the symbols are as defined above.

This expression compares to the slip ratio determined by Fauske (56) which involved the square root.

The model of Moody (59) is valid only for flow in long pipes where thermal equilibrium between the phases is possible.

The modeling of the critical flow from short pipes which includes the effect of upstream choking was presented by Henry (60). This model includes nonequilibrium conditions and the effects of entrance geometry. The equation for the critical mass velocity is (60):

$$G_c^2 = g_c \left[\frac{xv_g}{P} - (v_g - v_{\ell_o})N \frac{dx_E}{dP} \right]^{-1} \qquad L/D > 12 \tag{38}$$

where

$$N = 20x_E \qquad x < 0.05$$
$$N = 1.0 \qquad x \geq 0.05$$
$$x = Nx_E \{1 - \exp[-0.0523(L/D) - 12)]\}$$

$$x_E = \left(\frac{S_o - S_\ell}{S_g - S_\ell} \right)_E$$

$$\frac{dx_E}{dP} = - \left[\frac{(1-x)\frac{dS_\ell}{dP} + x\frac{dS_g}{dP}}{S_g - S_\ell} \right]_E$$

L = the length of the pipe
D = the diameter of the pipe
S = the entropy

subscript E indicates that properties and derivatives are evaluated at equilibrium conditions at local pressure and upstream stagnation condition, and all other symbols are as defined above.

The only weakness in the model of Henry is its dependence upon an empirical factor N which represents the equilibrium or metastable conditions within the flow stream. Since N has been derived from data that span the range of interest in loss of coolant accidents, this restriction is minor.

D. Computer Codes for Loss of Coolant Accident Analysis

Each reactor vendor and some of the national laboratories have their own version of a system of computer programs that calculate the response of the reactor to a loss of coolant accident. Each of the vendor models is considered to be proprietary and only the necessary information essential for licensing of power plants is included in documents that are available to the public. Aerojet-General at Idaho Falls, Idaho has developed a set of computer codes to predict the response of a reactor to a loss of coolant accident to use in the analysis of the experiments that are being performed at their facility.

The complete solution of the heat, mass, and momentum equations for the reactor coolant during the course of a loss of coolant accident is impractical and would require several of the largest and fastest computers to attempt a numerical solution. Rather, all of the vendors have adopted a lumped parameter approach where the primary coolant system is represented by a small number of nodes and conservation of mass, momentum, and energy are applied to each node. This results in considerable lack of detail and preciseness which is compensated by calibration to the available experimental data.

All of the systems of codes are different but their general characteristics are very similar. The development of Appendix K to 10 CFR 50 (44) has forced much similarity in the code systems since it specifies in detail many of the approaches to the calculation that are to be used and identifies specific correlations that are to be used and the ranges of applicability that are to be permitted for each correlation. For example, the critical flow rate model developed by Moody (59) is specifically cited and exit loss coefficient ranges are identified with the stipulation that the discharge coefficient shall be the one that gives the highest cladding temperature even if it is not in the range of coefficients that is specified.

All of the code systems consider multiple loop plants with each loop divided into several nodes with the nodes connected by junctions so that the calculation is based upon the transport of mass, momentum, and energy through each junction.

The primary system is divided into nodes which usually include the following regions (42):

1. The upper plenum and the core above the steam-water interface
2. The lower plenum and the core below the steam-water interface
3. The water refilled region of the downcomer
4. The steam filled region of the downcomer which also includes the piping between the pressure vessel and the rupture
5. The hot leg of the primary coolant loops
6. The steam generators
7. The intact cold legs of the primary coolant system
8. The portion of the ruptured cold leg between the rupture and the steam generator.

The NRC, in requirements as stated in Appendix K to 10 CFR 50, requires the use of the correlation of Moody (59) for the calculation of the mass flow rate of the coolant escaping from the ruptured pipe. Also, it is required to assume complete loss of any emergency core cooling water that is injected during the blowdown phase of the accident. It is customarily assumed that the nitrogen pressurizing gas in the accumulators expands adiabatically as the emergency core cooling water is injected into the primary coolant system. These assumptions introduce considerable conservatism into the analysis of the coolant inventory in the primary coolant system.

The heat transfer characteristics of the coolant system are analyzed by assuming that the steady state departure from nucleate boiling correlations may be used to predict the time of loss of cooling of the fuel rods. Appendix K of 10 CFR 50 further states that once departure from nucleate boiling has occurred the correlation developed by Groeneveld (49) shall be used for the post departure from nucleate boiling heat transfer. Some of the systems of codes go further than this and assume that the heat transfer after departure from nucleate boiling is adequately represented by insulation boundary conditions at the surface of the fuel rod and ignore any heat transfer to the steam. Once a point has entered into departure from nucleate boiling the assumed behavior indicates that return to nucleate boiling cannot occur until that area is rewetted as the quench front moves upward through the core during the reflood stage of the accident. A history of the peak cladding temperature during a calculated loss of coolant accident is presented in Fig. 13 where the effect of the heat-transfer coefficient after rewetting is shown. A heat-transfer coefficient of 15 Btu/hr-ft^2-°F is sufficient to halt the increase in cladding temperature and any larger value of h will begin the cooling process. Thus, the initiation of nucleate boiling is not required for the cooling of the cladding but its presence is instrumental in speeding the return of the cladding temperatures to low values.

The heat transfer calculations that model the thermal response of the fuel rods include the heat produced by the following phenomena:

1. Decay of radioactive fission products
2. Release of stored energy within the fuel
3. Energy released by the reaction of zirconium and water.

The energy released by decay of fission products is dictated by Appendix K of 10 CFR 50 (44) to be 20% larger than the ANS standard for decay heat (61). This ensures conservative temperatures of both the cladding and the fuel.

The reflooding phase of the accident is characterized by the uniform advancement of the quench front up through the core with no credit taken for any cooling in advance of the quench front and no credit taken for any non-uniformity of the velocity of the quench front. The velocity of the quench front is derived from experimental data from the FLECHT tests (62) where overheated fuel rods were flooded with water in an attempt to simulate the reflood phase of the accident. Similarly, the heat-transfer coefficients during the reflood phase are taken from the same data.

The general impression one gets of the computer models for the representation of the response to a loss of flow accident is one of considerable conservatism that is either dictated by the NRC or is self imposed by the reactor vendors in an attempt to assure the safety of the environment and population in the unlikely event of a loss of coolant accident.

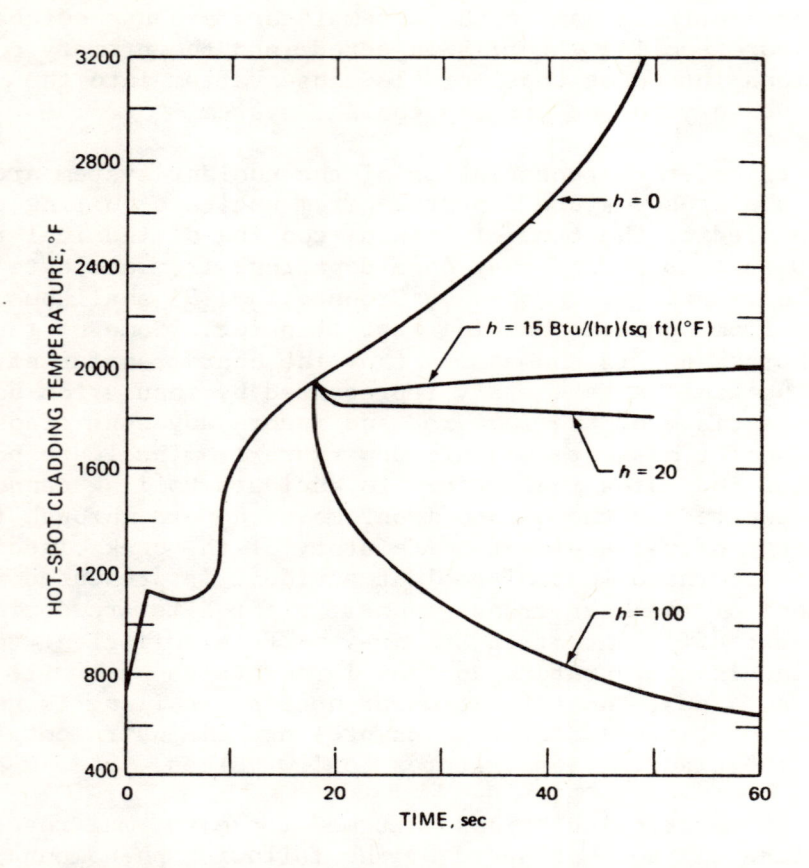

Fig. 13. Hot-spot Cladding Temperature vs Time for a 36-in.-ID Double-end Hot-leg Rupture and Variable Quench Coefficient (42)

IV. CONCLUSIONS

The principal conclusion that is apparent from the discussion of the principles of analysis is that conservative prediction of the consequences of a loss of coolant accident is possible. The models exist and agree with the experimental data in all of the important areas, but there is very little experimental data to assure the designer that the combination of the representations is realistic. For this reason the NRC, in their published requirements for the models and methods for employing the models, have introduced conservatism that results in conflicting assumptions. For example, the containment pressure is assumed to be higher than calculated when designing the containment building and at the same time the containment pressure is assumed to be lower than calculated when determining the escape of coolant and the termination of the blowdown phase of the accident.

It is hoped that continued experimentation will demonstrate the conservatism in the present modeling so that more realistic calculations may be utilized. The introduction of realism should permit the relaxation of some of the design criteria and a resultant reduction in the economic penalties associated with designing to accommodate the worst conceivable accident.

The continued development of the technology of safety should provide some small amount of public reassurance that the design of nuclear power plants will not cause any damage to the environment or the public health.

V. REFERENCES

1. A. SESONSKE, <u>Nuclear Power Plant Design Analysis</u>. Technical Information Center, U.S. Atomic Energy Commission (1973).
2. J. F. HOGERTON, "Background Information on Atomic Power Safety," Atomic Industrial Forum (1964).
3. "Sequoyah Nuclear Plant, Final Safety Analysis Report," Vol. 4, Tennessee Valley Authority (1974).
4. "Edwin I. Hatch Nuclear Plant Unit 2, Preliminary Safety Analysis Report,"Vol. 1, Georgia Power Company (1970).
5. J. A. CHRISTENSEN, R. J ALLIO, and A. BIANCHERIA, "Melting Point of Irradiated UO_2," WCAP-6065, Westinghouse Electric Corp. (1965).
6. M. M. EL-WAKIL, <u>Nuclear Heat Transport</u>, International Textbook Company, Scranton (1971).
7. F. KREITH, <u>Principles of Heat Transfer</u>, 3rd ed., Intext Educational Publishers, New York (1973).
8. D. A. REHBEIN and R. W. CARLSON, "The Effect of Variable Thermal Conductivity in Thermal-Hydraulic Calculations," to be published in <u>Nucl. Technology</u> (1976).
9. S. GLASSTONE and M. C. EDLUND, <u>The Elements of Nuclear Reactor Theory</u>, D. Van Nostrand Company, Inc., New York (1952).
10. C. G. PONCELET, "LASER--A Depletion Program for Lattice Calculations Based on MUFT and THERMOS," WCAP 6073, Westinghouse Electric Corp. (1966).
11. C. G. PONCELET, "Burnup Physics of Heterogeneous Reactor Lattices," WCAP-6069, Westinghouse Electric Corp. (1965).
12. E. DUNCOMBE, C. M. FRIEDRICH, and J. K. FISCHER, "CYGRO-3--A Computer Program to Determine Temperature, Stresses, and Deformations in Oxide Fuel Rods," WAPD-TM-961, Bettis Atomic Power Laboratory (1970); I. GOLDBERG and L. L. LYMANN, "FIGRO--FORTRAN-IV Digital Computer Program for the Analysis of Fuel Swelling and Calculation of the Temperature in Bulk-Oxide Cylindrical Fuel Elements," WAPD-TM-618, Bettis Atomic Power Laboratory (1966); F. J. HOMAN, W. J. LACKEY, and C. M. COX, "FMODEL-- A FORTRAN IV Computer Code to Predict In-Reactor Behavior of LMFBR Fuel Pins," ORNL-4835, Oak Ridge National Laboratory (1973).
13. L. W. DEITRICH, Personal Communication (1976).
14. BRIAN R. T. FROST, "Theories of Swelling and Gas Retention in Ceramic Fuels," <u>Nuclear Applications and Technology</u>, <u>9</u> (2), 128-140 (Aug. 1970)
15. F. H. NICHOLS, "Theory of Columnar Grain Growth and Central Void Formation in Oxide Fuel Rods," <u>J. Nucl. Mater.</u>, <u>22</u>, 214 (1967).
16. J. M. HELLMAN, Editor,"Fuel Densification Experimental Results and Model for Reactor Application," WCAP-8219, Westinghouse Electric Corp. (1973).
17. A. BIANCHERIA, "The Effect of Porosity on Thermal Conductivity of Ceramic Bodies," <u>Trans. Amer. Nuc. Soc.</u>, <u>9</u> (1), 15 (1966).

18. L. S. TONG and J. WEISMAN, *Thermal Analysis of Pressurized Water Reactors*, American Nuclear Society, Hinsdale, Illinois (1970).
19. J. H. RUST in *Elements of Nuclear Reactor Design*, Joel Weisman, editor, Elsevier Scientific, Amsterdam, Holland, to be published.
20. H. VON UBISCH, S. HALL, and R. SRIVASTAV, "Thermal Conductivities of Mixtures of Fission Product Gases with Helium and with Argon," *Proc. 2nd U.N. Intern. Conf. Peaceful Uses of Atomic Energy*, 7, 697 (1958).
21. R. A. DEAN, "Thermal Contact Conductance Between UO_2 and Zircaloy-2," CVNA-127, Westinghouse Electric Corp. (May 1962).
22. A. M. ROSS and R. L. STOUTE, "Heat Transfer Coefficient Between UO_2 and Zircaloy-2," AECL-1552, Atomic Energy of Canada Ltd. (June 1962).
23. S. McLAIN and J. H. MARTENS, "Fuel Element Design," in *Reactor Handbook, Vol. IV: Engineering*, Interscience, New York (1964).
24. W. R. PIERCE, "Temperature Distributions in Oxide Fuel Pins," Master's Thesis, The University of Virginia, Charlottesville, Virginia (1967).
25. L. S. TONG, "Boiling Crisis and Critical Heat Flux," AEC Critical Review Series, TID-25887 (1972).
26. R. B. BIRD, W. E. STEWART, and E. N. LIGHTFOOT, *Transport Phenomena*, John Wiley and Sons, Inc., New York (1960).
27. D. S. ROWE, "Crossflow Mixing Between Parallel Flow Channels During Boiling. Part I, COBRA--Computer Program for Coolant Boiling in Rod Arrays," BNWL-371, Pt. I, Battelle Northwest Laboratory (March 1967).
28. D. S. ROWE, "COBRA II: A Digital Computer Program for Thermal-Hydraulic Subchannel Analysis of Rod Bundle Nuclear Fuel Elements," BNWL-1229, Battelle Northwest Laboratory (February 1970).
29. J. T. RODGERS and N. E. TODREAS, "Coolant Interchannel Mixing in Reactor Fuel Rod Bundles Single-Phase Coolants," in *Heat Transfer in Rod Bundles* ASME, New York (1968).
30. N. KATTCHEE and W. C. REYNOLDS, "Hectic-II, An IBM 7090 Fortran Computer Program for Heat Transfer Analysis of Gas or Liquid Coolant Reactor Passages," IDO-28595, Aerojet-General Nucleonics, San Ramon, CA (December 1965).
31. S. F. ARMOUR and D. L. SMITH, "MANTA--Mixing Analyzed Nodal Thermal-Hydraulic Analyses," GEAP-4805, General Electric Company (February 1965).
32. P. A. LOWE, "A Two-Dimensional Turbulent Flow Mixing Model for Parallel Flow Rod Bundles," *ANS Transactions*, 10, 1 (June 1967).
33. D. E. SCHRAMM and R. F. BERLAND, "A Multi-Channel, Two-Dimensional Phase Flow Model and IBM 7090 Code," NAA-SR-TDR 9444, North American Aviation (December 1963).
34. I. E. OLDAKER, "Sheath Temperatures--A G-20 Computer Program for Fuel Bundles," AECL-2269, Atomic Energy of Canada Ltd. (May 1965).
35. J. P. SANDERS, "The Modified U-3 Code: A Thermal-Hydraulic Code for Axial Flow with Mixing in Fuel Bundles," ORNL-4016, Oak Ridge National Laboratory (November 1966).
36. R. NIJSING and W. EIFLER, "Heat Transfer Calculations of Organic Cooled Seven-Rod Cluster Fuel Elements," *Nucl. Engrg. and Design*, 4, 253-275 (1966); W. EIFLER, R. NIJSING, and J. AIROLA, "Description of IBM 360 Computer Program for the Calculation of Liquid Cooled Seven-Rod Cluster Fuel Elements," EUR 3733, Commission des Communautes Europeennes (1968).
37. L. S. TONG, "Pressure Drop Performance of a Rod Bundle," in *Heat Transfer in Rod Bundles*, ASME, New York (1968).

38. W. ZERNICK, H. B. CURRIN, E. ELYASH, and G. PREVITI, "THINC, A Thermal Hydrodynamic Interaction Code for a Semi-open or Closed Channel Core," WCAP-3704, Westinghouse Electric Corp. (February 1962).
39. D. S. ROWE, "COBRA IIIC: A Digital Computer Program for Steady State and Transient Thermal-Hydraulic Analysis of Rod Bundle Nuclear Fuel Elements," BNWL-1695, Battelle Northwest Laboratory (March 1973).
40. C. C. ST. PIERRE, "SASS Code 1 -- Subchannel Analysis for the Steady State," AECL-APPE-41, Atomic Energy of Canada Ltd. (September 1966).
41. R. W. BOWRING, "HAMBO, A Computer Programme for the Subchannel Analysis of the Hydraulic and Burnout Characteristics of Rod Clusters. Part 1, General Description," AEEW-R524, Atomic Energy Establishment, Winfrith (April 1967); R. W. BOWRING, "HAMBO, A Computer Programme for the Subchannel Analysis of the Hydraulic and Burnout Characteristics of Rod Clusters. Part 2, The Equations," AEEW-R582, Atomic Energy Establishment, Winfrith (1968).
42. G. YADIGAROGLU, K. P. YU, L. A. ARRIETA, and R. GRIEF, "Heat Transfer During the Reflooding Phase of the LOCA-State of the Art," EPRI 248-1, Electric Power Research Institute (1975).
43. L. J. YBARRONDO, C. W. SOLBRIG, and H. S. ISBIN, "The 'Calculated' Loss-of-Coolant Accident: A Review," _AIChE Monograph Series_, Number 7 (1972).
44. "ECCS Evaluation Models," Appendix K, _Licensing of Production and Utilization Facilities_, Title 10, Code of Federal Regulations, Part 50.
45. "Acceptance Criteria for Emergency Core Cooling Systems for Light Water Nuclear Power Reactors," Section 50.46, _Licensing of Production and Utilization Facilities_, Title 10, Code of Federal Regulations, Part 50.
46. "U.S. Atomic Energy Commission's opinion on the Acceptance Criteria for Emergency Core Cooling Systems for Light-Water-Cooled Nuclear Power Reactors," Docket RM-50-1, _Atomic Energy Clearing House_, _Vol. 19_, No. 53, pp. 2-62 (December 31, 1973).
47. P. W. MARRIOT, I. M. JACOBS, R. F. MURRAY, III, A. E. ROGERS, and W. A. WILLIAMS, "The Loss of Coolant Accident and the Environment -- A Probabilistic View," contributed by the Nuclear Engineering Division, American Society of Mechanical Engineers, Winter Annual Meeting, 72-WA/NE-9 (November 1972).
48. D. M. McELIGOT, P. M. MAGEE, and G. LEPPERT, "Effect of Large Temperature Gradients on Convective Heat Transfer in the Downstream Region," _Trans. ASME, J. Heat Transfer_, _87_, 67 (1965).
49. D. C. GROENEVELD, "An Investigation of Heat Transfer in the Liquid Deficient Regime," AECL-3281, Atomic Energy of Canada Ltd. (1969).
50. R. S. DOUGALL and W. M. ROHSENHOW, "Film Boiling on the Inside of Vertical Tubes with Upward Flow of the Fluid at Low Qualities," MIT Report 9079-26, Cambridge, Mass. (1963).
51. J. B. McDONOUGH, W. MILICH, and E. C. KING, "Partial Film Boiling with Water at 2000 psig in Round Vertical Tube," MSA Research Corp. Technical Report 62 (1958).
52. W. M. ROHSENHOW and H. Y. CHOI, _Heat, Mass and Momentum Transfer_, Prentice Hall, Inc., Englewood Cliffs, N.J. (1961).
53. R. SIEGEL and J. R. HOWELL, _Thermal Radiation Heat Transfer_, McGraw-Hill Book Co., Inc., New York (1972).
54. H. C. HOTTEL and R. S. EGBERT, "Radiant Heat Transmission from Water Vapor," _Trans. AIChE_, _38_ (1942).

55. J. R. HOWELL and M. PERLMUTTER, "Monte Carlo Solution of Thermal Transfer Through Radiant Media Between Gray Walls," ASME Jour. of Heat Transfer, 86 (1964).
56. H. K. FAUSKE, "Contribution to the Theory of Two Phase One Component Critical Flow," ANL-6633, Argonne National Laboratory (1962).
57. H. K. FAUSKE, "Two-Phase Critical Flow," paper presented at the M.I.T. Two-Phase Gas-Liquid Flow Special Summer Program (1964).
58. F. R. ZALOUDEK, "Steam Water Critical Flow from High Pressure Systems," HW-80535, General Electric Co., Hanford Atomic Products Operation (1964).
59. F. J. MOODY, "Maximum Flow Rates of a Single Component Two Phase Mixture," ASME Jour. of Heat Transfer, 87 (1965).
60. R. E. HENRY, "The Two Phase Critical Discharge of Initially Saturated or Subcooled Liquid," Nucl. Science and Engrg., 41 (1970).
61. Proposed American Nuclear Society Standards -- "Decay Energy Release Rates Following Shutdown of Uranium-Fueled Thermal Reactors," approved by Subcommittee ANS-5, ANS Standards Committee, October 1971.
62. "PWR FLECHT (Full Length Emergency Cooling Heat Transfer) Final Report," WCAP-7665, Westinghouse Electric Corp. (April 1971); "PWR Full Length Emergency Cooling Heat Transfer (FLECHT) Group I Test Report," WCAP-7435 (January 1970); "PWR FLECHT (Full Length Emergency Cooling Heat Transfer) Group II Test Report," WCAP-7544 (September 1970); "PWR FLECHT Final Report Supplement," WCAP-7931 (October 1972).

MANAGEMENT OF RADIOACTIVE WASTES

J. O. Blomeke
Oak Ridge National Laboratory
Oak Ridge, Tennessee 37830

INTRODUCTION

If nuclear fission is to provide a major share of U.S. energy requirements, we must provide technically, socially, and politically acceptable methods for handling the radioactive wastes that arise in a multiplicity of chemical and physical forms at every stage of the nuclear fuel cycle. Some of the radioisotopes in these wastes are so long-lived that they must be retained outside the environment for hundreds of thousands of years. They appear principally in the high-level liquid wastes from reprocessing spent fuels and in a wide assortment of solid refuse, contaminated with plutonium and other transuranium elements, that is generated in fuel reprocessing and mixed-oxide fuel fabrication.

Much work has been done over the past 20 years to develop suitable techniques for coping with these wastes; however, plant and field demonstrations are needed before much of the more recently developed technology can be reduced to practice.

This paper reviews the nature and characteristics of the wastes, the status of the work that has been done, the options that are available in waste management, and our plans and expectations for the near future.

PROJECTIONS OF THE NUCLEAR INDUSTRY

The projected growth of nuclear power in the U.S. through the year 2000 (Fig. 1) is based on ERDA's unpublished "Moderate Growth/High" case of February 1975. In this instance, the total installed nuclear capacity is estimated to rise from 82 to 1000 GW at the end of this century, and plutonium recycle in LWRs is assumed to begin in 1981. Larger, commercial HTGRs are introduced in 1983, and the commercial LMFBR in 1993. The fuel fabrication requirements for these reactors are shown in (Fig. 2). The amount of mixed oxide (LWR-Pu) and fast breeder (LMFBR) fuels fabricated is important because their fabrication is the source of 80 to 90% of the so-called "alpha" waste that must be managed. Thus, the first substantial amounts of alpha waste should appear in 1980, and proposed regulations (1) would allow this material to be stored on-site for five years before shipment to a federal repository.

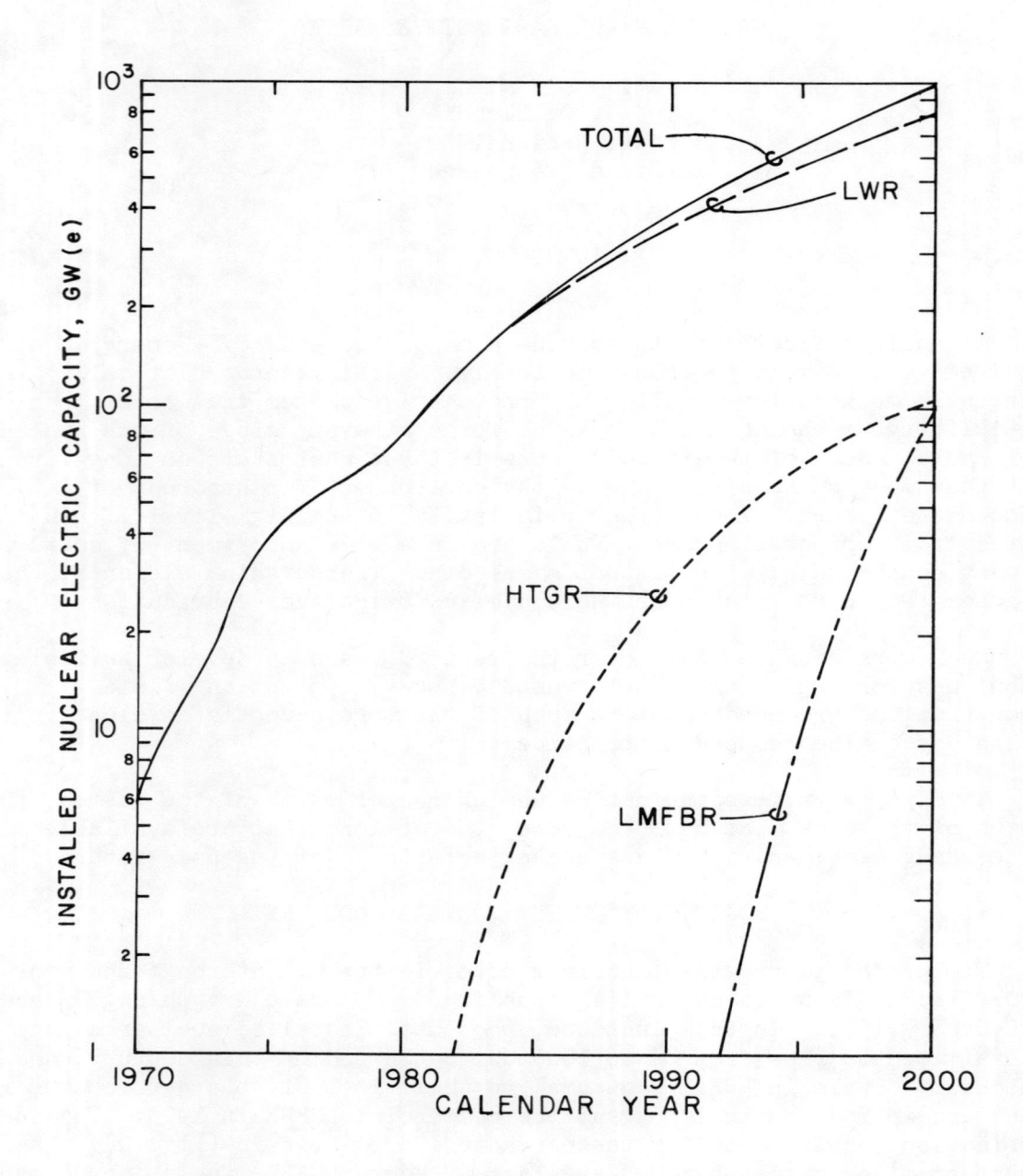

Figure 1. Projected Installed Nuclear Capacity in the U.S. (ERDA-33)

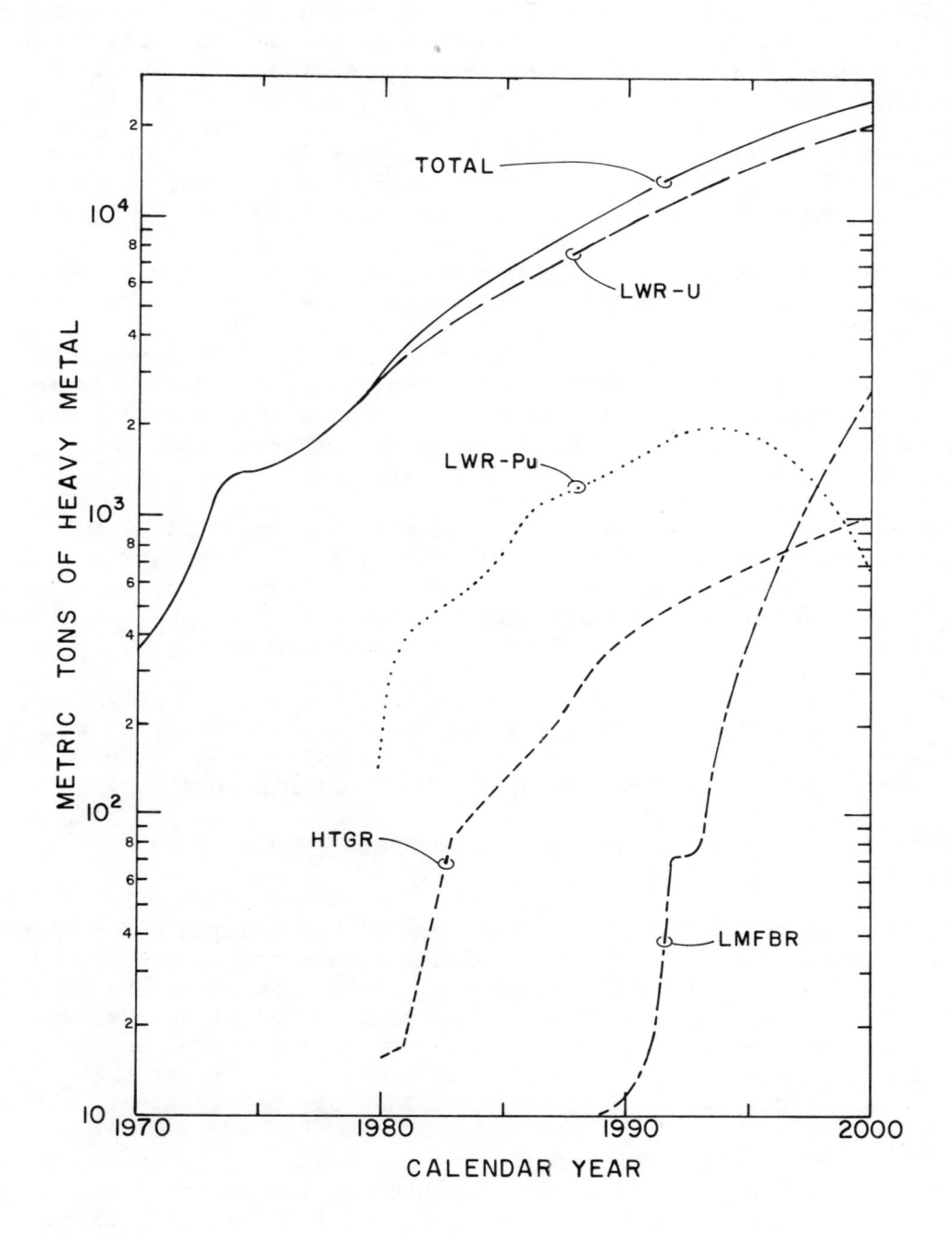

Figure 2. Fuel Fabrication Loads

The fuel reprocessing schedule (Fig. 3) is based on the assumption that Allied-General Nuclear Services starts operation in 1978, Nuclear Fuel Services in 1980, Exxon in 1983, and that additional capacity comes on-stream at a rate of 1000 metric tons per year beginning in 1986. Under present regulations (2), the so-called high-level wastes from fuel reprocessing may be stored in liquid form for no longer than five years, and then must be solidified and shipped to a federal repository within 10 years following its generation. This implies that a repository should be ready to receive high-level wastes no later than 1988.

We have categorized the principal wastes from the fuel cycle and have calculated annual rates of generation and shipment, as well as the total quantities in inventory (Table I). We also show here a measure of disposal of each waste that can be developed within the next 5 to 10 years. These estimates do assume the satisfactory completion, including demonstration, of selected R&D programs that are underway on chemical separations, methods of waste conditioning and packaging, suitable shipping systems, and repositories for long-term storage and disposal.

In the case of high-level wastes, we have assumed that all liquid process waste streams are combined and solidified into a glass or ceramic form, and that this waste contains 0.5% of the plutonium and uranium, 100% of all other actinides, and all nonvolatile fission products. After 10 years it is shipped to a mine and isolated in bedded salt.

Transuranium wastes, like high-level wastes, contain >10 µCi/kg of long-lived alpha activity, but they are not "heat-generating." We assume here that they are compacted, incinerated, or otherwise reduced in volume by factors of 3 to 10, and shipped 1 year after generation.

The beta-gamma wastes are separated, packaged, and shipped after 1 year. All have >10 µCi/kg of alpha activity.

In the case of ore tailings, 95% recovery of the uranium and plutonium from 0.2% uranium ore is assumed. The tailings contain the unrecovered 5%, as well as all the radioactive daughters. Note that while the _mass_ of actinides in the tailings is greater than that in the high-level wastes, the _toxicity_ is much less. The greater toxicity of high-level wastes is due to ^{90}Sr. When it decays after 10^3 years, the total toxicity of the tailings (on this basis) will be about twice that of the high-level wastes.

The ^{14}C and decommissioning wastes are not shown. We estimate that about 150,000 Ci of ^{14}C will have been generated by the end of this century. We can almost certainly separate the ^{14}C from gaseous effluents, but suitable methods for concentrating, fixing, and disposing of it haven't been developed. Decommissioning wastes are almost impossible to project because of uncertainties in plant decommissioning schedules, and in the levels to which these facilities must ultimately be decontaminated. We believe, however, that all facilities can, if necessary, be decontaminated to levels where perpetual surveillance will not be required.

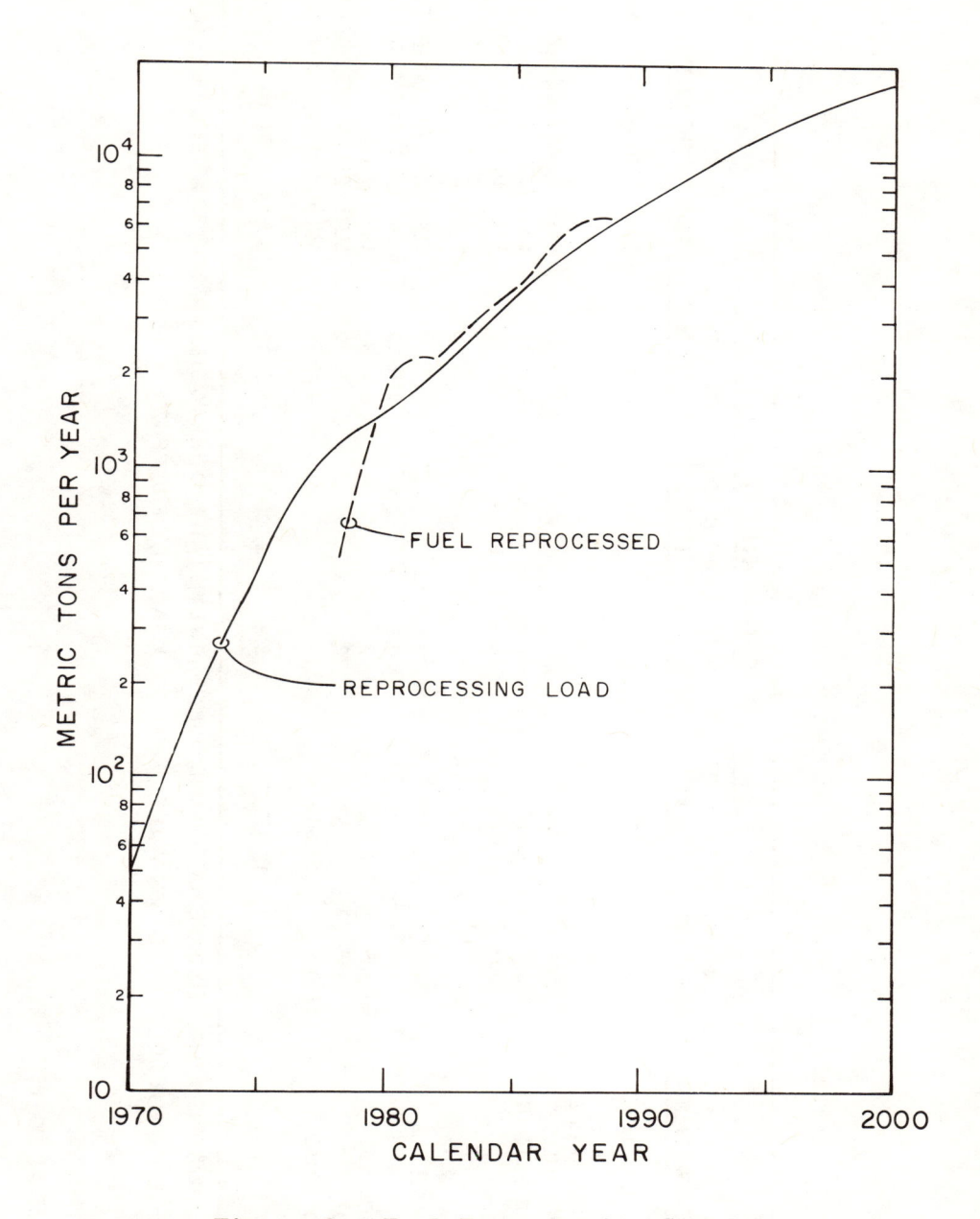

Figure 3. Fuel Reprocessing Schedule

Table I. Fuel Cycle Wastes Projected for the U.S. in the Year 2000

CATEGORY OF WASTE	NUMBER OF ANNUAL SHIPMENTS[a]	VOLUME (10^3 FT^3)	ACTIVITY (MCi)	ACTINIDES (TONS)	TOXICITY[b] (m^3 WATER)	POSSIBLE DISPOSITION
HIGH-LEVEL SOLIDIFIED	195[c]	370	126,000	1,110	4.1E 16	SALT (1500 ACRES COMMITTED)
TRANSURANIUM WASTES						
CLADDING HULLS	420[c]	364	587	75	3.1E 13	SALT (60 ACRES USED)
MISC. α, β, γ SOLID	2900[d]	2250	22	0.5	8.6E 12	SALT (210 ACRES USED)
ALPHA SOLID	650[c]	5960	110	13	1.5E 12	SALT (60 ACRES USED)
BETA-GAMMA WASTES						
NOBLE GASES	180[c,e]	17[e]	1,170		9.7E 10	SURFACE VAULTS (10^5 FT^3)
IODINE	16[d]	0.6	0.006		6.5E 8	SALT OR TRANSMUTATION
LWR TRITIUM	3750[d]	20,700	2		2.4E 10	DEEP WELLS
FP TRITIUM	730[d]	67	73		4.8E 11	SURFACE VAULTS (10^5 FT^3)
MISC. β, γ SOLID	11300[d]	51,000	16			BURIAL GROUNDS (1130 ACRES)
ORE TAILINGS		18,000,000	5.8	83,000	2.5E 13	SURFACE STORAGE (1600 ACRES)

[a] HIGH-LEVEL WASTE IS SOLIDIFIED IMMEDIATELY AND SHIPPED 10 YEARS LATER; OTHER WASTES ARE SHIPPED AFTER ONE YEAR.
[b] VOLUME REQUIRED FOR DILUTION TO RCG VALUES (APPENDIX B, TABLE II, OF 10 CFR 20).
[c] RAIL SHIPMENTS.
[d] TRUCK SHIPMENTS.
[e] PRESSURIZED AT 2000 PSI.

CHARACTERISTICS OF HIGH-LEVEL WASTES

Of all the nuclear fuel cycle waste, the high-level wastes from reprocessing spent fuels are by far the most important, and successful management of them will offer the key to management of all the other wastes. As generated, these wastes are complex aqueous mixtures containing radioactive isotopes of about 35 fission product elements and about 18 actinide and daughter elements. Corrosion products from fuel cladding and plant equipment, and chemical reagents that are added during reprocessing are also present. The properties of all the high-level wastes projected to have been accumulated by the year 2000 are summarized in Table II. In this table, only 0.1% of the iodine is assumed to be present in the high-level waste.

As mentioned previously, current regulations call for these wastes to be solidified within five years after they are generated and for the resultant stable solids to be shipped to a federal repository within ten years after the liquids are generated. Under these ground rules, part of the wastes to be accumulated by the year 2000 will be stored as solids in one or more repositories, and the remainder will be stored in liquid and solid forms at perhaps eight or ten reprocessing plants that may then exist in this country.

A convenient measure of the toxicity is the quantity of air or water that would be needed to dilute the waste to a level considered acceptable for inhalation or ingestion by the general population. This level is set by the U.S. Government's Radioactive Concentration Guide, and is known as the "RCG" level. For long-term storage in geologic structures, the ingestion hazard is the more relevant because a breach of containment would most likely result in contact between waste and ground water. The potential hazard is dominated initially by two fission-product radioisotopes: ^{90}Sr and ^{137}Cs. After about 350 years, however, the hazard decreases by more than three orders of magnitude and from that time on is controlled by long-lived actinides and, to a lesser extent, by three fission-product radioisotopes ^{99}Tc, ^{135}Cs and ^{93}Zr.

OPTIONS FOR MANAGEMENT OF HIGH-LEVEL WASTES

The principal options available for handling high-level wastes are seen schematically in (Fig. 4). The high-level wastes, with or without a prior period of storage as liquids to permit the decay of short-lived and medium-lived isotopes, are either sent to a chemical separations plant for removal of selected constituents or converted directly to a chemically inert solid form suitable for further storage or for disposal. A third, more remote, possibility is direct disposal as liquids in specially selected geological formations under the reprocessing plant.

Although storing wastes indefinitely in manmade surface structures is in principle possible, the need for continuous, virtually perpetual surveillance with this storage option relegates it either to use in conjunction with a permanent disposal method or, if necessary, as a fall-back position until an acceptable method of disposal can be developed.

Table II. Properties of High-Level Wastes to be Accumulated
in the U.S. Through the End of This Century

Property	Accumulated by year 2000	Time Elapsed Following Year 2000 (Years)			
		10^3	10^4	10^5	10^6
Radioactivity, MCi	126,000	62.6	25.0	4.5	1.5
Total fission products	123,000	4.45	4.24	3.25	0.69
^{90}Sr	10,800[a,b]	–	–	–	–
^{93}Zr	0.32	0.32	0.32	0.30	0.15[a,b]
^{99}Tc	2.44	2.44[a,b]	2.40[a,b]	1.79[a,b]	0.09
^{137}Cs	15,800	–	–	–	–
Total actinides	3,300	58.1	20.8	1.24	0.80
^{226}Ra	–	0.0001	0.004	0.028[b]	0.005[b]
^{229}Th	–	0.0005	0.004	0.031	0.068[a]
^{239}Pu	0.38	0.91	2.86	0.44[a]	–
^{240}Pu	2.75	9.6[a]	3.8[a]	–	–
^{241}Am	60.9	16.7[b]	0.39	–	–
^{243}Am	13.9	14.2	6.3[b]	0.001	–
^{244}Cm	2,600[a,b]	–	–	–	–
Thermal power, kW	634,000	1,500	485	35	19
Fission products	524,000	5.0	4.6	2.9	0.18
Actinides	110,000	1,500	480	32	19
Inhalation toxicity, m³ air[c]	1.2×10^{22}	3.4×10^{20}	1.5×10^{20}	1.0×10^{19}	4.1×10^{18}
Fission products	5.6×10^{20}	1.8×10^{15}	1.7×10^{15}	1.3×10^{15}	2.0×10^{14}
Actinides	1.1×10^{22}	3.4×10^{20}	1.5×10^{20}	1.0×10^{19}	4.1×10^{18}
Ingestion toxicity, m³ water[c]	4.1×10^{16}	1.0×10^{13}	3.4×10^{12}	1.6×10^{12}	5.9×10^{11}
Fission products	4.1×10^{16}	1.8×10^{10}	1.7×10^{10}	1.3×10^{10}	2.3×10^{9}
Actinides	4.3×10^{14}	1.0×10^{13}	3.4×10^{12}	1.6×10^{12}	5.9×10^{11}

[a] These nuclides control the inhalation toxicity at the indicated time.

[b] These nuclides control the ingestion toxicity at the indicated time.

[c] Cubic meters of air or water for dilution to RCG (Appendix B, Table II, 10 CFR 20).

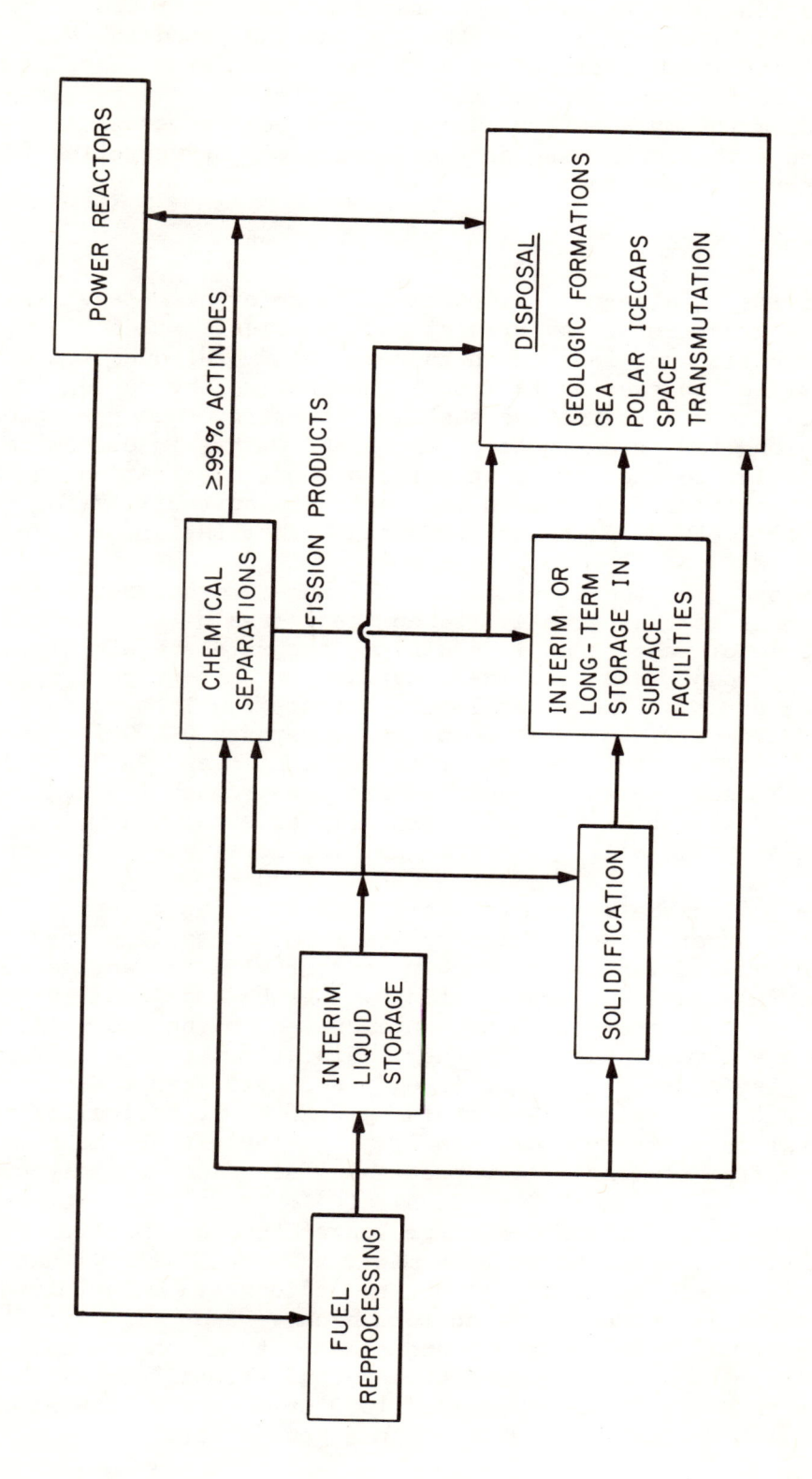

Figure 4. Options for Management of High-Level Wastes

Constituents can be removed from the wastes to reduce the volumes, heat-generation rates or hazards of the residuals that are to be stored or disposed of by one or more of the methods indicated in (Fig. 4). For example, the actinide elements control the hazard after several hundred years, and the disposal problem might be mitigated by separating the actinides from the much larger quantities of heat-generating fission products and either disposing of them separately or recycling them to the power reactor with new fuel.

Actinide Separations

The merits of alternate disposal schemes for high-level wastes are clouded to a great extent by present deficiencies in technology, probably much higher costs and, in some instances, by a moral question of whether we are justified in passing to future generations the burden of ultimately disposing of the wastes that we shall be creating. One possible way to reduce the long-term risk is to separate the wastes into chemical components that are similar to each other in terms of heat dissipation, duration of their potential hazard to the environment, and suitability for elimination by such exotic methods as neutron-induced transmutation or disposal in space.

On the basis of the duration of the long-term hazard, there is potential merit in processes that greatly reduce the losses of uranium and plutonium to the high-level waste and, in addition, separate the high-level waste into fission-product and actinide fractions. The required separation processes have not yet been developed, but apparently secondary treatment processes could permit overall recovery of perhaps 99.999% of the uranium, 99.995% of the plutonium, 99.95% of the neptunium and 99.9% of the americium and curium. Also, about 99.9% of the fission-product iodine, containing the long-lived isotope ^{129}I, will probably be separated — on the basis of its high relative volatility — from the high-level wastes during reprocessing.

Figure 5 illustrates the possible merits of these separations by comparing the hazard index of the wastes from conventional processing of fuel from a typical light-water power reactor with the wastes resulting from the postulated secondary treatment. The hazard index is defined as the volume of water that could be contaminated to the RCG — allowable value by one volume of waste or other radioactive material. For reference, the figure also shows the hazard index associated with the mineral pitchblende and with a uranium ore containing 0.2% U_3O_8, typical of the large deposits that occur in the Colorado Plateau. Note that the hazard index of the waste from conventional reprocessing decreases rapidly over the first thousand years (due to the decay of ^{90}Sr and ^{137}Cs) but remains equivalent to that of pitchblende for longer than one million years; the waste resulting from secondary treatment, however, falls within the range of widely dispersed, naturally occurring radioactive materials after only several hundred years, a time span for which we can reliably extrapolate the effects of geologic, climatic and other natural phenomena. Apparently then, this type of waste could be buried (and effectively diluted with nonradioactive species by a factor of 1000 or more) in natural geologic formations with negligible risk to future generations.

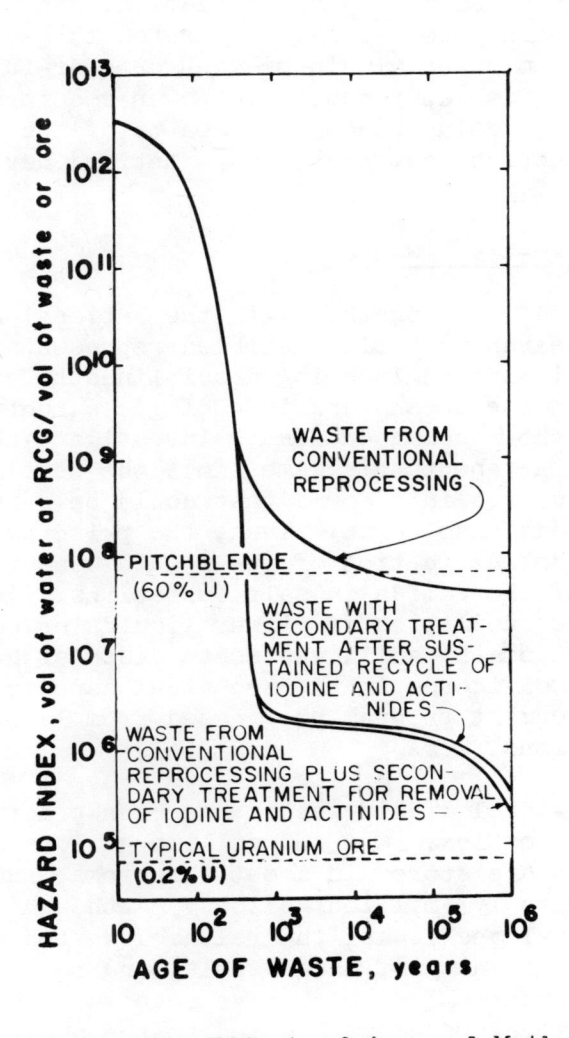

Figure 5. The Effect of Age and Method of
Treatment on the Hazard Index
of High-Level Wastes from LWRs

There are several prerequisites for this concept to become a realistic option, however. Exceptionally sharp separations of the six-or-more actinides from the fission products on a routine, day-to-day plant operating basis must be achieved without materially complicating the management of the remaining fission-product effluents. This means that all side streams from the processes must be recyclable, and that the final processed high-level wastes must be of such a nature that they can be solidified and handled by techniques which are currently being developed. We have been emphasizing high-level wastes, but it must be recognized that, if we are going this route, the actinides must be removed from <u>all</u> the wastes in which they appear. For example, about 1% of the total plutonium recycled is lost in the wastes from fuel preparation and fabrication, and with the cladding hulls. Finally, assuming that we have separated the actinides, we must be able to dispose of them. And within present or near-future technology this means burning them to fission products in power reactors. This can be done, but the implications of doing so on other parts of the fuel cycle (i.e., reactor operation, fabrication, and transportation) have not as yet been fully explored.

Solidification of High-Level Wastes

In the mid-1950's, AEC together with the National Academy of Sciences and the National Research Council, assembled representatives of the appropriate scientific and engineering disciplines to consider the disposal of wastes in the oceans and in geologic structures (<u>3</u>). From these deliberations the recommendation to investigate natural salt formations evolved. At about this same time, the development of processes for solidifying the wastes into forms that could be safely shipped and stored for long periods also began. Thus, the principal approach to waste management has been solidification of the liquid wastes followed by permanent disposal of the solids in salt formations. Interim storage of the wastes in surface facilities as either liquids or solids, or both, and the eventual shipment of the solid wastes to disposal sites were an integral part of this idea. Confidence in the technical aspects of the concept mounted as the development program progressed from laboratory to pilot-plant and field demonstrations.

Storage of high-level liquid wastes has been practiced for nearly 30 years within the AEC complex. Over 80 million gallons of acid and alkaline solutions and sludges are stored in about 200 tanks, ranging in individual capacity from 30,000 to 1.3 million gallons. Although 20 of these tanks are known to have developed leaks, the released wastes have been contained within the immediate vicinity of the vessels, and no radiation exposures to personnel or to uncontrolled areas have resulted. The experience gained with existing tank storage systems has led to the design of newer systems that justify greater confidence in containment integrity; however, this experience has also added to the incentive to develop a suitable disposal method.

Four processes have been developed for converting high-level liquid waste to solid forms believed amendable to safe storage, shipment and disposal. AEC's fluidized-bed waste calcining plant in Idaho has been working for several years and has converted to granular solids more than two million gallons of the wastes generated by reprocessing highly enriched fuels (4). At Pacific Northwest Laboratories, solidification of wastes from reprocessing power-reactor fuels has been demonstrated (5) by other means. More than 50-million curies of radioactive fission products have been processed, yielding solid products with thermal-power densities ranging up to 320 watts per liter.

The products here consist primarily of the oxides, silicates or borosilicates of the original waste constituents and are chemically and radiolytically stable at temperatures as high as 800 to 1200 °C. Characteristics such as final volume, thermal conductivity, resistance to leaching by water and so on vary rather widely, with the limits of acceptability being dictated by anticipated future handling and disposal requirements. Work is continuing both in the U.S. and in Europe on materials with maximum resistance to those undesired natural processes that could carry radionuclides away from disposal sites (6).

Storing and Shipping Solids

Highly radioactive solid materials — including spent fuels, isotopic radiation and heat sources, and solid wastes — are already regularly stored and shipped. During storage, most of these materials are placed under water for efficient heat removal and biological shielding. However, conceptual designs of solid-storage facilities, in which either water or air removes the heat, show that such systems are simpler and cheaper than equivalently safe systems that store the wastes as liquids. As a contingency measure, pending the demonstration of a suitable disposal method, consideration has been given to constructing retrievable surface storage facilities for both high-level and alpha-bearing wastes. Containment of these materials for a century or more could be provided with conventional, engineered surface structures (7).

The favored concept for storing high-level wastes for a long period is one termed the "Sealed Storage Cask Concept" (Fig. 6). Each canister of waste is first sealed in a 2-to-3-in.-thick steel cask which is then placed inside a 3-ft-thick concrete gamma-neutron shield. Cooling is by natural convection through an annulus. If the SSC were adopted through the year 2010, we would have about 70,000 casks placed on 25-ft centers in an arid area as shown in (Fig. 7). About 1100 acres of land would be required. Alpha wastes would be stored in vaults requiring perhaps an additional 400 to 500 acres of land. There is every reason to believe that such facilities could provide competent containment of the wastes for as long as full surveillance and maintenance are supplied.

The development of safe shipping systems is one of the major challenges facing the nuclear industry, and considerable effort has been and will continue to be devoted to this problem. The proposed shipments of solidified high-level wastes, however, constitute only a part of the total. For example, we estimate that the number of annual shipments of spent fuel from reactors to reprocessing plants may alone number 20,000 to 50,000 by the year 2000, whereas the number of shipments of solidified wastes in that year will be only about a hundredth of that. Also, the radioactivity of the wastes

AIR OUT (110° F)

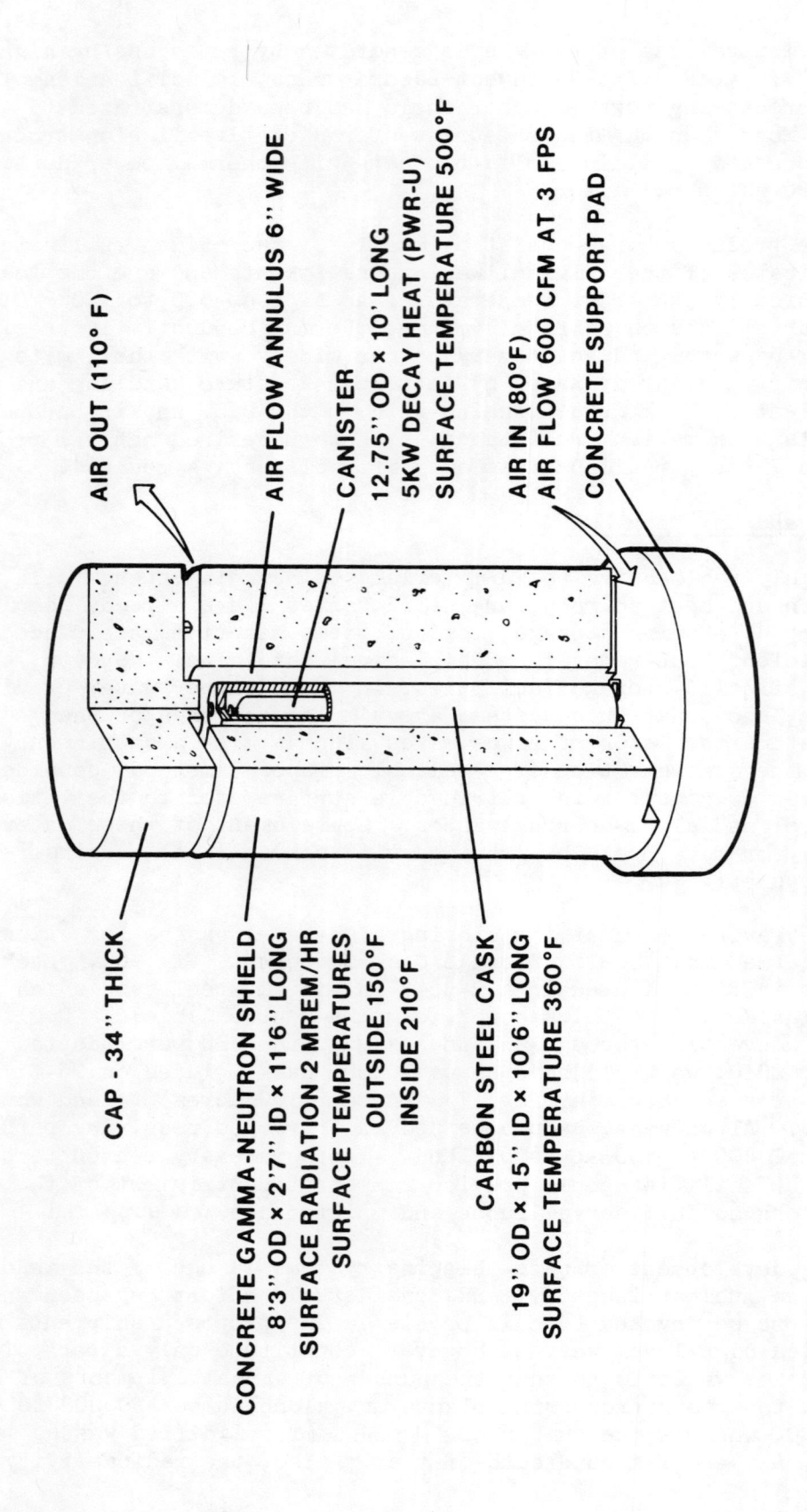

AIR FLOW ANNULUS 6" WIDE

CANISTER
12.75" OD × 10' LONG
5KW DECAY HEAT (PWR-U)
SURFACE TEMPERATURE 500°F

AIR IN (80°F)
AIR FLOW 600 CFM AT 3 FPS

CONCRETE SUPPORT PAD

CAP - 34" THICK

CONCRETE GAMMA-NEUTRON SHIELD
8'3" OD × 2'7" ID × 11'6" LONG
SURFACE RADIATION 2 MREM/HR
SURFACE TEMPERATURES
 OUTSIDE 150°F
 INSIDE 210°F

CARBON STEEL CASK
19" OD × 15" ID × 10'6" LONG
SURFACE TEMPERATURE 360°F

Figure 6. Sealed Storage Cask Concept

Figure 7. Sealed Storage Cask Concept Storage Area

will be only about a tenth of that of the spent fuels. Thus, in devising the methods and hardware needed to ship fuels safely, we shall be developing the technology needed for shipping wastes.

A conceptual cask for shipping high-level solidified waste is shown in (Fig. 8). This cask would carry nine, 12-in.-diam canisters of waste, would have overall dimensions about 10 ft in diameter by 14 ft long, and would weigh about 90 tons. Note the external fins for heat dissipation and the water jacket for neutron shielding.

Disposal in Bedded Salt

The designation by the NAS-NRC of natural salt deposits as the most promising place to dispose of waste was based on a number of considerations. Perhaps the most important of these is that salt deposits have remained free of circulating ground water since they were formed several hundred million years ago. Several conditions contribute toward maintaining this natural isolation unchanged in future years: 1) Most salt deposits are located in very stable regions typified by slow and gradual deformations. Therefore, they are less subject to earthquakes and other rapid geologic processes that could breach their protective covering of sedimentary rock. 2) Salt is one of the few geologic materials capable of rapid plastic deformation. Any fractures that might develop in the salt for any reason would "heal" rapidly, maintaining the integrity of the formation. 3) Salt deposits of such depth and extent exist that it would be all but impossible for any wastes buried inside to be disinterred by natural forces operating from the surface.

Other advantages of salt are its great abundance and wide distribution in the U.S.; the ease with which it can be mined; its resistance to damage by heat and radiation; its high thermal conductivity.

Following several years of laboratory and field experiments, Oak Ridge National Laboratory carried out a large-scale experiment called "Project Salt Vault," with high levels of radioactivity, in the Carey Salt Company mine at Lyons, Kansas (8). This experiment was successful in all respects, and in June 1970 AEC announced the tentative selection of an area adjacent to the Carey mine for the first waste repository. The selection was "tentative" because it was contingent on the outcome of additional work, mainly of a geologic nature, required to confirm the suitability of that particular site (9).

Subsequent geologic investigations revealed two areas of concern, both bearing on the long-term geologic integrity of the proposed Lyons site and both related to man's recent activities in that area. These were the potential for dissolution of the salt formation by ground water flowing down abandoned, improperly plugged oil and gas wells into permeable aquifers underlying the salt, and the past, present and future operations of a nearby mining company, which include hydraulic as well as mechanical mining of the salt for commercial purposes. Although our ongoing

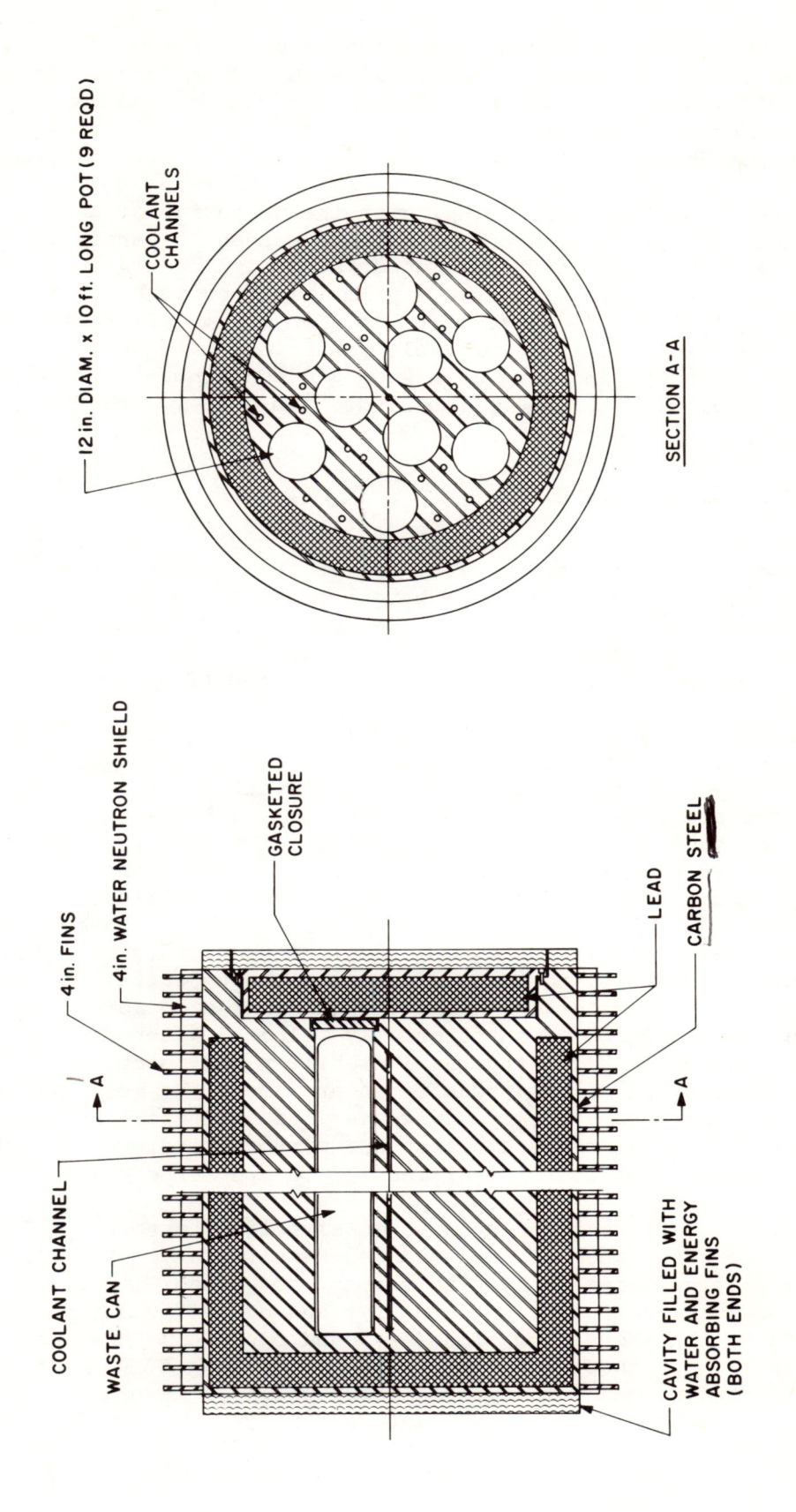

Figure 8. Shipping Cask for High-Level Solidified Waste

investigations yielded promise that these concerns could be successfully resolved from a technical point of view, a rather formidable array of public officials, including the Governor of Kansas and many members of the Kansas congressional delegation and state assembly, opposed the use of the Lyons mine, and it became necessary to search for alternative sites.

Work on salt is continuing, with the objective of obtaining all the supplementary information and experience needed to verify its suitability. A new salt-mine repository site under study is located in southeastern New Mexico, and we hope to have a pilot plant built by the early 1980's (Fig. 9). During this phase of pilot operations, all radioactive wastes will be kept in a fully retrievable condition while we gather information on operational safety and long-term geologic containment. We envision the eventual conversion of the pilot plant into a fully operational, permanent repository.

Other Geologic Environments

Although bedded salt formations were chosen as the most favorable site for the permanent isolation of high-level wastes for the reasons mentioned previously, other possibilities have been acknowledged. Among these are argillaceous (clay-containing) formations, limestone or crystalline-type deposits, nonsedimentary salt deposits (salt domes) ocean disposal and polar ice caps. What are the features of each of these alternatives?

Shales, clays, claystones and related argillaceous formations appear to be the second-best (after sedimentary bedded salt) general type of geologic environment for disposal of radioactive waste. These argillaceous formations are essentially impermeable (unless fractured or jointed), are insoluble, have a relatively high ion-exchange capacity, and in certain circumstances exhibit significant plasticity. However, there are two principal areas needing investigation: First, argillaceous materials usually contain significant quantities of moisture — both as pore fluid and as constituents of various clay minerals — that could be released when they are heated. We must, therefore, estimate the potential for thermal alteration of the mineralogy, and the quantities of water that would result. Then the consequences of any thermally released water must be assessed from the standpoints of structural stability and movement of fluids. Second, shales and related rock types are among the most difficult materials in which to perform underground excavations. Although existing technology can usually control mine-stability and support problems, the need for doing so would contribute significantly to operating costs.

Calcareous formations such as massive limestones and dolomites, and crystalline rocks such as granites and basalts, have many similar characteristics, especially with respect to suitability for waste disposal. These rocks are competent structurally, and no difficulty would be expected in the excavation of underground openings. However, that same structural competence, or more precisely the lack of plastic deformation characteristics, means that these rocks probably have appreciable natural permeability through existing fractures or joints. Even if rocks of this type could be found adequately isolated from circulating ground water, it would be difficult to guarantee their integrity over the very long hazardous lifetime of the wastes.

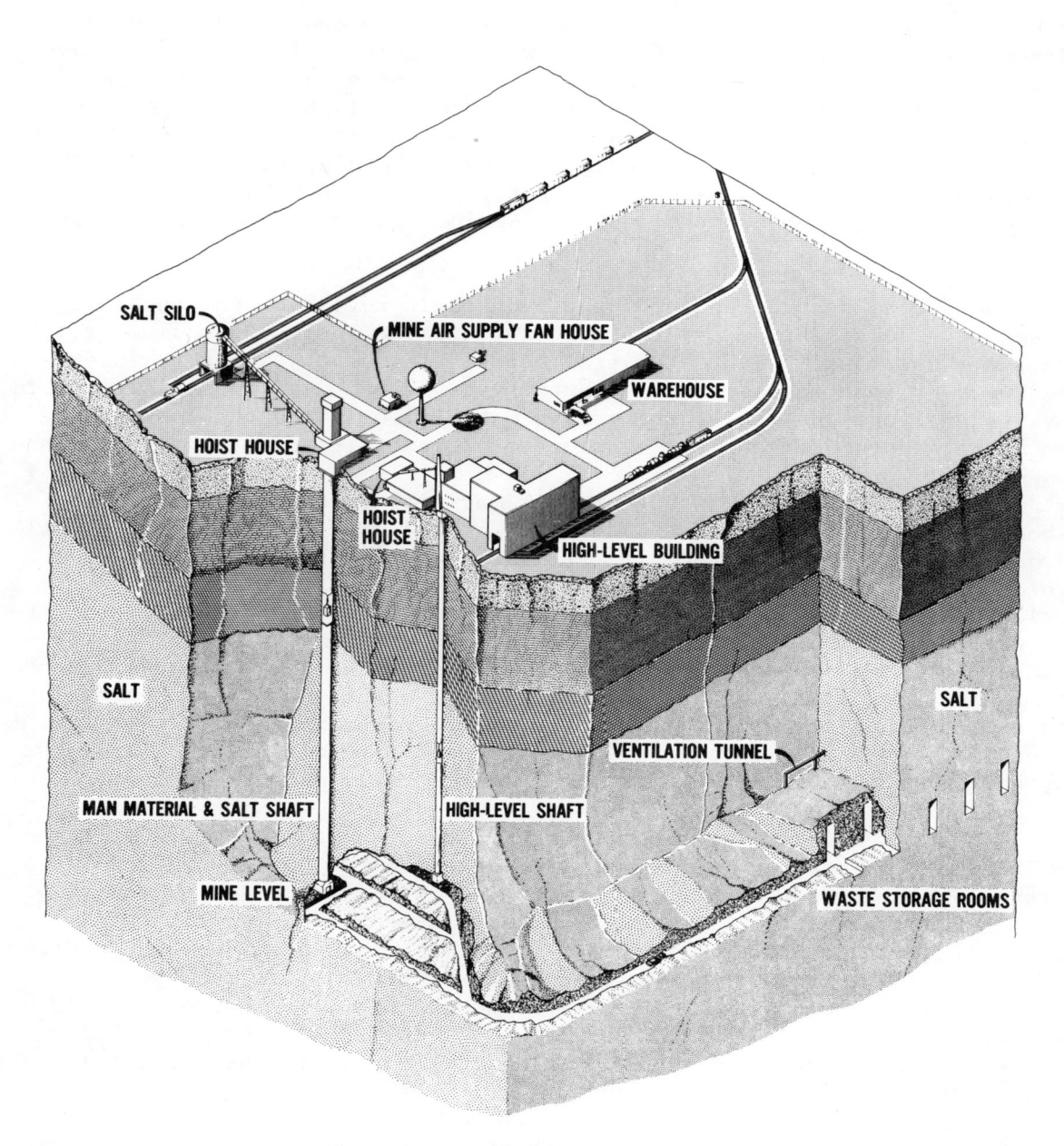

Figure 9. Pilot Plant Repository

Nearly all work on the feasibility of radioactive waste disposal in salt has been directed toward bedded salt deposits and has not encompassed the nonsedimentary domes and other mass-flow structures such as salt anticlines. Although there is no fundamental reason for disqualifying domes, there are special problems related primarily to selecting a specific dome and demonstrating its suitability. The first and most difficult of these problems concerns the tectonic stability of the structure, that is, its resistance to horizontal movements of sections of the earth's crust. Domes result directly from a tectonic instability, where massive volumes of salt have been intruded upward from deeply buried source beds, penetrating the overlying rock sequence. Before a particular dome could be considered suitable for waste disposal it must be shown that the dome is not now active, that rejuvenated movement (possibly triggered by waste disposal operations themselves) would be impossible and that past movements had not resulted in residual stress concentrations that would interfere with the excavation operations.

The second problem concerns the hydrologic regime around the salt structure. Because they are intruded through the overlying sediments, salt domes are not usually protected from circulating ground water by thick sequences of impermeable rocks. Investigations of any proposed site would have to analyze, in detail, the ground-water flow around the dome, and demonstrate future extensive dissolution of the salt to be very improbable.

Deep ocean trenches, where crustal material is being subducted and reabsorbed into the mantle, appear to offer the significant philosophical advantage of permanently removing the radionuclides not only from the biological environment but also from the earth's crust. However, a more serious consideration of the idea reveals some fundamental weaknesses. For example, subduction zones are characterized by extreme tectonic instability. Evidence at some zones suggest that a portion of the sediments are scraped off the sinking plate and are piled up at the margin of the continental slope. At other zones, the lighter, low-melting-point materials of the descending plate are vented through volcanoes. Therefore, removal of the wastes from the earth's crust may be neither complete nor permanent.

Perhaps a more acceptable approach to disposal at sea would be to emplace the canisters of waste in bored holes on the North Pacific's abyssal plain (10). This is reported to be one of the most tectonically stable areas on earth and new developments in deep ocean drilling and retrieval operations would appear to be applicable under these circumstances. Any sea disposal technique requires ocean shipping of the wastes, however, which certainly means increased hazard and probably, because of transloading, an increased cost. And retrieval of the wastes, should that ever become desirable for whatever reasons, would be effectively impossible within the present capabilities of technology.

The advantages of ice caps are their remoteness from inhabited regions of the earth and the simplicity of allowing the waste containers to bury themselves by melting their way down through the ice (11). This concept suffers from the obvious practical difficulties of routine operations with highly hazardous materials in extremely harsh arctic climates. It also has certain philosophical deficiencies, principally that, in the geologic sense, ice caps are very transient features. The existence of vast accumulations of ice is probably due to the concurrence of unusual geographic and climatic conditions, with a distinct possibility for drastic changes in the present ice caps with only fairly minor changes in the world mean climate. And, the conditions at the base of ice caps, which would be the final resting place for the wastes, are poorly known. In some cases, the ice may be floored with a layer of liquid water. Of course, it is known that even very slowly moving ice is very efficient at grinding rocks.

Extraterrestrial Disposal

Preliminary studies indicate that disposal of certain separated fractions of the high-level wastes to outer space may become feasible. Disposal of the *unseparated* high-level wastes into space is now impractical because of their very large mass and the need for massive radiation shields. We estimate costs in the range of one to three mills per kilowatt hour (electrical) for disposal of separated fractions of strontium and cesium, or of actinides from which 99 to 99.5% of the fission products have been removed.

The system being considered uses a space shuttle to place a payload of about 1000 pounds of waste (about 17,000 pounds total, including shields) in a low earth orbit, from which the waste package and a portion of the shield are ejected into a minimum earth escape, solar orbit. Ejection from our solar system, or solar impact, although potentially more acceptable, needs appreciably more energy. Reliable systems can probably be developed to permit safe recovery of the waste packages in the event of credible accidents during liftoff, ascent or orbiting phases of such a mission.

The primary advantage of disposal in space is complete removal of certain of the most long-lived components of the high-level waste from man's environment; the principal disadvantage is the obviously high cost. The questions that remain to be resolved are those of technical feasibility, reliability, cost and, of course, sociopolitical-acceptability.

PRESENT PLANS AND EXPECTATIONS

We believe that much of the technology that has already been developed can be applied to resolve the waste management problem. As we see it, the principal work that remains to be done within the next 10 years consists of engineering and plant demonstrations of techniques that are based on presently available technology. The development of promising longer-term alternatives, or back-up methods should proceed, but with a lower priority. A basic waste management flowsheet showing the principal direction of current work is presented in (Fig. 10). ERDA's waste management programs are being revised to emphasize the development and demonstration of the technology that is needed to prepare wastes for disposal. These demonstrations are scheduled to take place within the next 5 to 10 years. In addition, the construction of a "pilot" disposal facility for high-level and alpha wastes in bedded salt in New Mexico is scheduled for funding in 1978. The evaluation of other types of geologic formations than bedded salt (e.g., salt domes, shales, carbonates, volcanics, crystalline rocks, several permeable formations) will proceed with the expectation of developing 6 or 8 of the most promising combinations of waste type emplacement technique, geological site, and geological location.

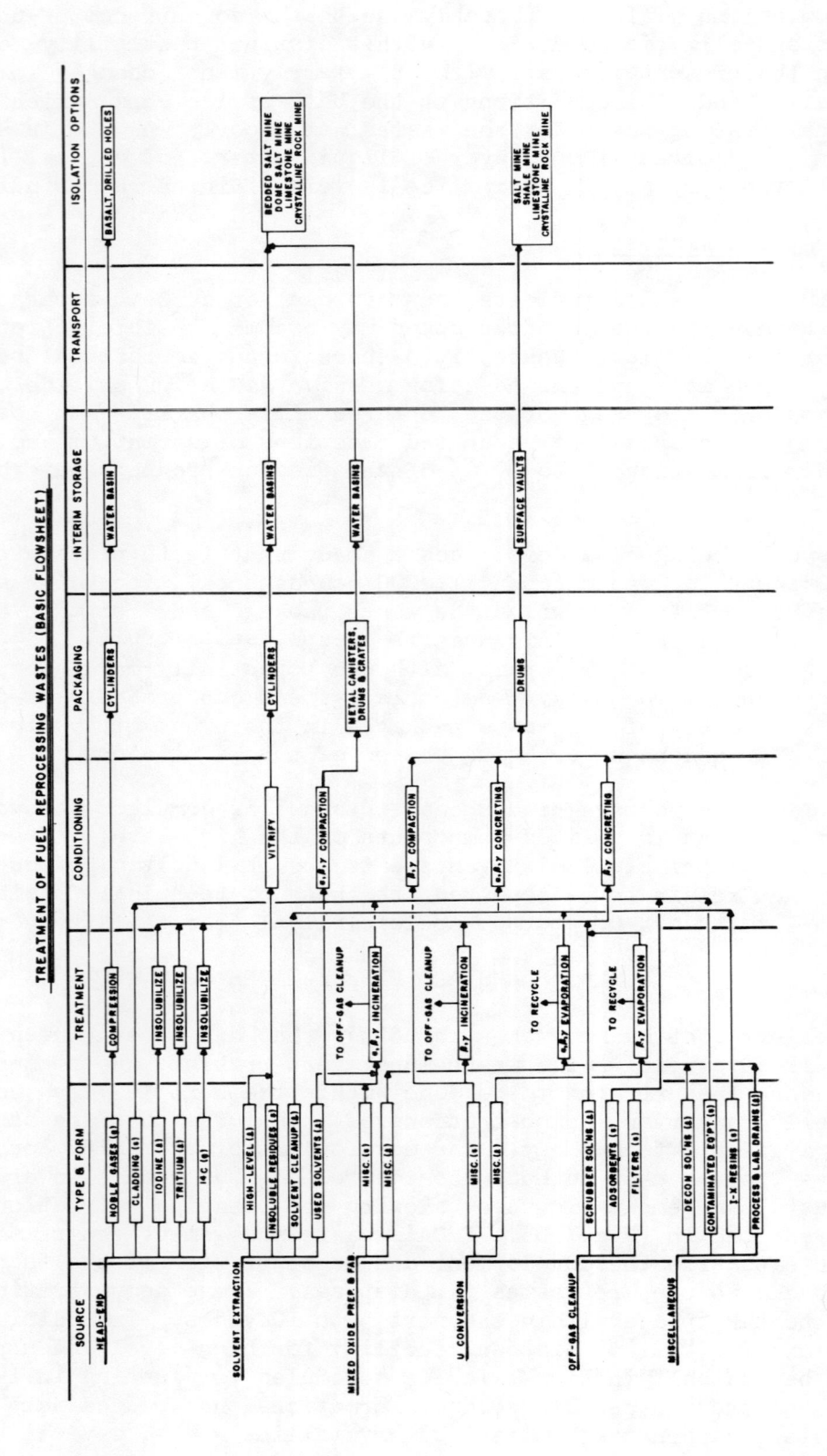

Figure 10. Basic Flowsheet for Radioactive Waste Management

ERDA has withdrawn the draft waste management environmental statement (WASH-1539) and plans to issue a new draft with an expanded scope concerning all options of waste management in 1976. With the delays that are contingent on resolution of the plutonium recycle question, it seems likely that the accelerated geologic disposal program might eliminate the need for federal retrievable surface storage facilities, and that by the mid to late 1980's we will have "closed" the LWR fuel cycle.

REFERENCES

(1). "Transuranic Waste Disposal. Proposed Standards for Protection Against Regulations," Federal Register 39 (178), 32921 (Sept. 12, 1974).

(2). Code of Federal Regulations Title 10, Part 50, Appendix F, "Policy Relating to the Siting of Fuel Reprocessing Facilities and Related Waste Management Facilities," Federal Register 35, 17350.

(3). "The Disposal of Radioactive Wastes on Land," Publ. 519, National Academy of Sciences — National Research Council, Washington, DC (April 1957).

(4). J. A. Wielang, G. E. Lohse, M. P. Hales, "The Fourth Processing Campaign in the Waste Calcining Facility," ICP-1004 (Dec. 1971).

(5). J. L. McElroy, et al., "Waste Solidification Summary Report, vol. 11, Evaluation of WSEP High-Level Waste Solidification Processes," BNWL-1667 (July 1972).

(6). Proceedings of a Symposium on the Management of Radioactive Wastes from Fuel Reprocessing, OECD/IAEA, Paris (Nov. 1972).

(7). U.S. Atomic Energy Commission, Management of Commercial High-Level and Transuranium-Contaminated Radioactive Waste, Draft Environmental Statement, WASH-1539 (Sept. 1974).

(8). "Project Salt Vault: Demonstration of the Disposal of High Activity Solidified Wastes in Underground Salt Mines," (R. L. Bradshaw, W. C. McClain, eds.), ORNL-4555 (March 1971).

(9). F. L. Culler, "Technical Status of the Radioactive Waste Repository — A Demonstration Project for Solid Radioactive Waste Disposal," ORNL-4680 (April 1971).

(10). W. P. Bishop and C. D. Hollister, "Seabed Disposal — Where to Look," Nuclear Technology, 24, 425 (1974).

(11). E. J. Zeller, D. F. Saunders, E. E. Angino, Science and Public Affairs, 29, 1, 4 (1973).

ADDITIONAL BIBLIOGRAPHY

Siting of Fuel Reprocessing Plants and Waste Management Facilities, ORNL-4451 (July 1970).

U.S. Atomic Energy Commission, High-Level Waste Management Alternatives, WASH-1297 (May 1974); also see K. J. Schneider and A. M. Platt (eds.), Advanced Waste Management Studies, High-Level Radioactive Waste Disposal Alternatives, BNWL-1900, 4 vols (May 1974).

Nuclear Power Growth 1974-2000, WASH-1139 (74) (Feb. 1974).

H. C. Claiborne, "Neutron-Induced Transmutation of High-Level Radioactive Waste," ORNL-TM-3964 (Dec. 1972).

P. J. Macbeth and W. W. Hickman, "ITSA — Above-Ground Retrievable Storage Method for Low-Level Transuranic Wastes," Nuclear Technology, 24, 383 (Dec. 1974).

Management of Plutonium Contaminated Solid Wastes (proceedings of the NEA seminar), OECD Nuclear Energy Agency, Paris (1974).

GENERAL FEATURES OF EMERGENCY CORE COOLING SYSTEMS

R. W. Shumway
Aerojet Nuclear Company
Idaho National Engineering Laboratory
Idaho Falls, Idaho

I. INTRODUCTION

Emergency core cooling (ECC) systems are one of a number of safeguard systems to help assure that radiation levels in the vicinity of light water power reactors are kept to acceptably low values. In particular, ECC systems are designed to protect the reactor core from becoming overheated should a loss-of-coolant accident (LOCA) occur. Rupture of a primary system pipe would allow the high pressure water to escape from the reactor system and could require emergency cooling water to remove the thermal energy stored in the core.

A reactor core consists of uranium fuel pellets packed in 12-ft long zirconium tubes (fuel cladding) which are about 1/2-in. in diameter. The "average" pressurized reactor contains about 40,000 tubes oriented vertically. The total weight of the tubes is about 20 tons and the weight of the uranium oxide is about 100 tons. During normal operation the surface temperature of a tube may be $650^{\circ}F$ while the centerline temperature of the fuel pellet may be as high as $4000^{\circ}F$. Water is pumped through the core at a rate of 18 tons per second and, although the transient time to pass through the core is less than 1 second, the water temperature increases $66^{\circ}F$. These numbers are presented to give the reader an impression of the amount of thermal energy generated in a reactor core. Again, the purpose of the ECC system is to remove the stored energy rapidly enough to prevent overheating of the fuel cladding to assure that the fuel stays in place if the primary water is lost.

The United States Nuclear Regulatory Commission (NRC) defines $2200^{\circ}F$ as the design upper limit on cladding temperature for all types of hypothesized pipe ruptures. Complete severance of a large primary pipe is unlikely, but protecting the core against such large breaks is, nevertheless, a requirement. Since $2200^{\circ}F$ is about $1100^{\circ}F$ less than the clad melting temperature, core meltdown will not occur and the fuel pellets will be maintained in a "coolable" geometry. One reason for keeping the allowable temperature so much lower than the melting temperature is that an exothermic reaction between the zirconium and water becomes significant at temperatures above $2200^{\circ}F$. For a variety of specified loss-of-coolant situations, the electrical utilities present to the NRC calculations of the rate of fluid flow through the core and the rate of energy removal. If calculations show that the temperature limit could be exceeded, the maximum operating power is reduced accordingly. Another restriction is that cladding oxidation must not exceed 17% of the cladding thickness. Exact calculations are difficult to make because the depressurization transient involves both steam and water flowing at possibly different velocities and temperatures. The flow is generally described mathematically as a homogeneous equilibrium mixture. Conservative numbers are used for such items as the heat generation rate to compensate for uncertainty caused by simplifying the calculations.

There are two types of water cooled reactors in the United States. Pressurized water reactors (PWR's) operate at a pressure of 2250 psia and have steam generator heat exchangers outside the reactor vessel. Boiling water reactors (BWR's) operate at 1040 psia and generate steam by boiling the primary water within the reactor vessel. The initial versions of power reactors did not use emergency core cooling systems. In 1962, work was begun by the Atomic Energy Commission (AEC) on a loss-of-fluid test (LOFT) to gain data on the potential extent of core meltdown and damage to the reactor vessel should an accident occur. By 1966 the concensus of opinion among government authorities was that reactor core meltdowns were not acceptable and emphasis shifted to core cooling systems. The LOFT (1) system began a complete redesign patterned after a large PWR. The LOFT system is now undergoing tests without a core in place. Tests with an active core are scheduled for mid-1977. Two small-scale electrically powered systems are currently producing test data. The Semiscale (2) facility is adjacent to the LOFT facility at the Idaho National Engineering Laboratory (INEL) and is scaled to LOFT. The BWR type test system, known as TLTA (3), is operated by the General Electric Company (GE) in San Jose, California, and is sponsored by the NRC, Electric Power Research Institute (EPRI), and GE. Comparisons of experimental data to mathematical predictions of the thermal hydraulic processes are used to evaluate mathematical techniques. Besides the LOFT, Semiscale, and TLTA tests there are other tests sponsored by NRC, EPRI and reactor vendors which concentrate on specific aspects of core cooling. Other countries also have research programs directed at studying the loss-of-coolant accident.

The following sections discuss PWR and BWR ECC systems, present test data from a few experiments, and compare some of the test data with calculations.

II. PWR SYSTEMS

System Description

Pressurized water reactors operate at a pressure of 2250 psig which is 600 psia above the saturation pressure. Thus no bulk boiling exists in the primary system. The piping from the steam generators and pumps connects to the reactor vessel above the elevation of the core as shown in Fig. 1; thus pipe ruptures will not necessarily result in a loss of water in the core. Water is circulated by the pumps from the vessel to the steam generators and back to the vessel. Three sections of piping connect the vessel to the steam generator and pumps. These pipes are about 2.5 ft in diameter and 20 ft long. Smaller pipes, including the emergency core cooling pipes, connect either to the large pipes or to the vessel. Core energy heats the water from a core inlet temperature of 544°F to an outlet value of 610°F. The steam generator secondary water absorbs this energy to create steam to drive steam turbines. Reactors have two, three, or four steam generators. The pipes carrying the water to the steam generator are known as hot leg pipes. The pipes connecting the steam generators to the pumps are called pump suction pipes and those connecting the pump to the vessel are known as cold leg pipes. A pressurizer tank attaches to one hot leg. System pressure is maintained by heating or cooling water in this tank.

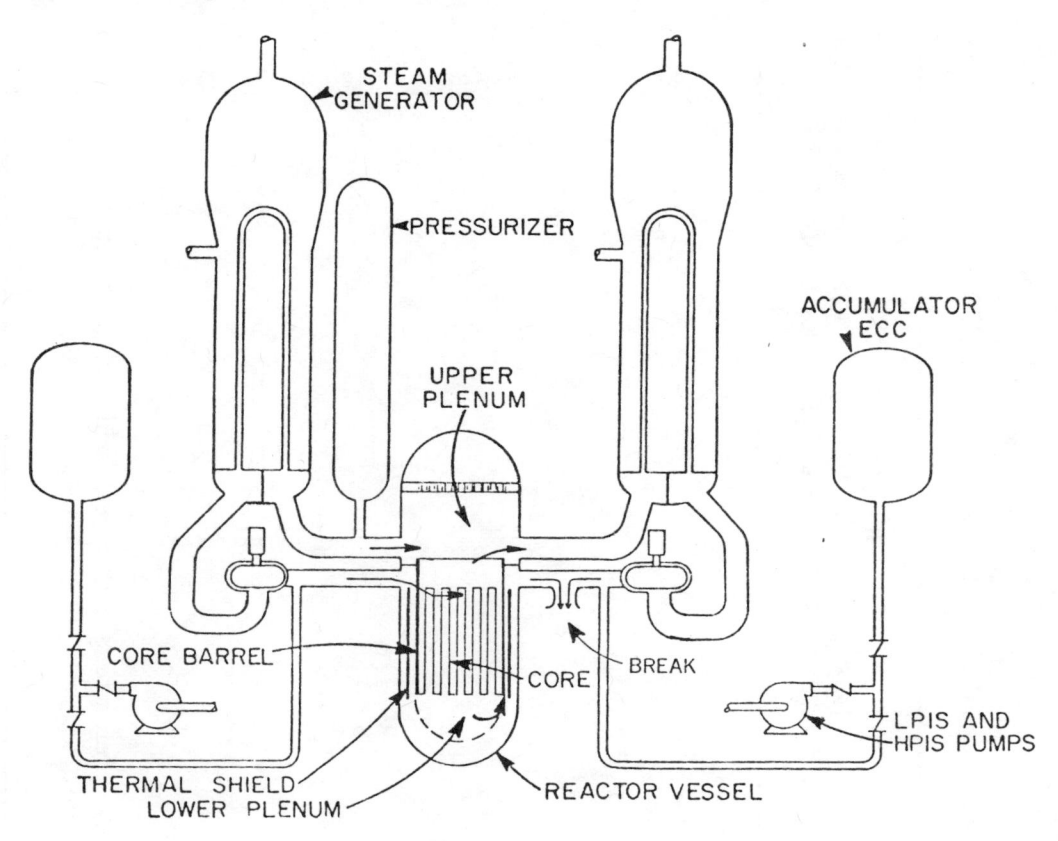

Fig. 1. PWR System During Blowdown

PWR ECC systems consist of both passive and active components. The passive components are tanks of water known as accumulators. A nitrogen gas dome above the accumulator water automatically pushes the water into the primary system whenever the system pressure falls below the nitrogen pressure. Accumulator lines attach to the cold legs or vessel as shown in Fig. 2 or to the vessel upper head. The ECC pumping systems have a lower capacity than the accumulators but cover a wider pressure range. Table I shows that the low pressure injection system (LPIS) pumps have a flow rate capacity about 1/10 of the accumulator rate and that the high pressure injection system (HPIS) pumps and charging pumps have a flow rate capacity about 1/10 of the LPIS pumps. The rate values on Table I are given in terms of the rate at which the systems could raise the water level in the core region of the vessel if the core were not hot. Steam generated by a hot core retards core reflooding as will be discussed later.

Table I. Typical Delivery Capability for the Various PWR-ECC Systems

ECC SYSTEM	INITIATING PRESSURE (PSID)	TOTAL NUMBER OF COMPONENTS	NUMBER ASSUMED AVAILABLE	DESIGNED COLD VESSEL CORE FLOODING RATE (in./sec)
Charging Pumps	1000 to 2750	2	1	0.1
HPIS Pumps	1500	2	1	0.1
Accumulators	200 to 750	2 or 3 or 4	1 2 3	10.0
LPIS Pumps	120 to 600	2	1	1.0

PSID = pressure difference between pressure vessel and containment atmosphere in pounds per sq in.

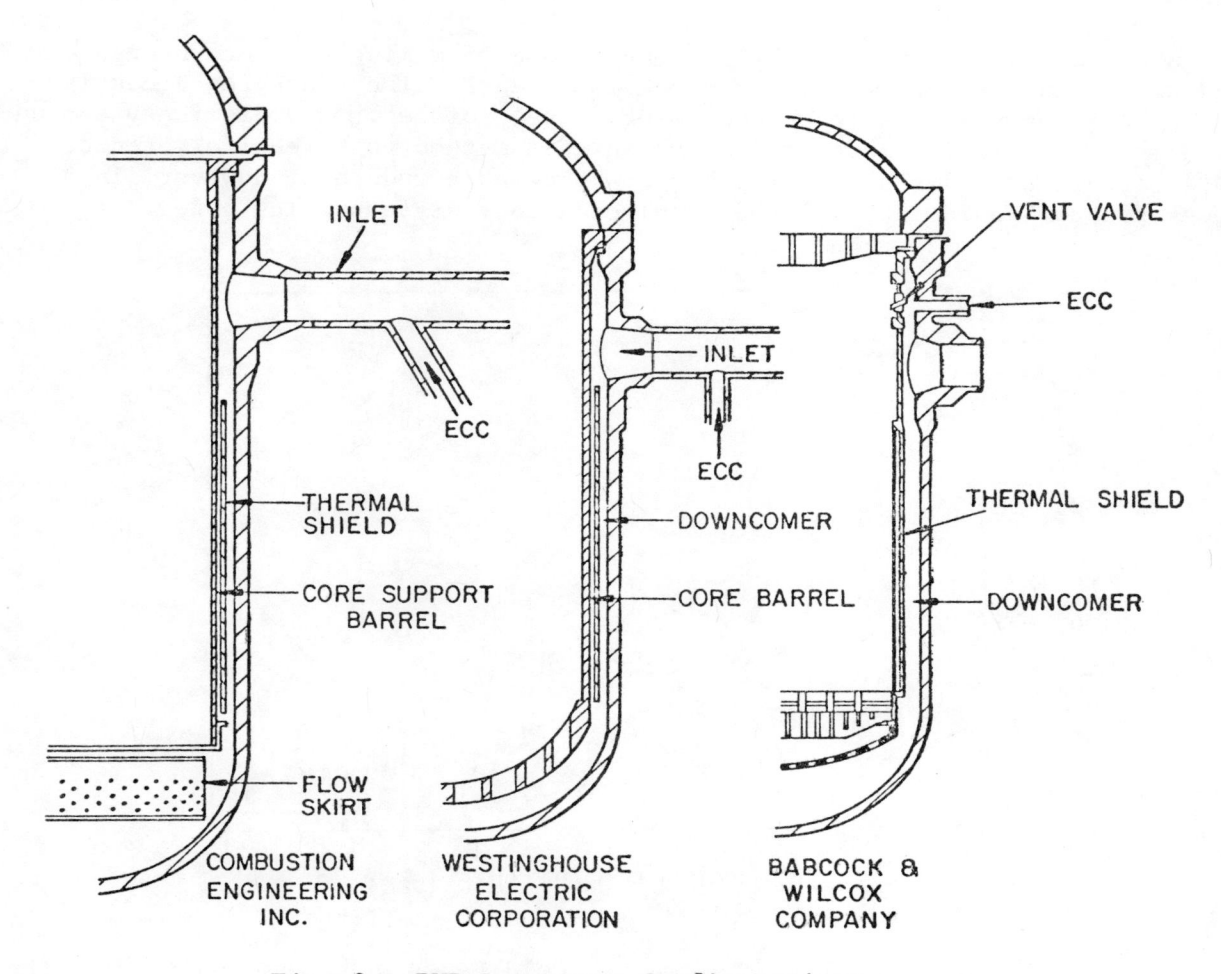

Fig. 2. PWR Downcomer Configurations

LOCA Hydraulics

The most widely studied LOCA is a rupture of a cold leg pipe. Figure 1 depicts a cold leg rupture causing the system to depressurize (blowdown). Initially the water flowing from the break has a temperature less than the saturation temperature, and water exits the break at a high rate. When the pressure falls to a value corresponding to the fluid saturation temperature, steam bubbles form and the flow rate is greatly retarded by two-phase flow choking. Figure 3 shows flow data at the break from Semiscale Test S-02-2 (4), a cold leg break test. The initial subcooled blowdown lasts for about 2.5 seconds longer on the vessel side of the break than on the pump side. This is because of the large supply of subcooled water in the downcomer, lower plenum, and unbroken piping loops compared to the subcooled water in the broken loop. The vessel side flow draws fluid from the bottom of the core and can cause flow stagnation within the core. This causes the mode of energy removal from the cladding to change rapidly from nucleate boiling to film or dispersed flow cooling and the temperatures of the cladding rise. Break flow rates for a hot leg break, shown in Figure 4, are more evenly divided and core cooling is much better for a large hot leg break than for a large cold leg break (5). The measured fluid density at the core inlet is shown in Fig. 5 to be significantly less for Semiscale Cold Leg Break Test S-02-2 than for the Hot Leg Break Test S-02-1 (5). One method of making the core hydraulics and heat transfer for a cold leg break more nearly like a hot leg break is to place area reducers in the cold leg vessel nozzles. A preliminary evaluation of recent Semiscale tests has shown this method to effectively reduce core temperatures for a cold leg break. Flow area reducers, however, in a power reactor would increase the required steady-state pumping power.

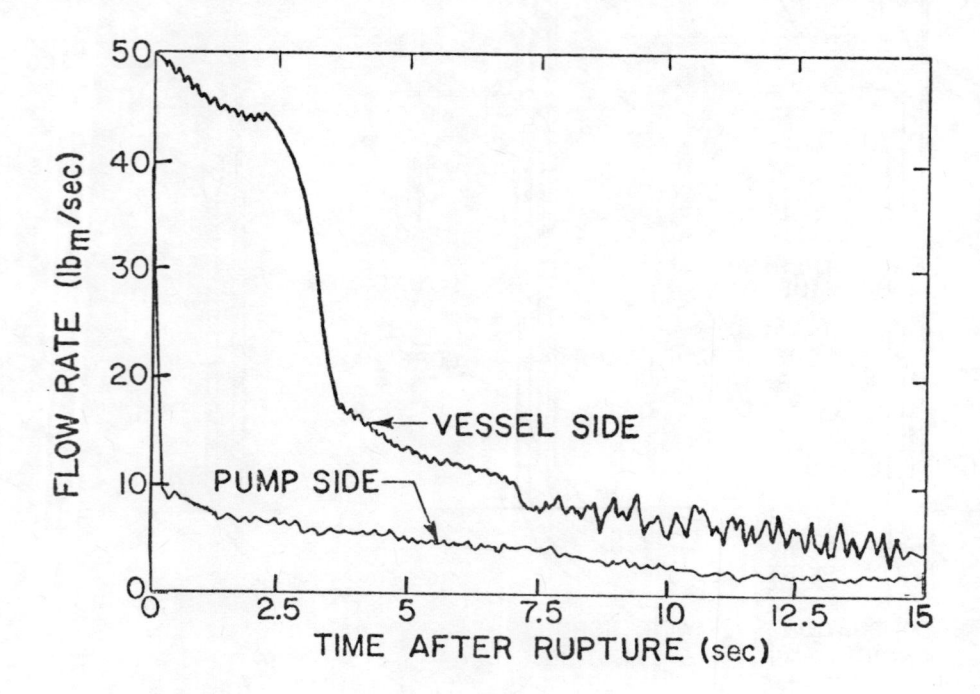

Fig. 3. Semiscale Break Flow Rates, Cold Leg Break, Test S-02-2

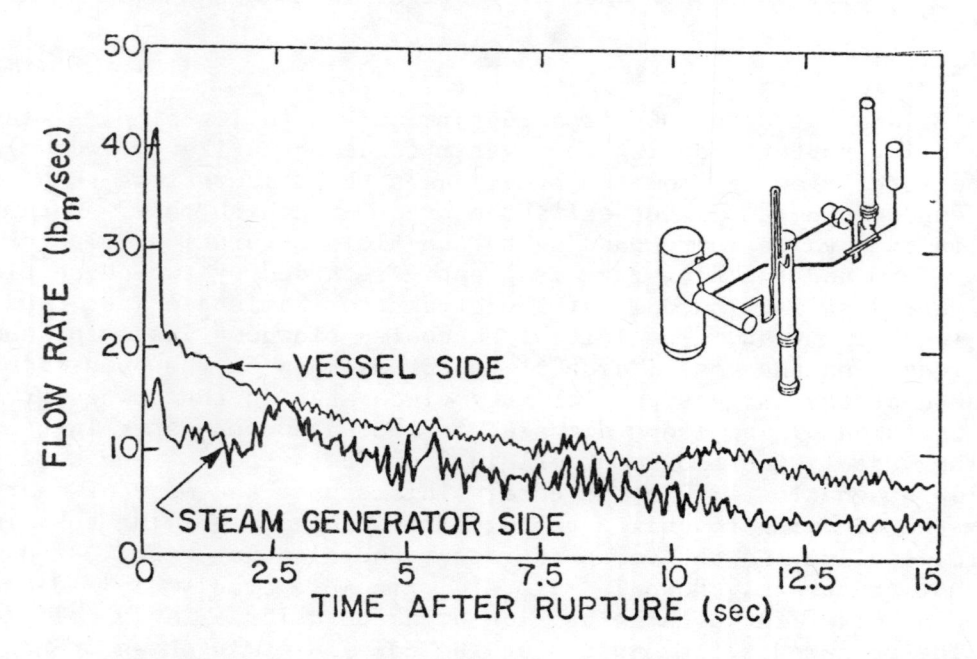

Fig. 4. Semiscale Break Flow Rates, Hot Leg Break, Test S-02-1

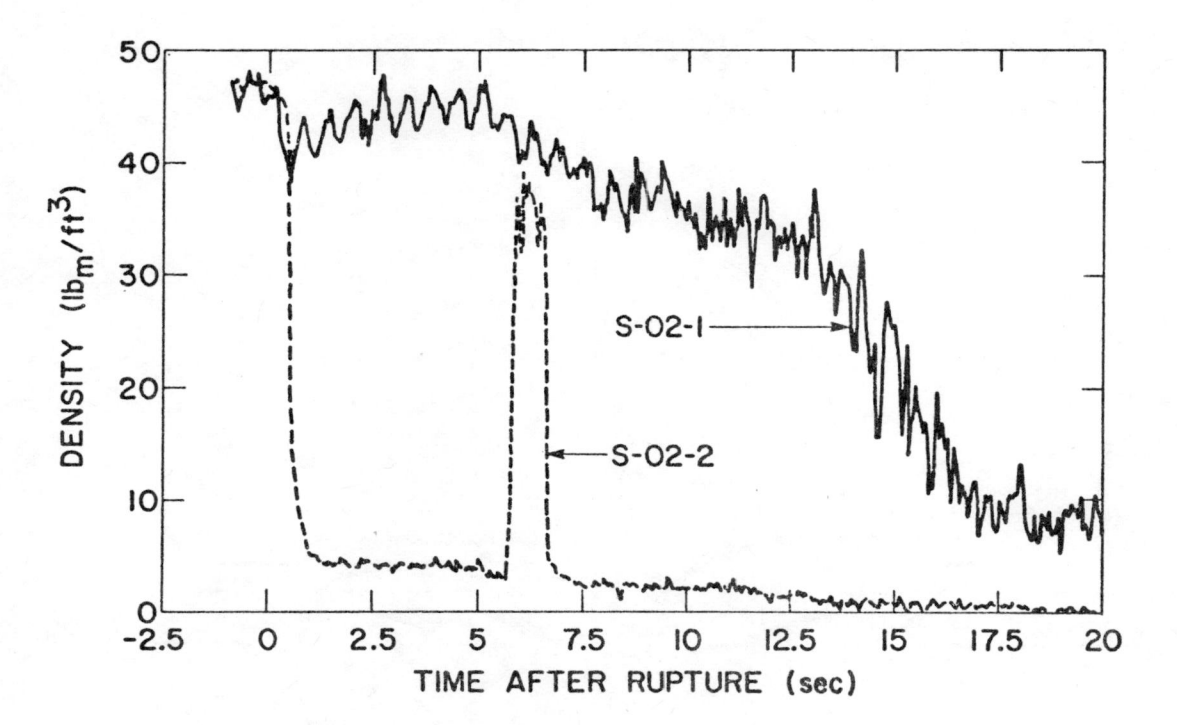

Fig. 5. Semiscale Core Inlet Density

The preceding paragraph indicates that emergency core coolant is needed more for a large cold leg break than for a large hot leg break. Moreover, the ECC could have more difficulty traversing from cold leg accumulators to the core following a cold leg break than following a hot leg break. Figure 6 illustrates a cold leg break with fluid upflow in the vessel annulus. For ECC injected into the cold legs or annulus to get to the core it must penetrate downward against the upflow. If the upflow is too large the ECC will flow around the annulus and out the break. This process is known as ECC bypass caused by countercurrent flow. An expression (6) which describes the velocity of the liquid downflow, J_ℓ, against the velocity of the gas flow, J_g, is:

$$J^{*1/2}_g + M J^{*1/2}_\ell = C \qquad (1)$$

where

$$J^*_g = J_g \left[\rho_g / (\Delta\rho \, gD) \right]^{1/2}$$

$$J^*_\ell = J_\ell \left[\rho_\ell / (\Delta\rho \, gD) \right]^{1/2}$$

ρ_ℓ = liquid density

ρ_g = gas density

$\Delta_\rho = \rho_\ell - \rho_g$

g = acceleration due to gravity

D = downcomer hydraulic diameter

M,C = slope and intercept constants.

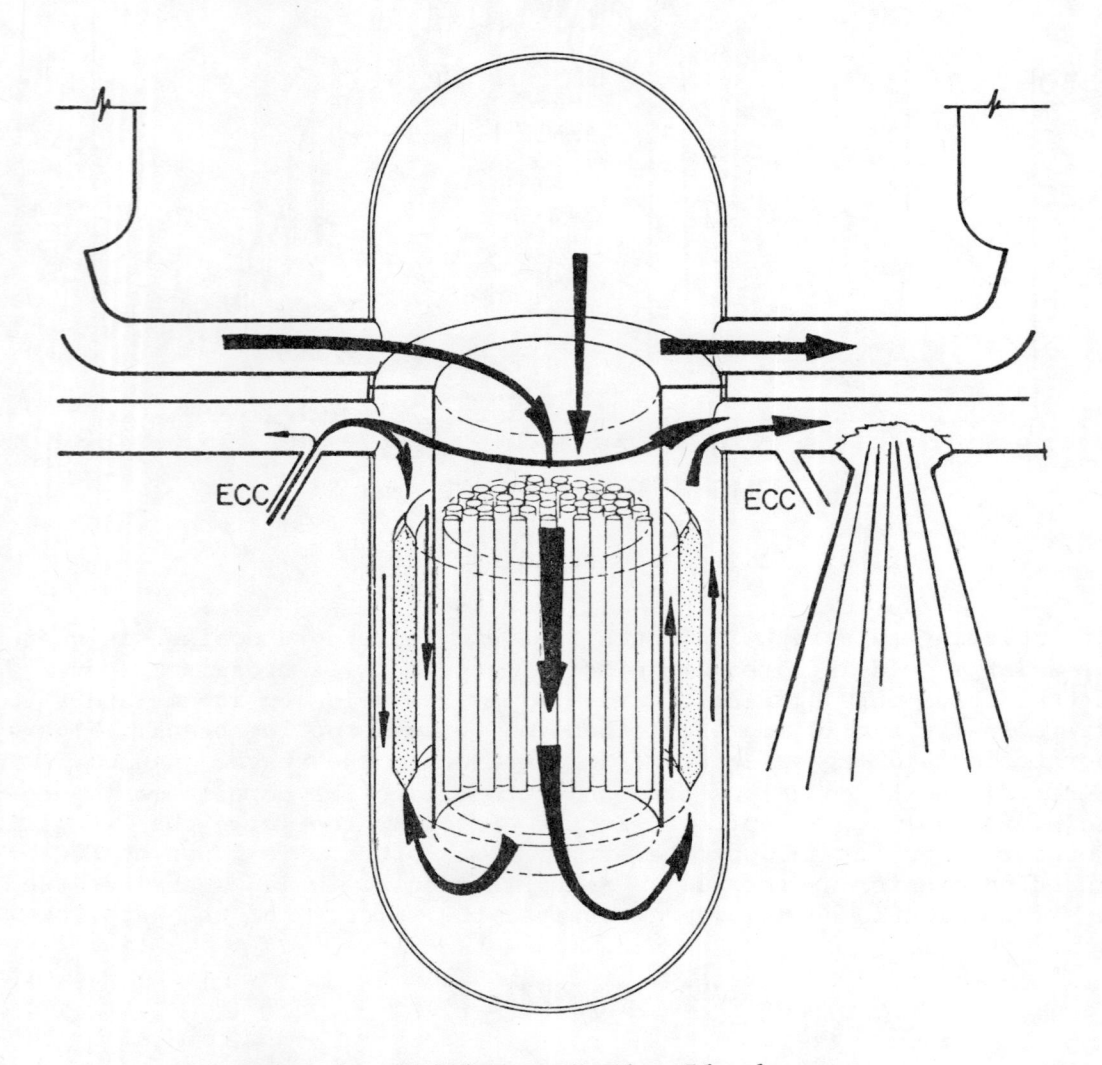

Fig. 6. ECC Bypass During Blowdown

To obtain data on countercurrent flow in an annular geometry, a plexiglass model of Semiscale was built and air-water data obtained (7) as shown in Fig. 7. The data follow the linear form given by Equation (1). When the air velocity was large enough that $J^*_g{}^{1/2} D^{1/4}$ was greater than 0.5, no water could flow down the annulus to the lower plenum. Steam-water data (8), also shown in Fig. 7, shows a similar trend until $J^*_g{}^{1/2} D^{1/4}$ is reduced to below 0.35. Below this value steam condensation resulted in all the water reaching the lower plenum. Semiscale blowdown data show that the fluid upflow is too large during blowdown for ECC water to reach the lower plenum. Reactor calculations do not take credit for water reaching the lower plenum until the end of blowdown. The NRC is sponsoring large vessel countercurrent tests at Pacific Northwest Laboratories (PNL) to obtain data in reactor size annular gap since potential three-dimensional effects in large gaps could allow ECC downflow earlier than predicted.

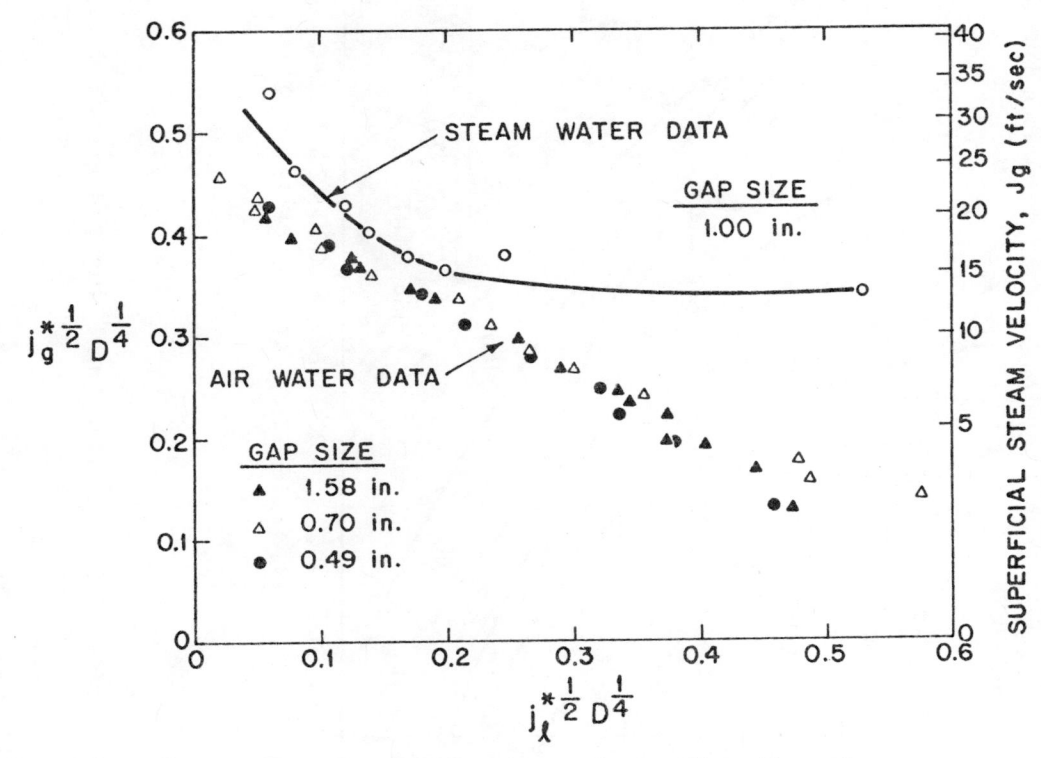

Fig. 7. Semiscale Countercurrent Flow Test Data

Following blowdown there is a refill period where the pumps and accumulators raise the water level to the bottom of the core. Reflooding the core from the bottom is an effective method of cooling the core provided the flooding rate is as large as 2 in./sec. Steam is generated in the nucleate boiling region and, when the water inlet velocity is higher than the nucleate boiling region velocity, water is entrained in the steam and carried up through the core. The water helps cool the steam and improves the heat transfer in the upper regions of the core. The axial variation in the reflood heat transfer and flow regimes is depicted in Fig. 8. A downward pressure force on the core water is created as the steam escapes from the primary system to the containment building. The core pressure pushes water up the annulus and core flow oscillations occur as the annulus water level counterbalances the steam pressure drop around unbroken piping loops. The Babcock and Wilcox Company manufactures plants with vent valves between the upper plenum and annulus (shown in Fig. 2) which allow the steam to escape from the upper plenum to the break without going through the piping loops. Vent valves increase the reflood rate.

Most of the reflood data have been generated at the Full Length Emergency Cooling Heat Transfer (FLECHT) (9,10,11) facility near Pittsburgh, Pennsylvania. Currently, FLECHT tests are sponsored by NRC, EPRI, and Westinghouse Electric Corporation. These data have been used to construct the ECC performance map (12) applicable for FLECHT type nominal reactor conditions shown in Fig. 9. For a given hot spot cladding temperature abscissa at the start of reflood, this map shows the flooding rate necessary to stay below 2200°F

Fig. 9. PWR-ECC Performance Map Using PWR-FLECHT Data

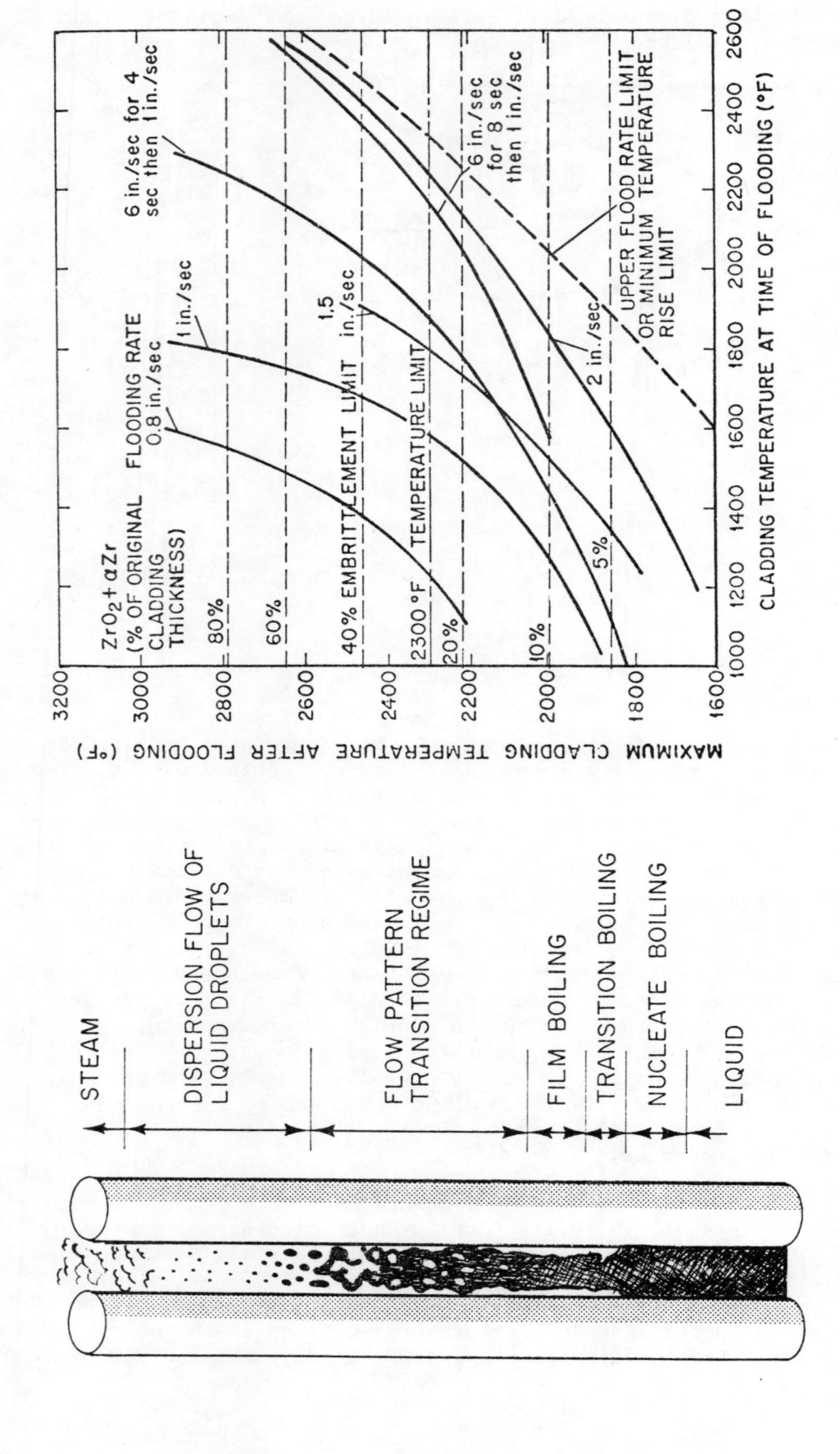

Fig. 8. Core Reflood Heat Transfer and Flow Regimes

temperature limit. Note that the temperature limit is always exceeded prior to the oxidation or cladding embrittlement limit. The 17% zirconium oxide, ZrO_2, limit corresponds approximately to a 40% ZrO_2 plus zirconium condition. PWR reflooding rates are predicted to be from 1 to 2 in./sec.

Calculations Compared to Data

The most widely used computational method used for LOCA calculations is known as RELAP4 (13). The reactor system is modeled as up to 75 fluid control volumes and 50 energy conduction surfaces. Transient equations of momentum, mass and energy conservation are solved with the plant operating conditions as the calculational starting conditions. The fact that this procedure works fairly well is shown in the following figures. Figure 10 shows the predicted upper plenum pressure versus data from Semiscale Test S-02-7 (14). Core inlet flow for a cold leg break is one of the most difficult parameters to predict because a stagnation point moves through the core as the flow reverses. Figure 11 shows core flow data versus prediction for the first Semiscale Heated Core Cold Leg Break Test S-02-2.

Cladding temperature time is also difficult to predict because the fluid condition and heat transfer logic both must be correct. Figure 12 shows that the excellent cooling for a hot leg break, discussed earlier, was predicted well by RELAP4. However, for the Cold Leg Break Test S-02-2 the heat transfer logic had to be altered slightly to improved the accuracy of the calculation. The results are shown in Fig. 13.

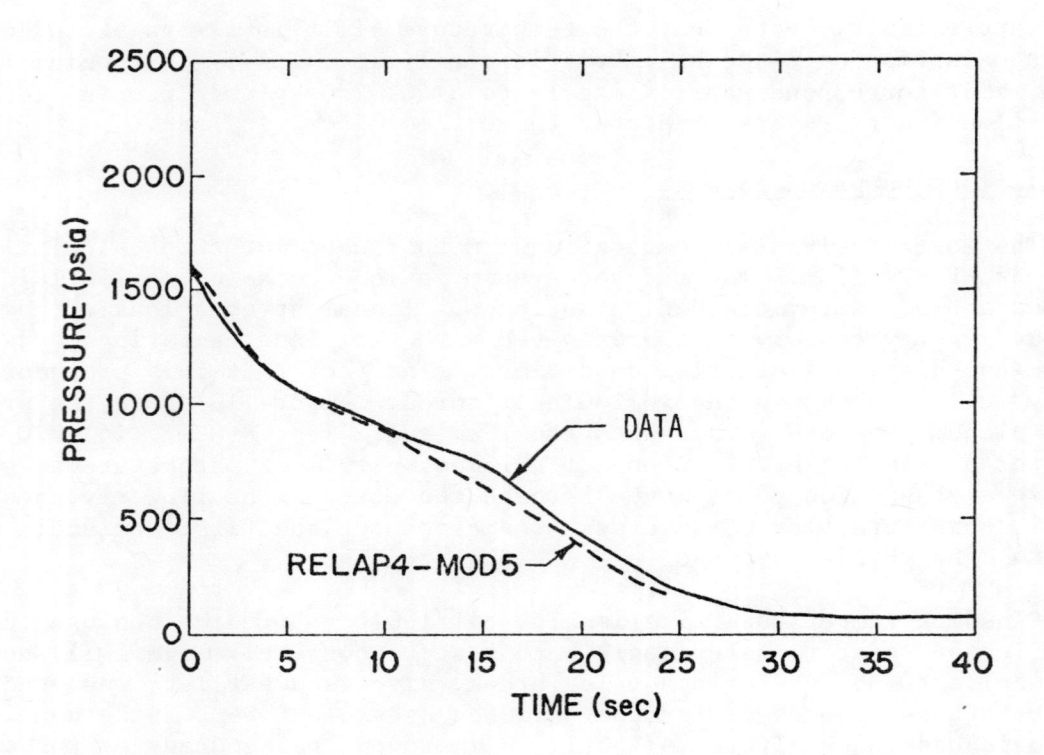

Fig. 10. Semiscale Vessel Upper Plenum Pressure, Test S-02-7

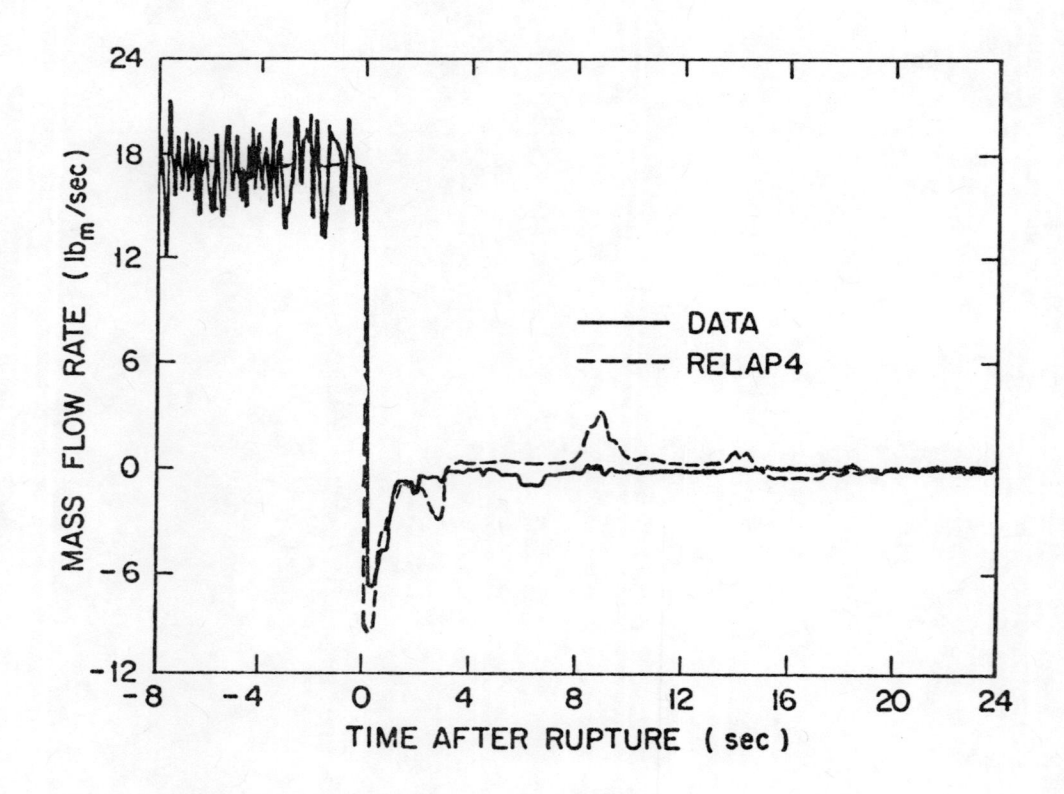

Fig. 11. Semiscale Core Inlet Flow Rate, Test S-02-2

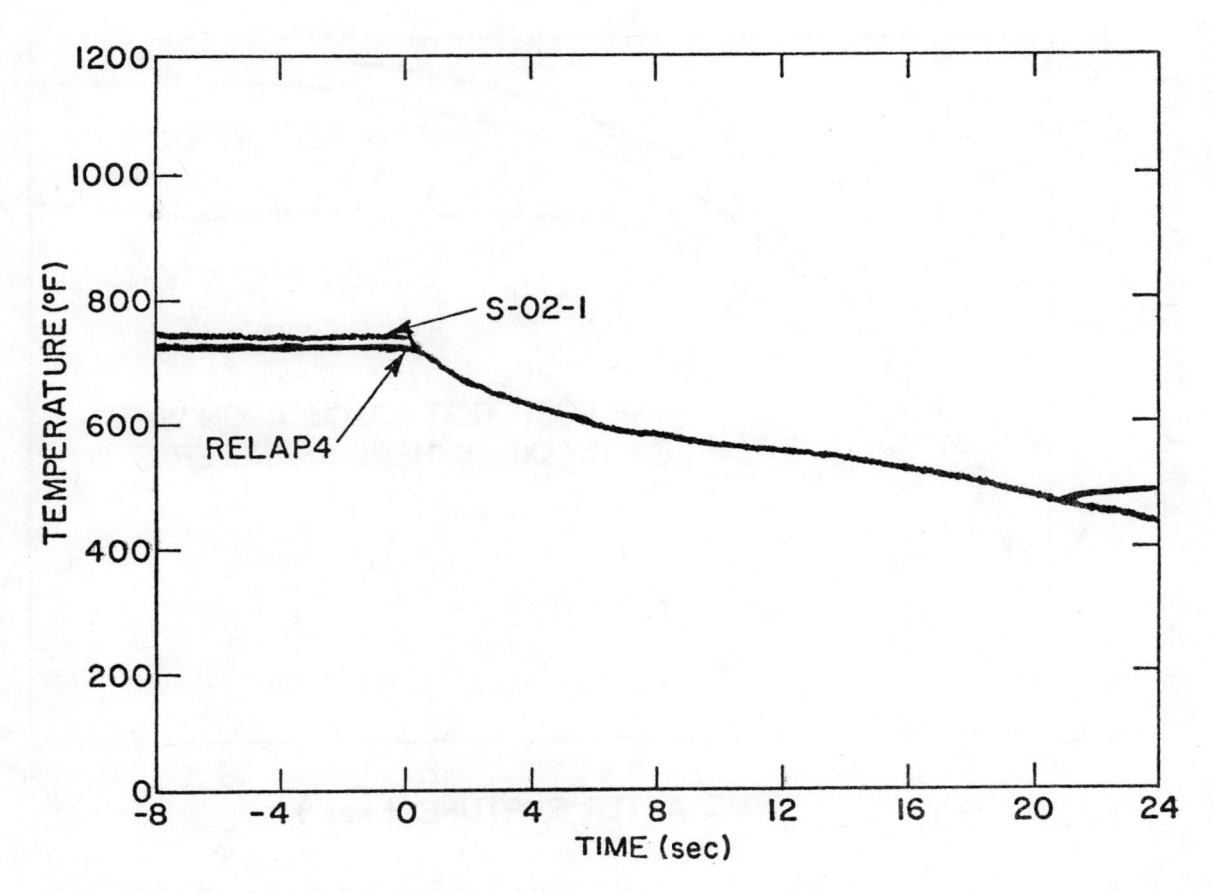

Fig. 12. Semiscale Hot Spot Rod Temperature, Hot Leg Break

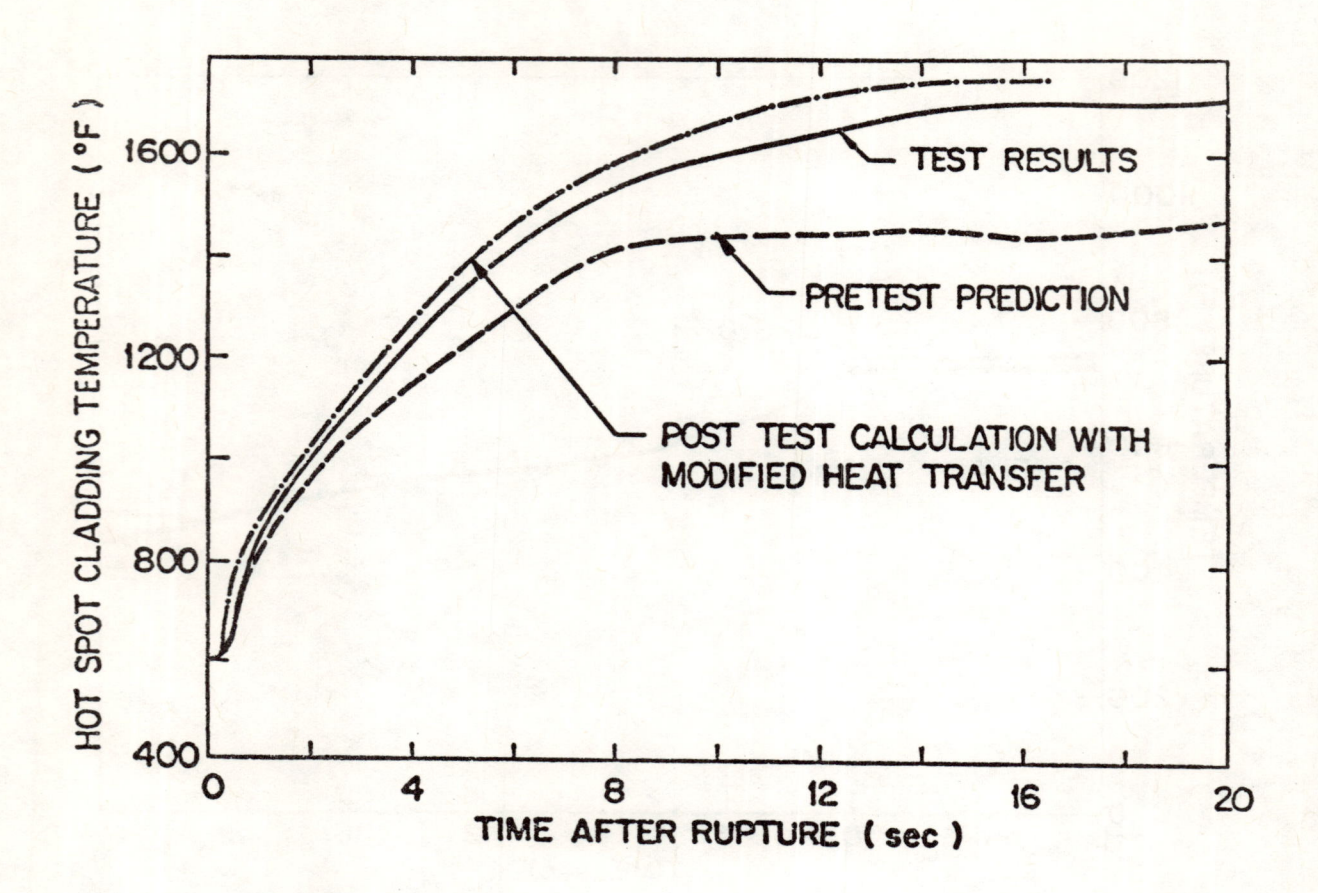

Fig. 13. Semiscale Rod Hot Spot Temperature, Cold Leg Break

III. BWR SYSTEMS

System Description

Boiling water reactors operate at pressure and temperature conditions (about 1040 psia and 550°F) that allow bulk boiling inside the core. As the steam escapes through the upper plenum it passes through baffles to remove moisture prior to exiting to the steam turbine. Recirculation pumps outside of the vessel draw water from an annular region and force it through vertical contraction sections inside the annulus which jet the water into the lower plenum. The water then proceeds upward through the core and falls back into the annulus. Figure 14 shows the general arrangement of the hardware.

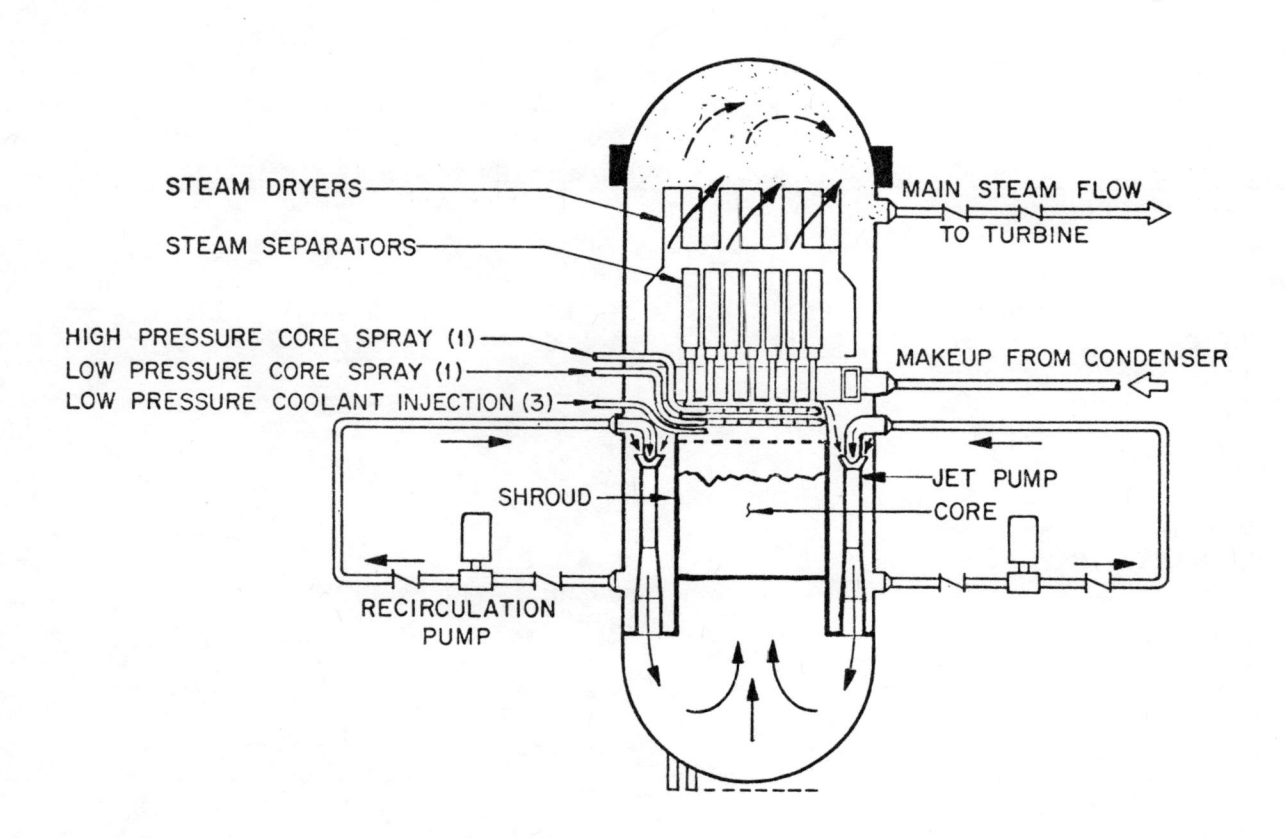

Fig. 14. Typical BWR System

There are four types of systems which contribute to core cooling in the event of a LOCA. These systems are activated by signals of a low reactor vessel water or by high pressure signals from the drywell region surrounding the vessel. There are two spray systems; a high pressure coolant spray (HPCS) and low pressure coolant spray (LPCS). At low reactor vessel pressures, a low pressure coolant injection (LPCI) helps reflood the core. If the leak rate from the vessel is high enough that the HPCS cannot keep the core covered, automatic depressurizing valves attached to the steam lines open to allow a more rapid depressurization rate. Table II shows that the amount of coolant injected into the system is increased as the pressure decreases. Electrical power for driving ECC pumps is assumed to be lost at the time the emergency cooling systems are needed. Auxiliary power furnished by diesel driven generators is available after they are activated. About 30 seconds are required for diesels to come to full speed.

TABLE II

STANDARD BWR PLANT CORE STANDBY COOLING SYSTEM DELIVERY CAPABILITY

FUNCTION	NUMBER INSTALLED	CAPACITY %	DESIGN FLOW FOR EACH SYSTEM	PRESSURE (PSID)	PRESSURE RANGE (PSID)
HPCS	1	100	1300 gpm	1130	1130-0
			4625 gpm	200	
AUTOMATIC DEPRESSURIZING VALVES	6	20	800,000 lb/hr	1125	1130-0
LPCS	1	100	4625 gpm	119	265-0
LPCI	3	33-1/3	4970 gpm	20	225-0

PSID = Pressure difference between pressure vessel and containment atmosphere in pounds per sq. in.

LOCA Hydraulics

Core temperatures are calculated to attain their highest values should the recirculation line break (see Fig. 14). Thus a recirculation line break is called the design basis accident. Figure 15 shows the sequence of events for a recirculation line break. Loss of pump power causes the core flow rate to decrease as the recirculation pumps begin to coast down. Valves in the main steam lines are closed within 3 seconds and system pressure increases until about 6 seconds. When the liquid in the annulus (downcomer) starts to flash, both core flow and pressure increase. At about 10 seconds the vessel water level falls below the jet pump nozzles and core flow rapidly decreases. As the system pressure starts its next rapid decrease, core cooling is aided by bubble formation in the lower plenum which swells some of the lower plenum water up through the core. This core flow increase, shown in Fig. 15, reaches its peak at about 17 seconds. Core flow and core cooling are then negligible until the ECC spray starts coming into the rod bundles at 30 to 40 seconds. Bundles of 49 or 64 rods are housed in a metal structure known as a canister. Because the canister is cooler than the rods the liquid from the spray nozzles can quench the canister before it can quench the rods. Following canister quench, radiation from the rods to the canister improves the rod cooling rate. The temperature calculations given in Fig. 16 (12) show that the sudden decrease in the canister midplane temperature at 160 seconds improved cooling on rods 1 and 9, but rod 17 is closer to the center of the bundle and shows no immediate effect of canister quench. Before rod to rod radiation cooling became significant on rod 17, the LPCI system had filled the lower plenum and core reflooding began. BWR reflooding rates are relatively high (3 in./sec) since the flow resistance between the core and the break does not create a significant steam backpressure problem.

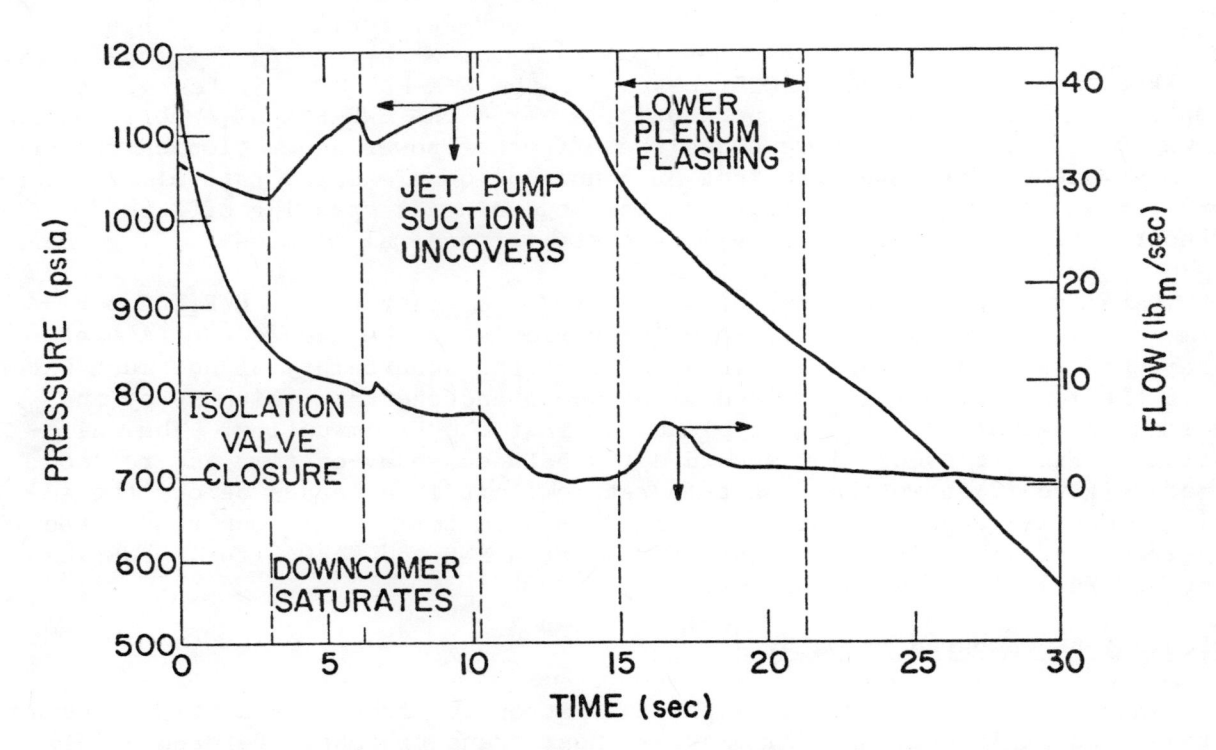

Fig. 15. BWR-Recirculation Line Break, Pressure and Core Flow

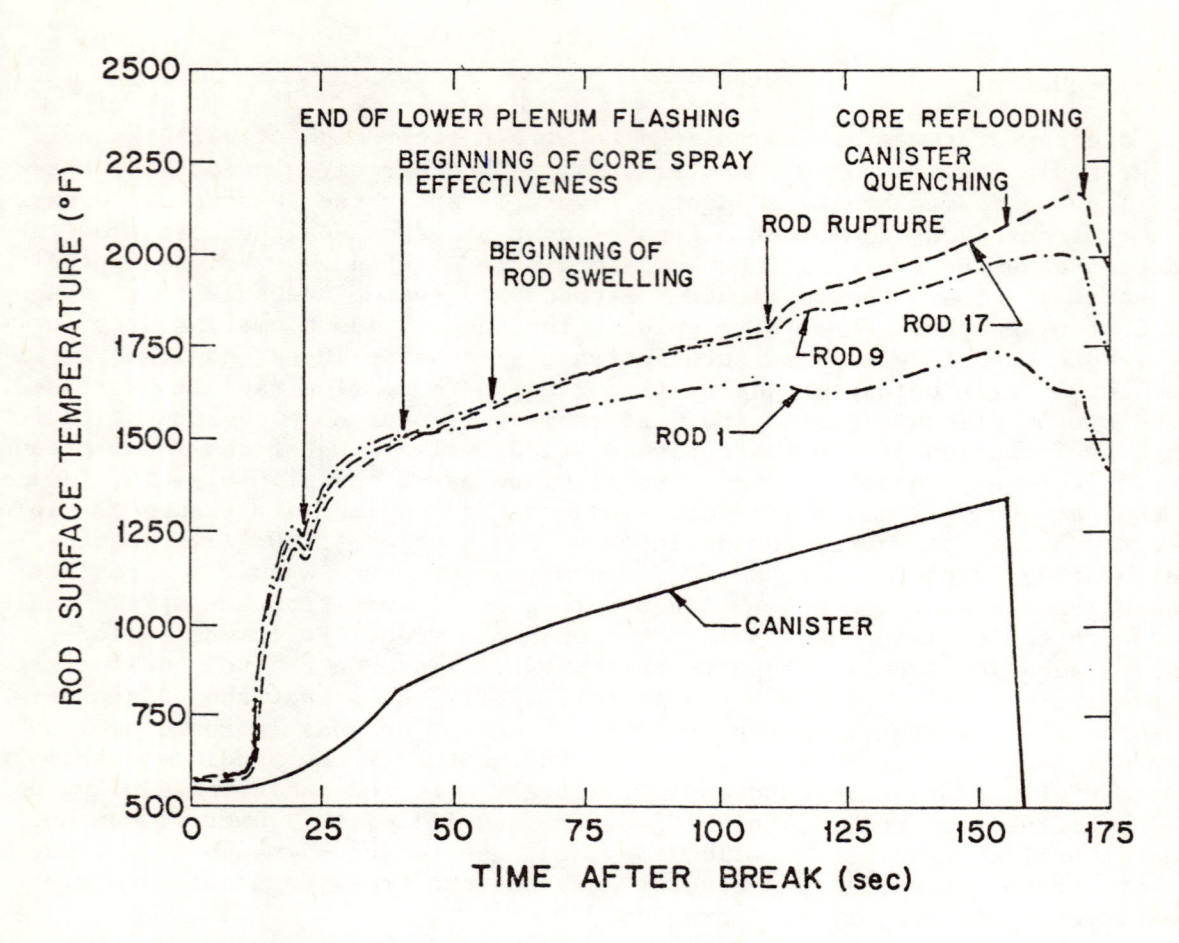

Fig. 16. Calculated Rod Temperatures for a BWR Recirculation Line Break

Peer tests to the PWR-FLECHT tests are the BWR-FLECHT (15) tests. A sample of experimental rod temperature data from the BWR-FLECHT experiment is shown in Fig. 17. The figure shows the effect of power generation on the bundle midplane temperature transient at a nominal bundle spray rate of 2.45 gallons per minute and a temperature at the beginning of spraying of 1250°F. Without bottom flooding the transient lasts for several minutes.

Data from BWR-FLECHT have been used (12) to generate the BWR ECC performance map shown in Fig. 18. A major variable on the map is the time at which core reflooding begins. This reflects the uncertainty of how much water is in the lower plenum at the end of blowdown and the rate which spray and LPCI can penetrate to the lower plenum against the steam upflow. The calculation in Fig. 16 used a 130 second delay between spray-on time and reflood time. Figure 17 shows that the temperature limit is exceeded before the embrittlement is reached. This same trend existed for a PWR. One reason the embrittlement limit is reached at 2870°F for a BWR and 2470°F for a PWR is that BWR fuel cladding is thicker.

Calculations Compared to Data

The BWR test which receives the most attention from a prediction versus data viewpoint is the TLTA (3) blowdown heat transfer test. Reference (3) presents the information shown in Figs. 19 and 20 for the core flow and rod

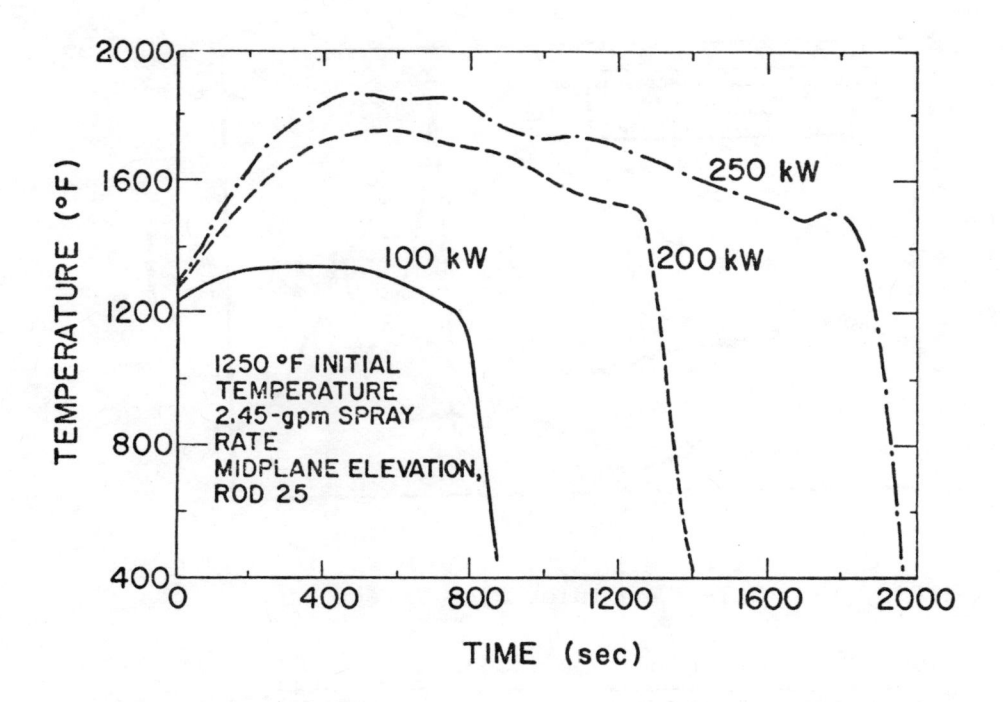

Fig. 17. BWR-FLECHT Measured Effect of Rod Bundle Power on Rod Temperature

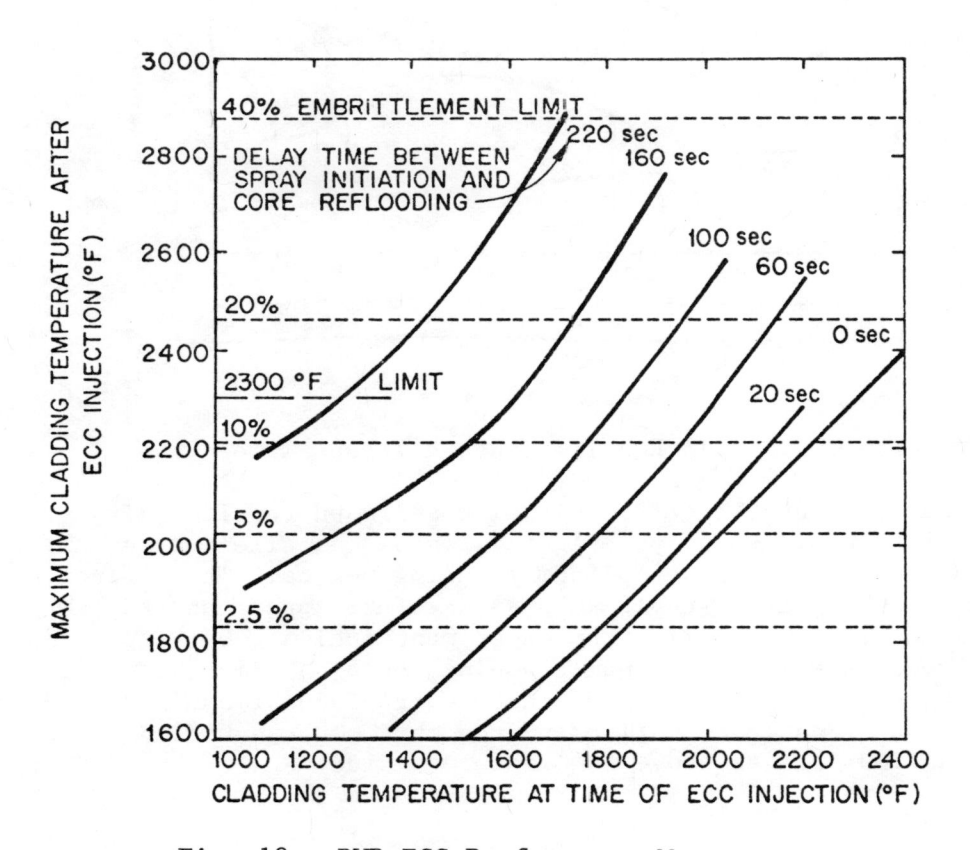

Fig. 18. BWR ECC Performance Map

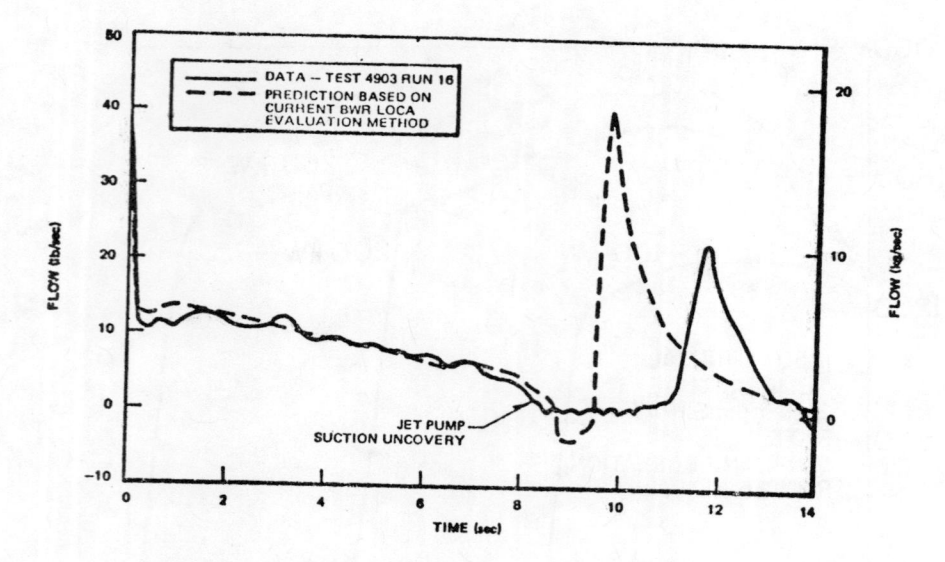

Fig. 19. TLTA Core Inlet Flow Versus Time

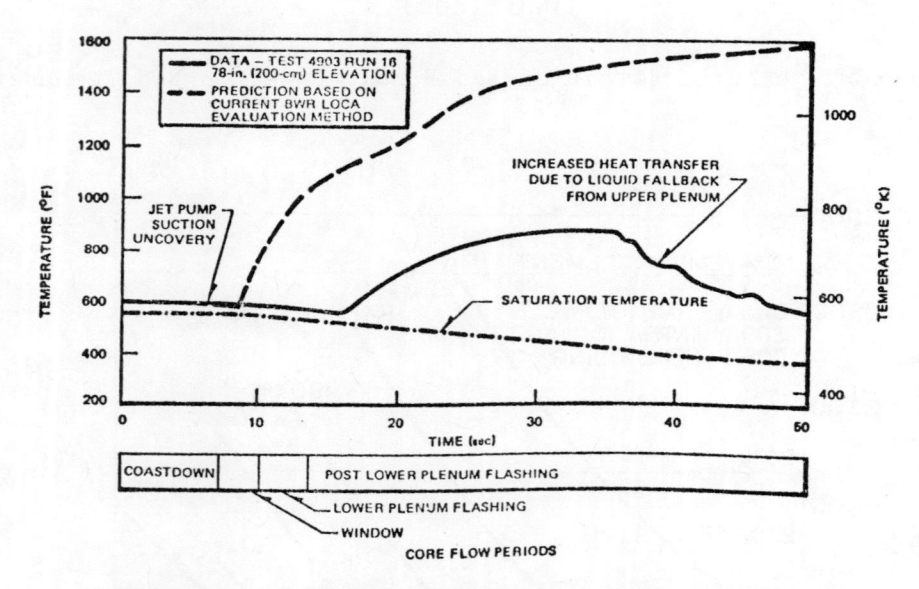

Fig. 20. TLTA Rod Temperature Versus Time

temperature data versus the calculational model used for licensing evaluation by the General Electric Company. Figure 19 shows excellent agreement between data and prediction until lower plenum flashing begins. The influence of lower plenum flashing at 10 to 12 seconds does not appear on the temperature curves shown in Fig. 20. During the experiment nucleate boiling still existed at 12 seconds so lower plenum flashing could not improve cooling. The calculation conservatively predicted that lower plenum flashing would not cause transition or nucleate boiling to reoccur. The peak clad temperature is several hundred degrees above the measured value.

IV. CONCLUSIONS

1. Ample evidence exists to show that LOCA behavior is sufficiently understood to perform conservative type safety calculations for large breaks. Small break experiments have not currently been extensively performed or evaluated.

2. Additional experimental work is underway to improve calculations in areas such as:

    a) boiling transition,
    b) countercurrent flow in the annulus, upper plenum and core,
    c) pump behavior.

3. Design changes which could improve ECC performance are:

    a) vent valves in the upper plenum,
    b) flow limiters in the cold leg,
    c) ECC injection into the lower plenum, pump suction, or upper plenum.

V. REFERENCES

1. L. J. YBARRONDO, P. GRIFFITH, S. F. ABIC, and G. D. MCPHERSON, "Examination of LOFT Scaling," ASME 74-WA/HT-53 (1974).

2. E. M. FELDMAN and D. J. OLSON, "Semiscale Mod-1 Program and System Description for the Blowdown Heat Transfer Tests," ANCR-1230 (August 1975).

3. G. L. SOZZI, R. J. MUZZY, and G. W. BURNETTE, "Blowdown Heat Transfer Phenomena in the Scaled BWR Test System," GEAP-21126 (January 1976).

4. H. S. CRAPO, M.F. JENSEN, and K. E. SACKETT, "Experiment Data Report for Semiscale Mod-1 Test S-02-2," ANCR-1232 (August 1975).

5. H. S. CRAPO, M.F. JENSEN, and K. E. SACKETT, "Experiment Data Report for Semiscale Mod-1 Test S-02-1," ANCR-1231 (July 1975).

6. G. B. WALLIS, <u>One-Dimensional Two-Phase Flow</u>, McGraw-Hill Book Co., New York (1969).

7. D. J. HANSON, "Experiment Data Report for Semiscale Transparent Vessel Countercurrent Flow Tests," ANCR-1163 (October 1975).

8. S. A. NAFF, R. S. ALDER, and R. A. PINSON, "Experiment Data Report for Semiscale System Countercurrent Flow Tests," ANCR-1151 (June 1974).

9. F. F. CADEK, et al., "PWR FLECHT Final Report Supplement," WCAP-7931 (October 1972).

10. E. R. ROSAL, et al., "FLECHT Low Flooding Rate Cosine Test Series Data Report," WCAP-8651 (December 1975).

11. J. P. WARING and L. E. HOCKREITER, "PWR FLECHT-SET Phase B1 Evaluation Report," WCAP-8583 (August 1975).

12. G. F. BROCKETT, et al., "Loss of Coolant: Control of Consequences by Emergency Core Coolings," Proceedings of 1972 Inter. Conf. on Nuc. Solutions to World Energy Problems, Washington, D. C., November 13-17, 1972. American Nuclear Society (1973).

13. K. V. MOORE and W. H. RETTIG, "RELAP4: A Computer Program for Transient Thermal-Hydraulic Analysis," ANCR-1127 (December 1973).

14. H. S. CRAPO, M. F. JENSEN, and K. E. SACKETT, " Experiment Data Report for Semiscale Mod-1 Test S-02-7," ANCR-1237 (November 1975).

15. J. D. DUNCAN and J. E. LEONARD, "Emergency Cooling in BWR's Under Simulated Loss-of-Coolant Conditions (BWR-FLECHT Final Report)," GEAP-13197 (June 1971).

A REVIEW OF ACCIDENT RISKS IN LIGHT-WATER-COOLED NUCLEAR POWER PLANTS

Lynn E. Weaver
School of Nuclear Engineering
Georgia Institute of Technology
Atlanta, Georgia 30332

INTRODUCTION

No treatment of nuclear power safety would be complete without including a discussion of WASH-1400, "Reactor Safety Study," sponsored by the United States Nuclear Regulatory Commission. Considering the length of the study and the limited space available for this presentation, an attempt is made to summarize the underlying methodology and conclusions. Most of the material is taken directly from WASH-1400 and the reader is referred to that document for more detailed and in-depth information.

CONCEPTS OF RISK

Risk is a commonly used word that can convey a variety of meanings to different people. A dictionary definition is "the possibility of loss or injury to people and property." Here estimates are made of potential fatalities and injuries to people and of property damage resulting from both nuclear power plant and non-nuclear accidents.

Particular emphasis is placed on the risk to the health and safety of the general public.

To provide a basis for the quantitative comparison of societal risks from accidents, the following technical definition of risk is used:

$$\text{Risk} \left\{ \frac{\text{consequence}}{\text{unit time}} \right\} = \text{Frequency} \left\{ \frac{\text{events}}{\text{unit time}} \right\} \times \text{Magnitude} \left\{ \frac{\text{consequences}}{\text{event}} \right\}$$

As a quantitative example of the use of such an equation, in 1971 about 15,000,000 auto accidents occurred in the U.S., and one in 300 accidents resulted in a fatality. Thus, the societal risk of death from auto accidents can be approximately calculated as:

$$15 \times 10^6 \frac{\text{accidents}}{\text{year}} \times \frac{1 \text{ death}}{300 \text{ accidents}} = 50,000 \frac{\text{deaths}}{\text{year}}$$

Further, if U.S. society consists of 200,000,000 people, the average individual risk can be expressed as:

$$\frac{50{,}000 \text{ deaths/year}}{200{,}000{,}000 \text{ persons}} = \frac{2.5 \times 10^{-4} \text{ deaths}}{\text{person-year}}$$

The final term expresses the individual risk as probability of death per person per year. This mode of expression is frequently used in the mathematical analysis of risks.

In the aforementioned example pertaining to fatalities in auto accidents, note that the risk is expressed both in terms of risk to society and risk to an individual. Additional risks due to auto accidents result from injuries and property damage. In the U.S. about 30 times as many people are seriously injured as are killed in auto accidents. Thus, on the average one person is injured in every 10 accidents. This yields, for societal risk,

$$15 \times 10^6 \frac{\text{accidents}}{\text{year}} \times \frac{1 \text{ injury}}{10 \text{ accidents}} = 1{,}500{,}000 \frac{\text{injuries}}{\text{year}}$$

and, for individual risk,

$$\frac{1{,}500{,}000}{200{,}000{,}000} \left(\frac{\text{injuries/year}}{\text{persons}}\right) = 7.5 \times 10^{-3} \left(\frac{\text{injuries}}{\text{person-year}}\right)$$

The cost of injuries and property damage due to auto accidents can be similarly calculated. In this case, the recorded statistic is the total dollar value of injuries and property damage due to auto accidents for each year. This value represents the societal risk and is $15.8 billion dollars for 1971. A reasonable measure of the average individual risk is the cost per registered driver per year. For 1971, this can be computed as follows from available data:

$$\frac{\$15.8 \times 10^9 \text{ per year}}{114 \times 10^6 \text{ registered drivers}} = \$140 \text{ per driver per year}$$

Historical data for risks commonly encountered by many, if not most, people in the U.S. are collected by many organizations.

Table I presents the accident data for 1967 and 1968 in terms of individual risk, i.e., the probability of death per person per year. In using numbers such as those displayed in Table I, all the factors associated with them, whether expressed or implied, must be known to avoid misinterpretation and misuse of the data. For example, consider the fatalities from falls, as listed in this table. A person looking at the data at the end of 1967 might have concluded that his risk, as a member of the U.S. population, of suffering a fatal fall in the next year was one chance in 10,000. Since the number 1×10^{-4} was derived from the number of fatalities counted during 1967, using it to predict future risk involves the reasonable assumption that the rate will remain the same (in this case, it did remain the same in 1968). Another assumption involved is that all members of the U.S. population are equally exposed or susceptible to the risk. This is rarely true in human events.

Table I. Some U.S. Accident Death Statistics - 1967, 1968

Accident	Total Deaths 1967	Total Deaths 1968	Probability of Death per Person per Year 1967	Probability of Death per Person per Year 1968
Motor Vehicle	53,100	55,200	2.7×10^{-4}	2.8×10^{-4}
Falls	19,800	19,900	1.0×10^{-4}	1.0×10^{-4}
Fires, burns	7,700	7,500	3.9×10^{-5}	3.8×10^{-5}
Drowning	6,800	7,400	3.4×10^{-5}	3.7×10^{-5}
Firearms	2,800	2,600	1.4×10^{-5}	1.3×10^{-5}
Poisoning	2,400	2,400	1.2×10^{-5}	1.2×10^{-5}
Cataclysm	155	129	8×10^{-7}	6×10^{-7}
Lightning	110	162	6×10^{-7}	8×10^{-7}

A breakdown of the 1967 data on falls by age group shows that almost 3 out of 4 (73%) fatal falls involved people of age 65 and over. Thus, the risk is much smaller for persons under 65 than for persons 65 and older, and it is not 1×10^{-4} for either group as shown in Table II.

Table II. 1967 Falling Deaths - by Age Group

Deaths by Falling 1967	Number per 100,000 in Age Group	Probability of Death per Person per Year (in age group)
19,800 total for all ages	10	1×10^{-4}
14,454 at age 65 and over	75	7.5×10^{-4}
5,346 at ages below 65	3	3×10^{-5}

The examples presented indicate that there are many factors that contribute to the quantification and evaluation of risk. The aforementioned examples of risks due to automobiles involve fatalities, injuries and property damage that can be quantified from commonly available data. The effects are principally short term ones (i.e., quickly measurable). There are undoubtedly other contributors to risks from automobiles that are not fully included in measured data. These would involve long term effects such as life shortening and decreased earning power due to injuries. These examples indicate that there are risk factors of interest both on a societal and individual basis. In addition, Table II brings the concept that risk is not always equally distributed in the population. Thus the measurement and evaluation of risk have many facets; these will be discussed more fully in later sections of this chapter.

ATTITUDES TOWARD RISK

An apparent consistency in public attitudes toward familiar risks, such as those listed in Table II, has been noted. Types of accidents with a death risk in the range of 10^{-3} per person per year to the general public are difficult to find. Evidently this level of risk is generally unacceptable, and when it occurs, immediate action is taken to reduce it.

At an accidental risk level of 10^{-4} deaths per person per year, people are less inclined to concerted action but are willing to spend money to reduce the hazard. Money is spent for traffic control, fire departments and fences around dangerous areas. Safety slogans for accidents with this risk level show an element of fear (e.g., "The life you save may be your own" as applied to automobile driving).

Risks of accidental death at a level of 10^{-5} per person per year are still recognized in an active sense. Parents warn their children about the hazards of drowning, firearms, poisoning, etc., and people accept a certain amount of inconvenience to avoid risks at this level. Safety slogans have a precautionary ring: "Never swim alone"; "Keep out of the reach of children."

Accidents with a probability of death of 10^{-6} or less per person per year are apparently not of great concern to the average person. He is aware of them, but feels they will not happen to him. He may even feel that such accidents are due partly to stupidity, e.g., "Everyone knows you shouldn't stand under a tree during a lightning storm." Phrases associated with these hazards have an element of resignation: "An act of God."

The concept that the degree of public acceptance of a risk is likely to be influenced by the perception of the associated benefits is presented in Fig. 1. It suggests a relationship between the benefits of an activity, expressed in arbitrary units, and the acceptable risk expressed as probability of death per year per exposed person. The highest level of acceptable risks has been taken as the normal U.S. death rate from disease; the lowest level for reference is taken as the risk of death from natural events (lightning, flood, earthquakes, insect and snake bites, etc.).

One of the obvious shortcomings of the approach in Fig. 1 is that it does not differentiate with respect to the magnitude of the consequences of accidents. This point is illustrated by considering two accidents with significantly different frequencies and consequences. The first occurs at a rate of once per year and results in one death per accident. The risk is

$$1 \frac{\text{accident}}{\text{year}} \times 1 \frac{\text{death}}{\text{accident}} = 1 \frac{\text{death}}{\text{year}}$$

The second type has a frequency of only once in 10,000 years but results in 10,000 fatalities per event. The risk is

$$\frac{1 \text{ accident}}{10,000 \text{ years}} \times \frac{10,000 \text{ deaths}}{\text{accident}} = 1 \frac{\text{death}}{\text{year}}$$

Although each of the accidents indicated above has the same _average_ annual risk, there is a factor of 10,000 in the size of the accidents. Society generally views the single large consequence event less favorably than the total of small events having the same average risk.

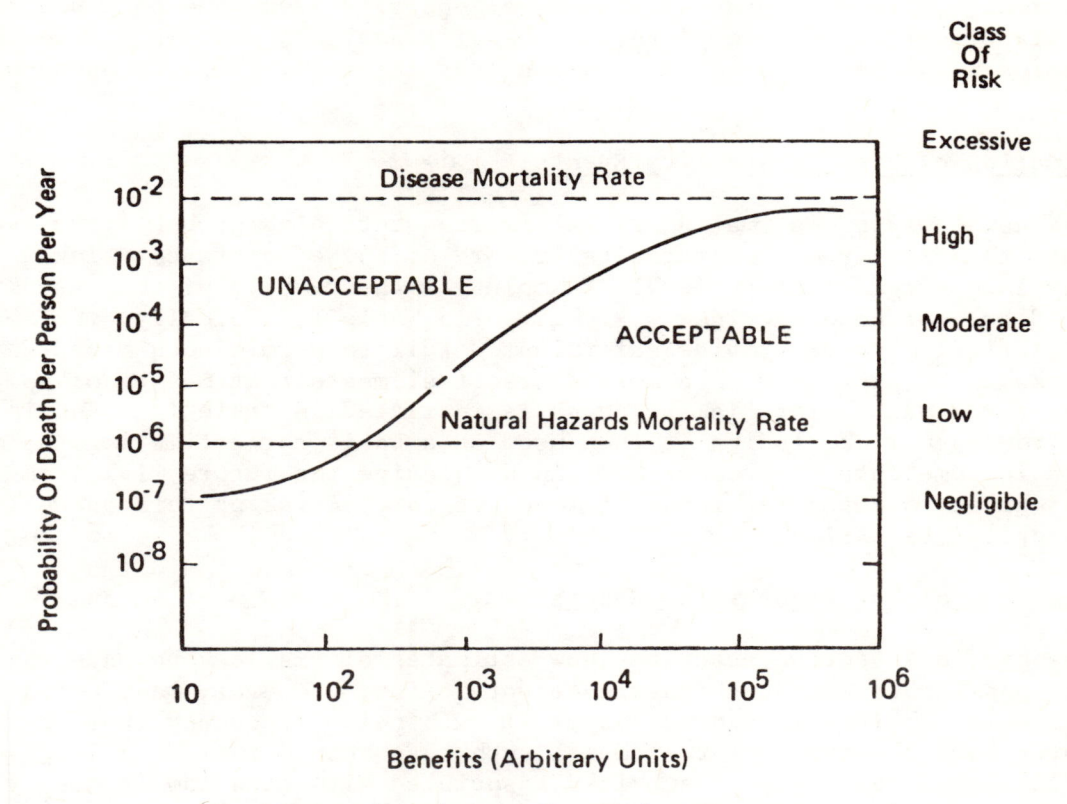

Figure 1. A Benefit-Risk Pattern

This attitude leads to the concept of "risk aversion." The term risk aversion is used to indicate, among other things, that accidents having the same average societal impact may be viewed differently depending on the sizes of the individual events. In general, single large accidents are viewed less tolerantly than multiple smaller accidents, even though the average annual consequences of the two are equal. In fact, the public appears to accept more readily a much greater societal impact from many small accidents than it does from the more severe, less frequent occurrences that have a smaller societal impact. One of the clear indications of this attitude is indicated by the public (and news media) attitude toward fatalities from automobile accidents in contrast to those from aircraft crashes. It appears that the public's aversion to large consequence events may be largely due to the view that, if such events are at all possible, they are likely, and their low probability is to be discounted.

The analyses referenced above are interesting because they represent early attempts to quantify the acceptability of the risks associated with a given activity in relation to the benefits gained from this activity; however, this field is still highly formative and much in need of development. These analyses are, therefore, of limited utility in this study. Explicit techniques for assessing the total cost of various risks and the total benefits derived from the activities causing them are still in the early stages of development even for measurable (fairly likely) risks. In the area of risks from low probability events that have not been observed, it is clearly not yet possible to perform a rigorous cost-benefit assessment. Decisions in the area of risk, as in many other areas, have generally been made on a qualitative basis with less than complete cost-benefit analyses available. Whether this approach can be improved upon in the near future is still an open question.

Determination of High Probability Events

The usual way of estimating risks for frequent (high probability) events is to use the data from the historical record of these events covering a suitably large segment of society. Examples of the results of such studies for broad categories of accidents are given in Table I. Normally, the available historical records provide sufficient detail to permit such broad category risks to be separated into more distinct elements that may be of special interest for a particular risk study as is indicated in Table II. The information from such risk studies is then used as a basis for estimating the risk expected in some future time period. In projecting the future risk, consideration is given to potential future influences in the risk pattern as well as the historical variation.

Determination of Low Probability Events

The previous section describes how estimates of risk can be made when directly applicable accident experience data exist. However, many potential risks to which society is exposed occur at such a low frequency that they have never been observed. For example, such cases could include a large meteor falling into a city. The risks associated with such low frequency events are more difficult to estimate and express in a meaningful way than those of more frequent events.

In some cases the probability of rare occurrences can be obtained by dividing the total occurrence into a series of events for which the individual probabilities of occurrence are known. A simple example of this is the chance of getting heads every time in fifty random flips of a fair coin. From experience we know the chance of getting heads in one flip is 0.5, the chance of getting heads both times in two flips is $(0.5)^2 = 0.25$ and in fifty flips the chance is $(0.5)^{50}$, which is about one chance in 10^{15}. Thus, although this event has undoubtedly never been observed, an estimate of its probability can be achieved. Similarly the chance of getting four-of-a-kind in two successive hands in a five card stud poker game can be calculated. Analysis of poker hands has shown that the probability of getting four-of-a-kind once is about one in four thousand. Thus, the chance of getting four-of-a-kind in two successive hands is about 10^{-7}. In these cases each rare occurrence was broken down into more likely events that were all the same; but this type of analysis is also applicable when the group of more likely events is of more than one type and/or frequency. The breaking up of a rare event into a series of more likely events is a basic principle of the event tree and fault tree techniques (discussed later) utilized for determining probabilities of reactor accidents. Application of the above technique has involved the determination of the failure probability of systems by combining the failure probabilities of their individual parts and components.

When an unlikely event cannot be described as a sequence of more likely events, it is sometimes possible to estimate its probability by extrapolation. Suppose, for example, the highest observed level of a river at some point was 35 feet above the normal level and an estimate of the likelihood of the river reaching the 40 foot level is desired. The historical frequency of floods versus their height can be plotted and extrapolated to predict the frequency of a flood height of 40 feet. Extrapolation requires that the physical factors affecting a particular situation remain constant. Thus, while one may be able to easily estimate the likelihood of a flood level of 40 ft, given a historical level of 35 ft, it may not be valid to extrapolate to the 50 foot flood, or to the 40 foot flood that might occur 1,000 years in the future. By using the principles previously discussed it is possible to make reasonable estimates of the probabilities of very unlikely events.

In the analysis of low probability events it is rather common to speak of the recurrence rate of an event, e.g., the 10,000 year flood, or the 10,000 year earthquake. This is another way of describing a rare flood or earthquake that has a probability of occurrence of 10^{-4}/year. Such estimates are usually extrapolations of limited experience and should not be interpreted too literally to mean the worst flood we expect in the next 10,000 years, since this would imply there would be no change in the factors (climate, local topography, etc.) affecting the frequency over the 10,000 year period. Such changes can, of course, occur in the long time periods involved. Also, just because an event is determined to have a probability of occurring only once in 10,000 years, doesn't mean that it will be 10,000 years before it occurs or that it will occur at that time. It means that the event would occur on the average of once every 10,000 years; however, although it is very unlikely, it could occur in this century.

Similar misinterpretations can easily be made with respect to the probabilities of reactor accidents. For example, suppose the probability of an accident involving melting of the nuclear core in today's reactors is 10^{-5} per reactor-year. Since about 1,000 reactors are expected to be in operation in the year 2000, there may be a tendency to say the probability of such an accident in the year 2000 will be $(10^{-5})(10^{3}) = 10^{-2}$. The error in this extrapolation is that it assumes the failure rate will remain constant at 10^{-5} for the next 25 years. As was demonstrated in other industries, it is expected that an increase in knowledge will improve safety. There will be a learning curve for the nuclear reactor industry where increasing attention is being devoted to safety both within the NRC and in the industry as a whole.

PRESENTATION OF RISK ESTIMATES

In view of the difficulties involved in expressing the consequences of various accidents in a common unit, four types of consequences for the determination and comparison of accident consequences have been selected. These are:

   a. Early fatalities,
   b. Early illnesses,
   c. Late health effects attributable to the accident,

d. Property damage.

With respect to the selection of these types, it is noted that data on the previous accident and natural event experience in the U.S. are frequently available for types a, b, and d. Further, information on type c is sometimes available in selected studies.

In WASH-1400 the major types of accidents have been identified and their probability of occurrence estimated on the basis of event tree and fault tree analyses of reactor operations. Each of these accidents has been analyzed to determine the range of consequences associated with it in terms of fatalities, illness, long-term health effects, and property damage. The results of these studies provide the probability-consequence relationships which will serve as the basic information in expressing reactor accident risks.

Similar determinations are made for low probability high consequence non-nuclear accidents that could result from other technological undertakings. Specifically, dam failures, aircraft crashes into large concentrations of people, and the release of large amounts of chlorine (a toxic chemical), have been studied and are compared with nuclear accidents.

To provide additional perspective on the significance of potential reactor accidents, the risks due to nuclear accidents will be compared to the more common societal risks resulting from man's technological activities and from natural events.

THE NATURE OF NUCLEAR POWER PLANT ACCIDENTS

Nuclear power plant accidents differ from those in conventional power plants because they can potentially release significant amounts of radioactivity to the environment. While very large amounts of radioactivity are generated by the fission process in the uranium dioxide fuel in a nuclear plant, the bulk of this radioactivity (about 98%) remains in the fuel as long as the fuel is adequately cooled. For large amounts of radioactivity to be released from the fuel, it must be severely overheated and essentially melt. Based on this knowledge, the major types of nuclear power plant accidents that have the potential to cause large releases of radioactivity to the environment have for some time been recognized. Attempting to prevent such accidents and to mitigate their potential consequences has been the primary objective of nuclear power plant safety design.

The safety design approach for nuclear power plants has often been described as consisting of three levels of safety involving (1) the design for safety in normal operation, providing tolerances for system malfunctions, (2) the assumption that incidents will nonetheless occur and the inclusion of safety systems in the facility to minimize damage and protect the public, and (3) the provisions of additional safety systems to protect the public based on the analysis of very unlikely accidents. The safety design approach has also been described as involving the use of physical barriers (fuel, fuel cladding, reactor coolant system, containment building) to attempt to prevent the release of radioactivity to the environment. The above descriptions are valid, but both are general statements covering the detailed concepts underlying reactor safety design. A definition of the locations and amounts of radioactivity in a nuclear power plant and an examination of the processes by which significant amounts of this radioactivity can be released from the fuel and transported to the environment outside the containment building, provide a somewhat more definitive view of the various elements that enter into reactor safety approaches.

All the places in which fuel is located in a nuclear power plant and the amount of radioactivity in each location are identifiable. The largest amount of radioactivity resides in the reactor core. A smaller, but still large amount of radioactivity is located in the spent fuel storage pool at the time refueling of the reactor is completed. In both these locations, the fuel is subjected to heating due to absorption of energy from the decay of radioactive materials. This continues even after reactor shutdown has terminated the fission process. This decay heat can be the source of overheating in fuel in a shutdown reactor or in fuel that has been removed from the reactor. Immediately following the shutdown of a reactor that has operated about a month or longer, the decay heat amounts to about 7% of the prior operating power level. While the heat has an initial rapid decrease after reactor shutdown, it constitutes a substantial heat source for some time, and continued cooling of the fuel is required.

Overheating of fuel occurs only if the heat being generated in the fuel exceeds the rate at which it is being removed. This type of heat imbalance in the fuel in the reactor core can occur only in the following ways:

a. The occurrence of a loss of coolant event will allow the fuel to overheat (due to decay heat) unless emergency cooling water is supplied to the core.

b. Overheating of fuel can result from transient events that cause the reactor power to increase beyond the heat removal capacity of the reactor cooling system or that cause the heat removal capacity of the reactor cooling system to drop below the core heat generation rate.

Item a identifies a class of accidents, called loss of coolant accidents (LOCA's), in which a rupture in the reactor coolant system (RCS) would lead to a loss of the normal coolant. The rupture would allow the high pressure, high temperature RCS water to flash to steam and blow down into the containment building. To cope with this type of potential event, a set of systems called engineered safety features (ESF's) is provided in each plant. A number of the engineered safety features, as well as physical processes, act to reduce the amount of radioactivity released to the environment should either a LOCA or transient event result in a significant release of radioactivity from the reactor core. For instance, a containment building is provided to contain the radioactivity released from the fuel and to delay and reduce the magnitude of release to the environment. Some of the radioactivity would be deposited on surfaces within the containment building or would be absorbed by water sprays, water pools or filters provided for this purpose.

Item b identifies a class of events called transients. A nuclear plant includes various electrical safety circuits and a system for rapid termination of the fission process to attempt to protect against damaging transients.

The ESF's also serve to mitigate consequences should the transients result in severe overheating of the fuel.

The spent fuel storage pool (SFSP) holds fuel that has been removed from the reactor and that is being stored until its heat generation rate decays to a low level at which time fuel is permitted to be shipped to a fuel reprocessing plant. The decay heat rate of the fuel in the storage pool is much lower than that of the fuel in an operating reactor core. At these low heat rates, the fuel is adequately cooled by the pool water, and significant releases of radioactivity can occur only in accidents involving essentially complete loss of water from the pool.

During normal reactor operation the bulk of the radioactivity remains trapped in the fuel pellets since the uranium dioxide, a ceramic of high melting point (~ 5,000°F), effectively retains the bulk of the radioactivity. A typical fuel rod is shown in Fig. 2. The gas plenum at the top of the fuel rod collects the small amount of gaseous radioactivity that normally leaks from the fuel pellets during operation. Figures 3 and 4 show the reactor coolant systems (RCS) for a typical PWR and a typical BWR plant, respectively. Figure 5 shows the BWR RCS inside the primary containment. The BWR primary containment completely encloses the RCS and is provided with a pressure suppression pool to prevent overpressurization of the containment by the initial stream release to the containment in the event of a LOCA. Figure 6 shows the BWR RCS and primary containment located in a reactor building. This building, sometimes called a secondary containment, is not really a containment, but a confinement building that provides a path by which radioactivity that leaks from the primary containment is discharged to the environment through filters and is discharged at an elevated level. Figure 7 shows the PWR RCS inside a containment building. A system to quench the steam released in a PWR LOCA is not needed to prevent initial overpressurization because of the large volume within the containment. However, systems are provided to remove heat and reduce the pressure in the containment building and to retain radioactivity that may be released from the core. Both PWR and BWR containments are designed to have low leakage rates in order to inhibit the release of radioactivity to the environment.

LOCATION AND MAGNITUDE OF RADIOACTIVITY

The fresh uranium dioxide pellets that serve as the fuel in the PWR and BWR reactors are only slightly radioactive. However, during reactor operation the fission process produces large amounts of radioactivity in the fuel. By far the largest fraction of the radioactivity is associated with the fission products resulting from the fission process. Some of the neutrons produced by the fission process are absorbed, to various degrees, by structural and coolant materials and thereby generate radioactivity. This radioactivity is generally referred to as induced radioactivity. The induced radioactivity is only a minute fraction of the total radioactivity that could potentially be released from the reactor in the event of a severe accident and is, therefore, not important.

While essentially all the radioactivity in the plant is initially created in the reactor core, transfer of spent fuel assemblies from the core results in considerable radioactivity being located in other parts of the

Typical Fuel Data

	PWR	BWR
Overall length, in.	149.7	~164
Outside diam., in.	0.422	0.563
Metal wall thickness, in.	0.0243	0.037
Pellet diam., in.	0.366	0.477
Pellet length, in.	0.600	0.5
Pellet stack height, in.	144	144
Plenum length, in.	4.3	16
Fuel rods in fuel assembly	204	49
Fuel rod pitch, in.	0.563	0.738
Fuel assemblies in core	193	764

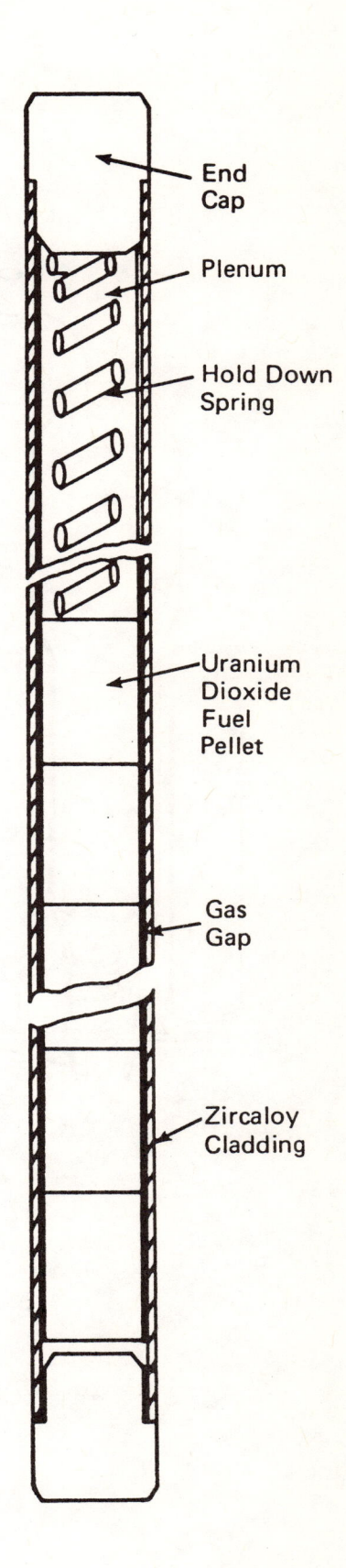

Figure 2. Cutaway of Fuel Rod Used for Commercial Water Cooled Nuclear Power Plants

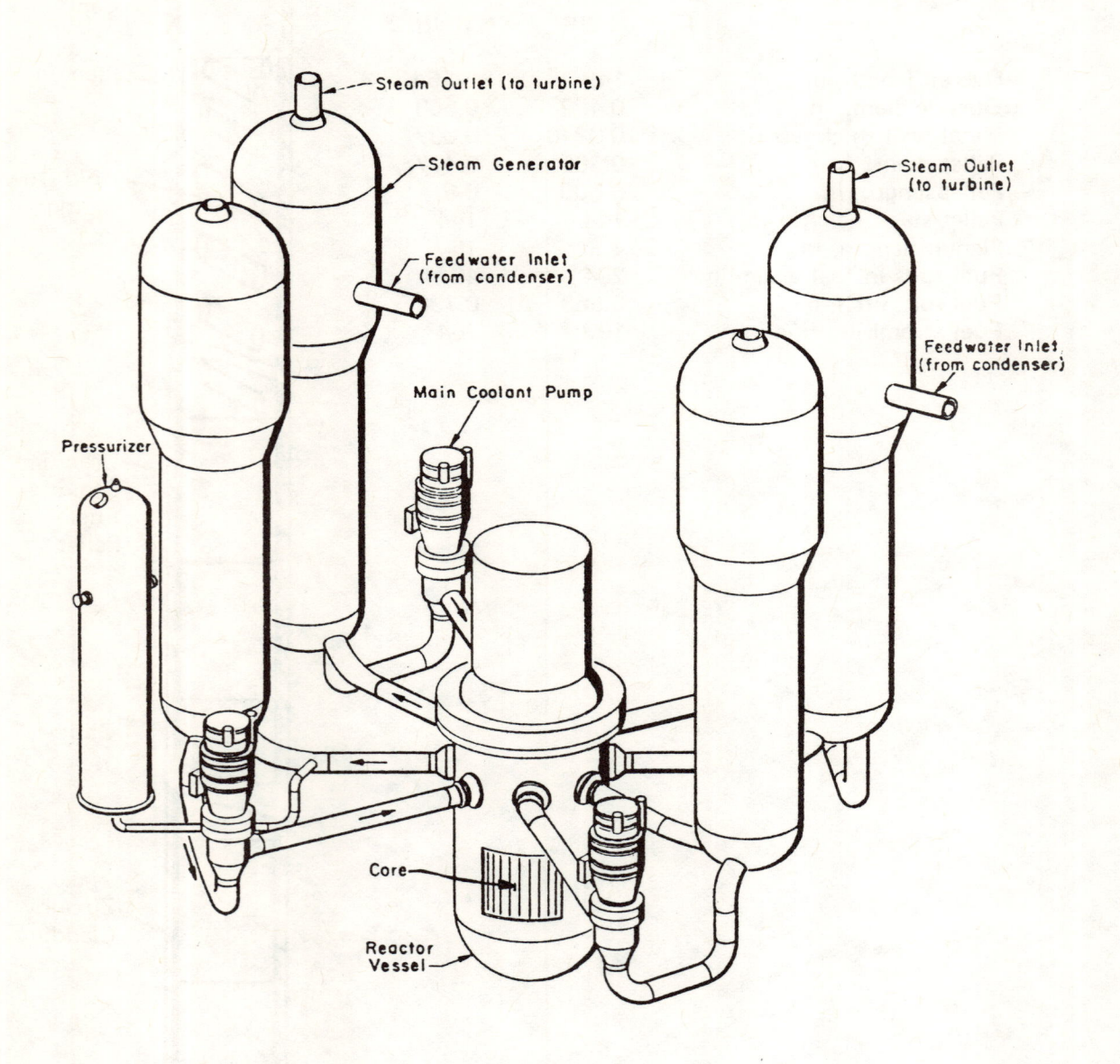

Figure 3. Schematic of Reactor Coolant System for PWR

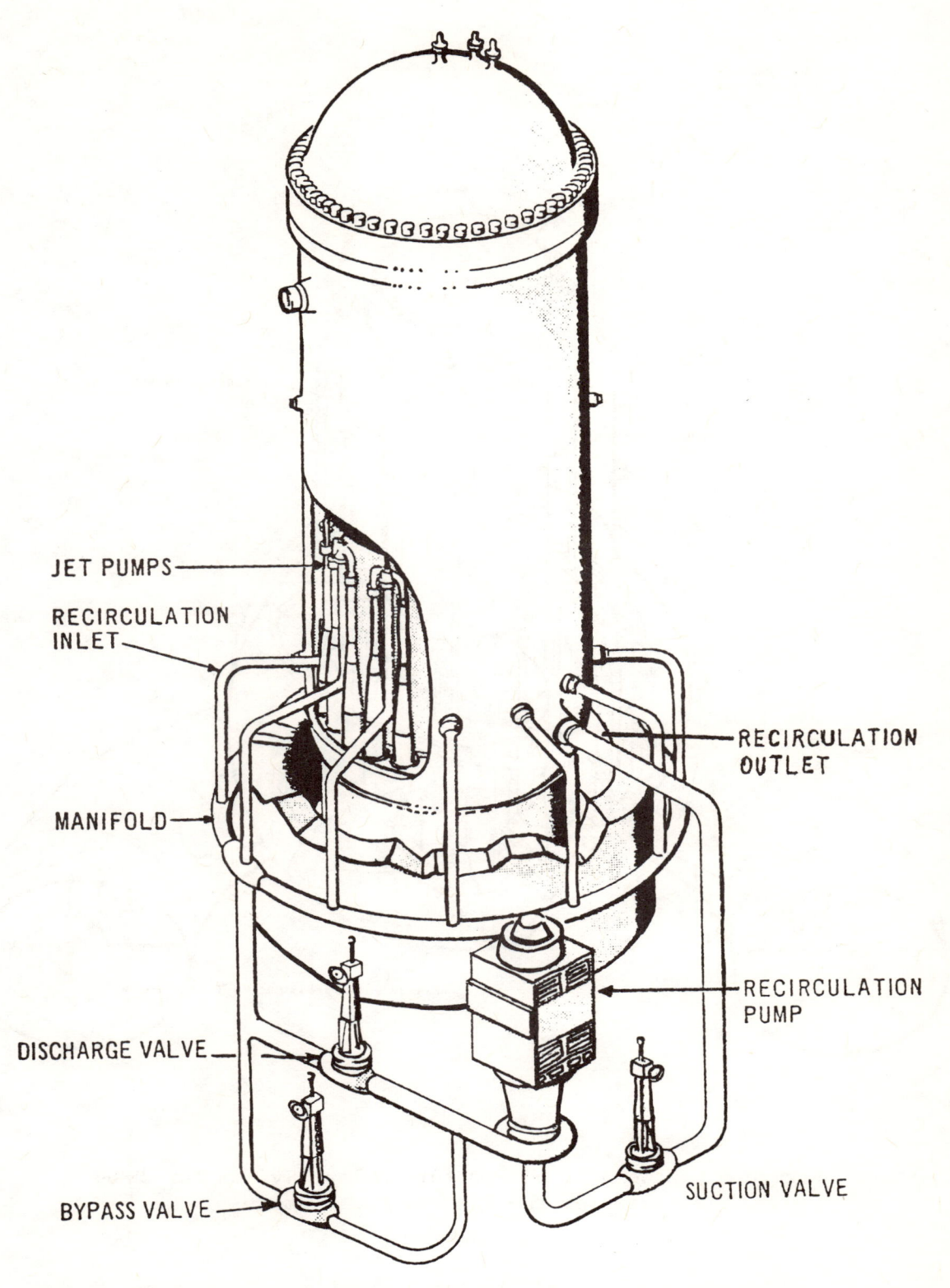

Figure 4. Schematic of BWR Reactor Coolant System

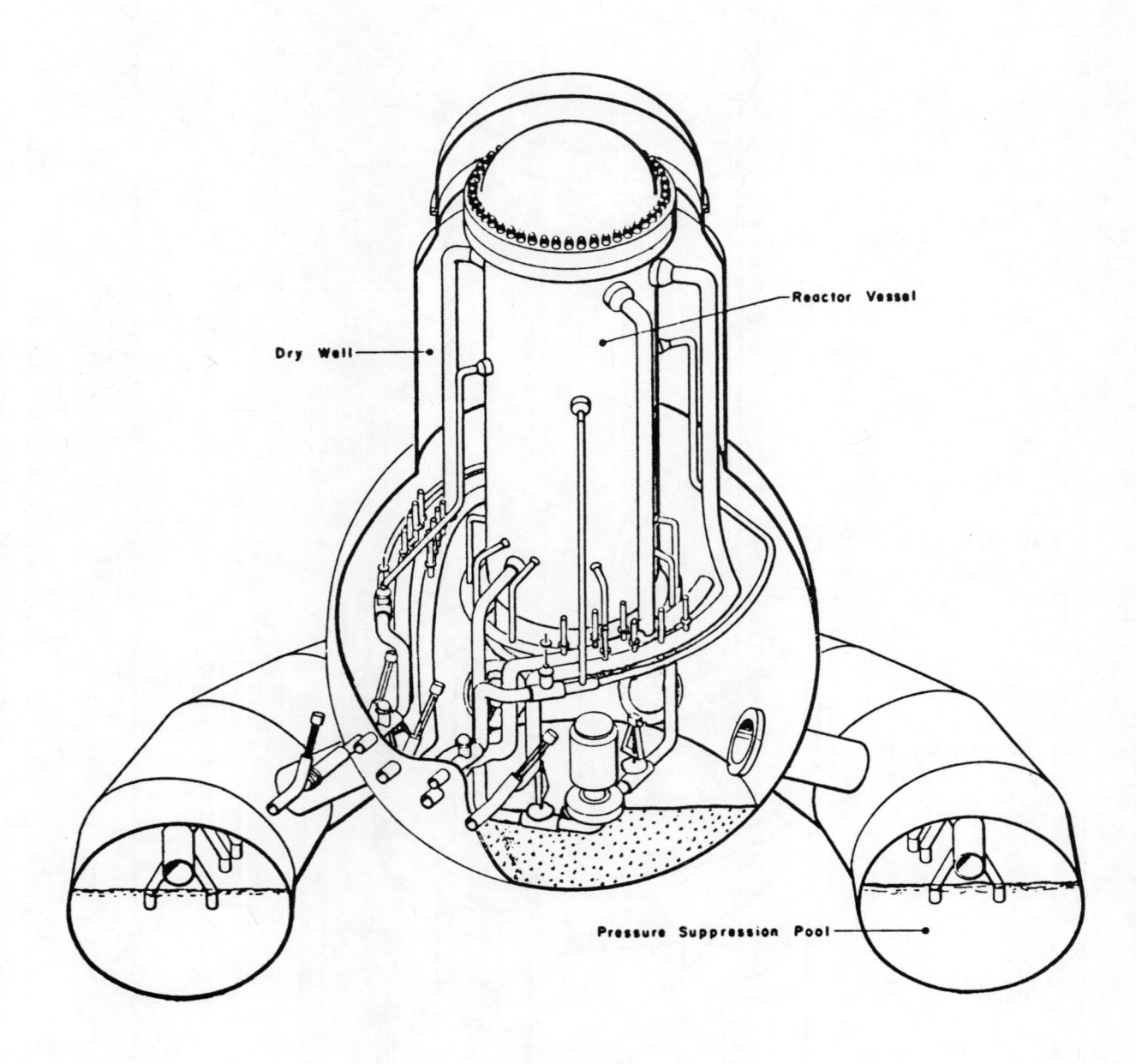

Figure 5. Schematic of Reactor Coolant System for BWR--
Inside of the Primary Containment

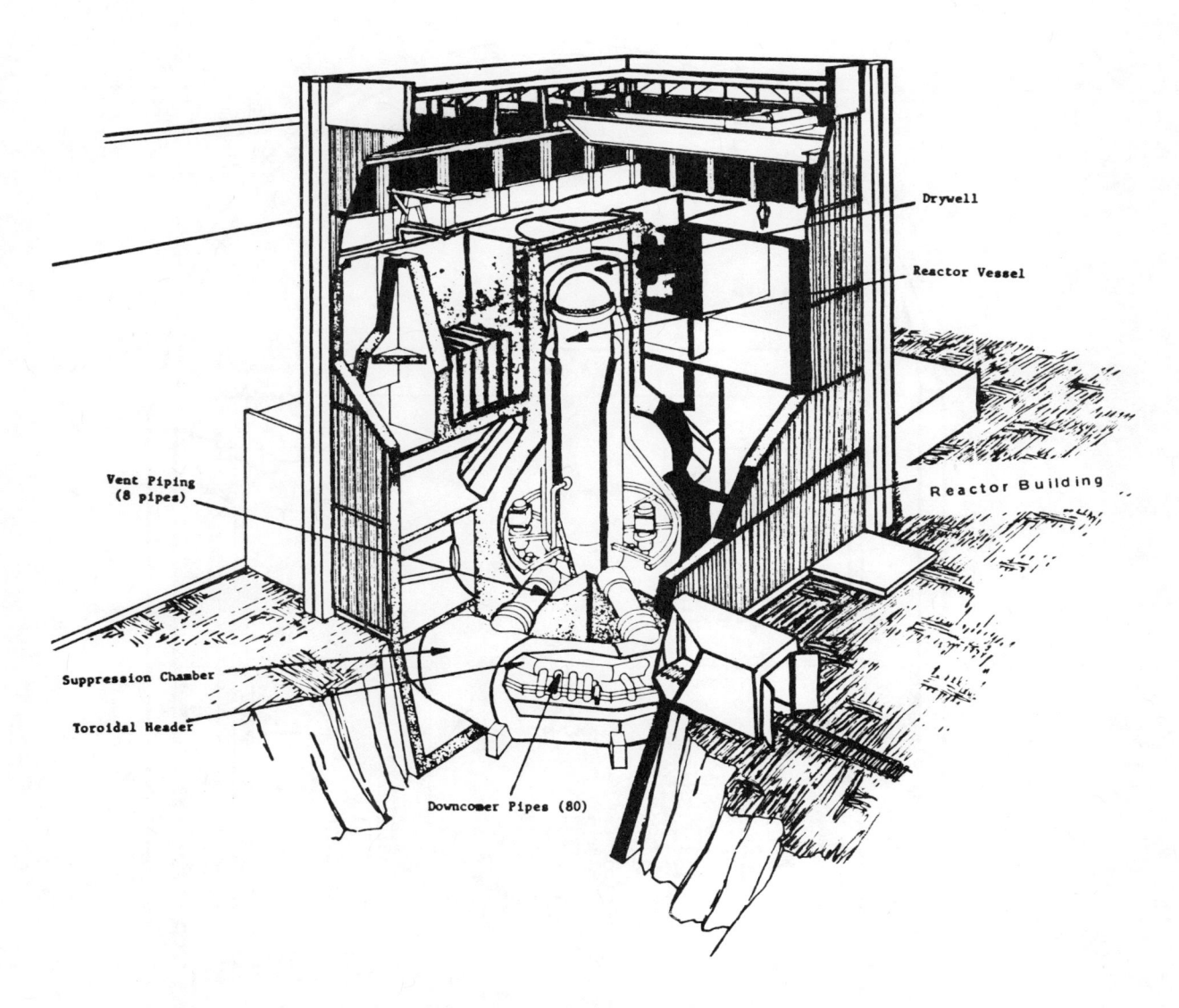

Figure 6. BWR Reactor Building Showing Primary Containment System Enclosed

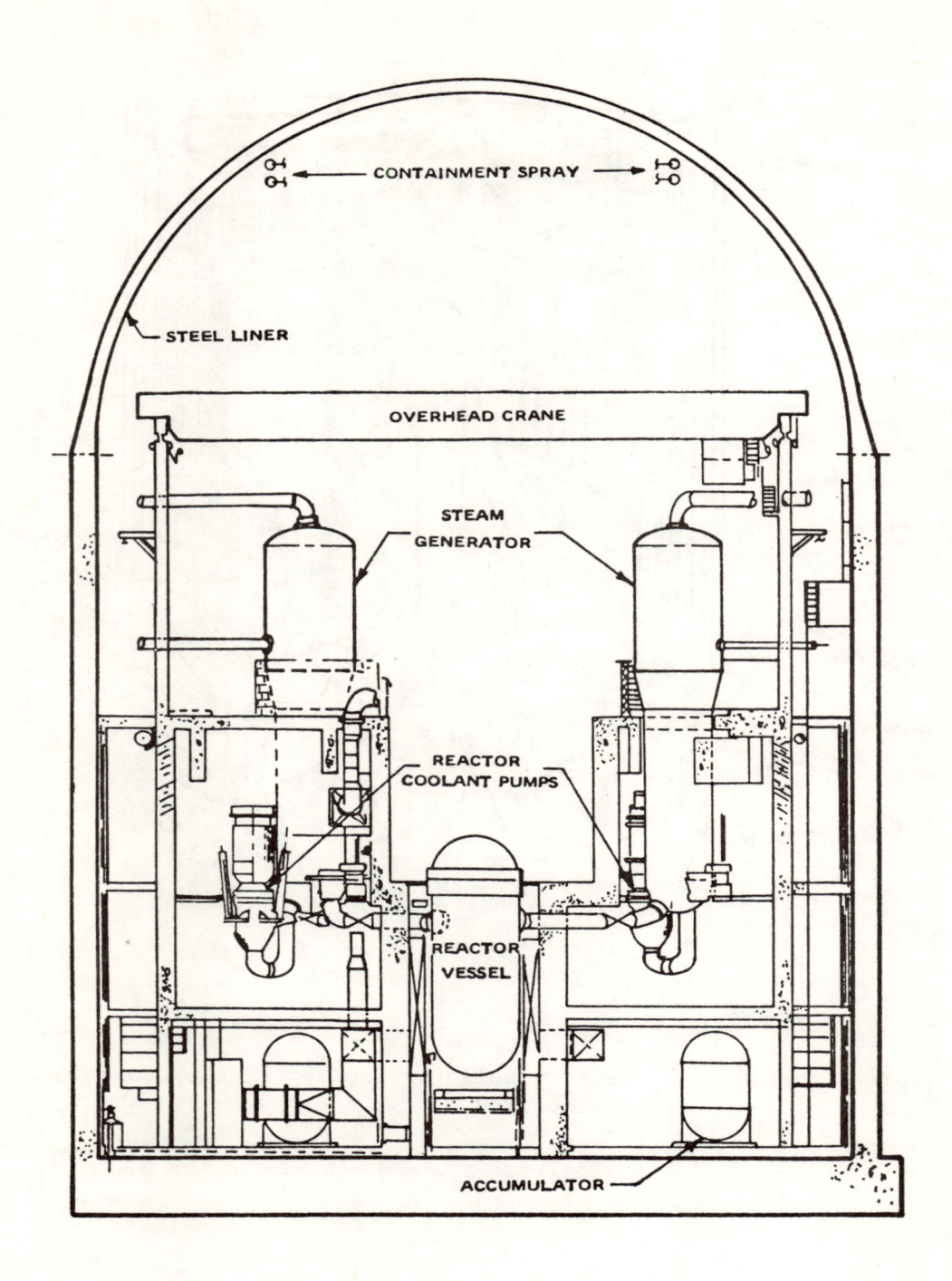

Figure 7. Typical PWR Containment

plant. The radioactivity inventory, second largest in amount compared to the reactor core, is located in the spent fuel storage pool (SFSP) which holds fuel that has been removed from the reactor and is awaiting shipment to an off site fuel reprocessing facility. The average number of fuel assemblies in the SFSP constitutes about half of a full reactor core loading. Radioactive fuel assemblies in the plant may also be located in the spent fuel shipping cask which holds up to about 10 fuel assemblies. The refueling transfer from the core to the SFSP involves only a single fuel assembly at a time. In addition to the above, smaller sources of radioactivity are normally present at the plant in the waste gas storage tanks (WGST) and the liquid waste storage tanks (LWST). These latter sources result, for example, from leakage of a small amount of radioactivity from the fuel rods during reactor operation, as well as radioactivity induced in impurities in the reactor cooling water. Typical magnitudes of the radioactive inventories in the above noted plant locations are shown in Table III.

The values given in Table III are typical for a 1,000 megawatt electric (MWe) plant operating at 3,200 megawatts thermal (MWt). In addition to the reactor power level, the plant radioactive inventory depends slightly on the length of power operation. For example, in the reactor core the total amount of radioactivity produced is directly related to the product of the power level and time at power. However, since the radioactivity decays to other isotopes, which are non-radioactive or less radioactive, an equilibrium amount of radioactivity occurs when the radioactive decay rate equals the production rate. For most of the radioactivity, equilibrium has occurred after several months of sustained operation. The reactor core radioactivity inventory shown in Table III is based on 550 days of sustained operation and represents the expected equilibrium radioactivity in an operating reactor. The inventory of radioactivity in the SFSP is based on a plant that has a common SFSP serving two 1,000 MWe reactors. The average number of spent fuel assemblies stored in the SFSP is based on assumed normal unloading and shipment schedules. The radioactive inventory in the shipping cask is based on a full load of fuel in the largest shipping cask currently licensed, and the shortest decay period (150 days) allowed for fuel shipped in the container. The refueling radioactivity represents that amount in a single fuel assembly at three days after reactor shutdown. This time is typical of the earliest time after shutdown that transfer of fuel from the reactor core to the SFSP begins.

Table III clearly shows that the reactor core contains by far the largest source of radioactivity in the plant. It also shows there is a relatively large inventory of radioactivity in the fuel in the SFSP and indicates that potential accidents of interest could result from melting of fuel initiated by a complete loss of water from the pool. While the spent fuel assemblies in a loaded shipping cask constitute a significant radioactive inventory, there is only a small potential for releasing a small fraction of this radioactivity in an in-plant accident. The radioactivity in shipping cask fuel has decayed long enough so that air cooling alone is sufficient to preclude fuel melting. However, the fuel clad temperatures reached may become high enough to cause cladding failures and the release of the small amount of gaseous radioactivity that collects in the fuel rod gap and plenum. The postulated accident related to refueling transfer is the inadvertent lifting of a fuel assembly completely out of the water-filled refueling canal or SFSP. Convective air cooling and heat radiation are also adequate to prevent fuel

Table III. Typical Radioactivity Inventory for a 1000 MWe Nuclear Power Reactor

Location	Total Inventory (Curies) Fuel	Total Inventory (Curies) Gap	Total Inventory (Curies) Total	Fraction of Core Inventory Fuel	Fraction of Core Inventory Gap	Fraction of Core Inventory Total
Core[a]	8.0×10^9	1.4×10^8	8.1×10^9	9.8×10^{-1}	1.8×10^{-2}	1
Spent Fuel Storage Pool (Max.)[b]	1.3×10^9	1.3×10^7	1.3×10^9	1.6×10^{-1}	1.6×10^{-3}	1.6×10^{-1}
Spent Fuel Storage Pool (Avg.)[c]	3.6×10^8	3.8×10^6	3.6×10^8	4.5×10^{-2}	4.8×10^{-4}	4.5×10^{-2}
Shipping Cask[d]	2.2×10^7	3.1×10^5	2.2×10^7	2.7×10^{-3}	3.8×10^{-5}	2.7×10^{-3}
Refueling[e]	2.2×10^7	2×10^5	2.2×10^7	2.7×10^{-3}	2.5×10^{-5}	2.7×10^{-3}
Waste Gas Storage Tank	–	–	9.3×10^4	–	–	1.2×10^{-5}
Liquid Waste Storage Tank	–	–	9.5×10^1	–	–	1.2×10^{-8}

(a) Core inventory based on activity 1/2 hour after shutdown.
(b) Inventory of 2/3 core loading; 1/3 core with three day decay and 1/3 core with 150 day decay.
(c) Inventory of 1/2 core loading; 1/6 core with 150 day decay and 1/3 core with 60 day decay.
(d) Inventory based on 7 PWR or 17 BWR fuel assemblies with 150 day decay.
(e) Inventory for one fuel assembly with three day decay.

melting in this case, but cladding failures and a relatively small release of radioactivity (from the fuel rod gap and plenum) could result. The radioactivity in the waste gas storage tanks (WGST) and liquid waste storage tanks (LWST) are very small compared to the other sources. Accidents postulated for release of radioactivity from these tanks include tank ruptures as well as malfunctions that could involve release of the contents of the tank.

Although accidents that involve release of radioactivity from the shipping cask fuel, the refueling process, the WGST and the LWST would be troublesome, particularly to in-plant personnel, none of these could result in public consequences nearly as serious as accidents involving melting of the fuel in the reactor core or in the SFSP. Thus, although the study treats accidents involving all the radioactive sources listed in Table III, the ensuing discussion is directed at potential accidents involving fuel in the reactor core and the SFSP.

LOSS OF COOLANT ACCIDENTS

A LOCA would result whenever the reactor coolant system (RCS) experiences a break or opening large enough so that the coolant inventory in the system could not be maintained by the normally operating makeup system. Nuclear plants include many engineered safety features (ESF's) that are provided to mitigate the consequences of such an event. A brief description of the LOCA sequence, assuming that all ESF's operate as designed, is as follows:

1. A break in the RCS would occur and the high pressure, high temperature RCS water would be rapidly discharged into the containment.

2. The emergency core cooling system (ECCS) would operate to keep the core adequately cool.

3. Any radioactivity released from the core would be largely retained in the low leakage containment building.

4. Natural deposition processes and radioactivity removal systems would remove the bulk of the released radioactivity from the containment atmosphere.

5. Heat removal systems would reduce the containment pressure, thereby reducing leakage of radioactivity to the environment.

If the ESF's were to operate as designed, the reactor core would be adequately cooled and only small consequences would result. However, the potential consequences could be much larger if ESF failures were to result in overheating of the reactor core. The public impact would depend on a large number of factors. Some of the more significant factors are discussed below.

LOCA Initiating Events

There are a number of ways in which a LOCA may be initiated. The most commonly considered initiating event would be a break in the RCS piping. Piping breaks that could cause a LOCA range in size from about the equivalent of a 1/2 inch diameter hole up to the complete severance of one of the main coolant loop pipes (about 3 feet in diameter). The maximum pipe diameter varies somewhat from plant to plant. The large pipe break is normally considered to be double-ended. This means that coolant from the RCS is expelled through both ends of the severed pipe, or the equivalent of two pipes that are about three feet in diameter.

The consequences of failure of pressure vessels (such as the reactor vessel and steam generators) have not normally been considered in the NRC's safety reviews since the high quality requirements applied in design, fabrication, and operation of these vessels have, in the past, been considered adequate to make the likelihood of failure of these vessels negligibly small. However, this study has considered both the likelihood and consequences of such failures in order to ascertain the extent to which they can potentially affect the overall risk from nuclear power plant accidents. The effects of steam generator failures and many types of reactor vessel failures as initiating events can be adequately controlled by existing ECCS systems. However, large disruptive reactor vessel failures could prevent adequate cooling of the core and can potentially cause failure of the containment building.

The specific LOCA initiating events analyzed in this study are:

a. Large pipe breaks (6" to approximately 3 feet equivalent diameter)

b. Small to intermediate pipe breaks (2" to 6" equivalent diameters)

c. Small pipe breaks (1/2" to 2" equivalent diameter)

d. Large disruptive reactor vessel ruptures

e. Gross steam generator ruptures

f. Ruptures between systems that interface with the RCS.

Effects of Engineered Safety Features

The basic purpose of the ESF's is the same for both PWR and BWR plants. However, the nature and functions of ESF's differ somewhat between PWR's and BWR's because of the differences in the plant designs. A number of the ESF's are included in a group termed the emergency core cooling system (ECCS) whose function is to provide adequate cooling of the reactor core in the event of a LOCA. Other ESF's provide rapid reactor shutdown and reduce the containment radioactivity and pressure levels that result from escape of the reactor coolant from the RCS. The following functional descriptions apply to current designs of BWR and PWR plants.

The ESF functions are illustrated in Fig. 8. The primary functions they perform are as follows:

a. Reactor trip (RT)--to stop the fission process and terminate core power generation.

b. Emergency core cooling (ECC)--to cool the core, thereby keeping the release of radioactivity from the fuel into the containment at low levels.

c. Post accident radioactivity removal (PARR)--to remove radioactivity released from the core to the containment atmosphere.

d. Post accident heat removal (PAHR)--to remove decay heat from within the containment, thereby preventing overpressurization of the containment.

e. Containment integrity (CI)--to prevent radioactivity within the containment from being dispersed into the environment.

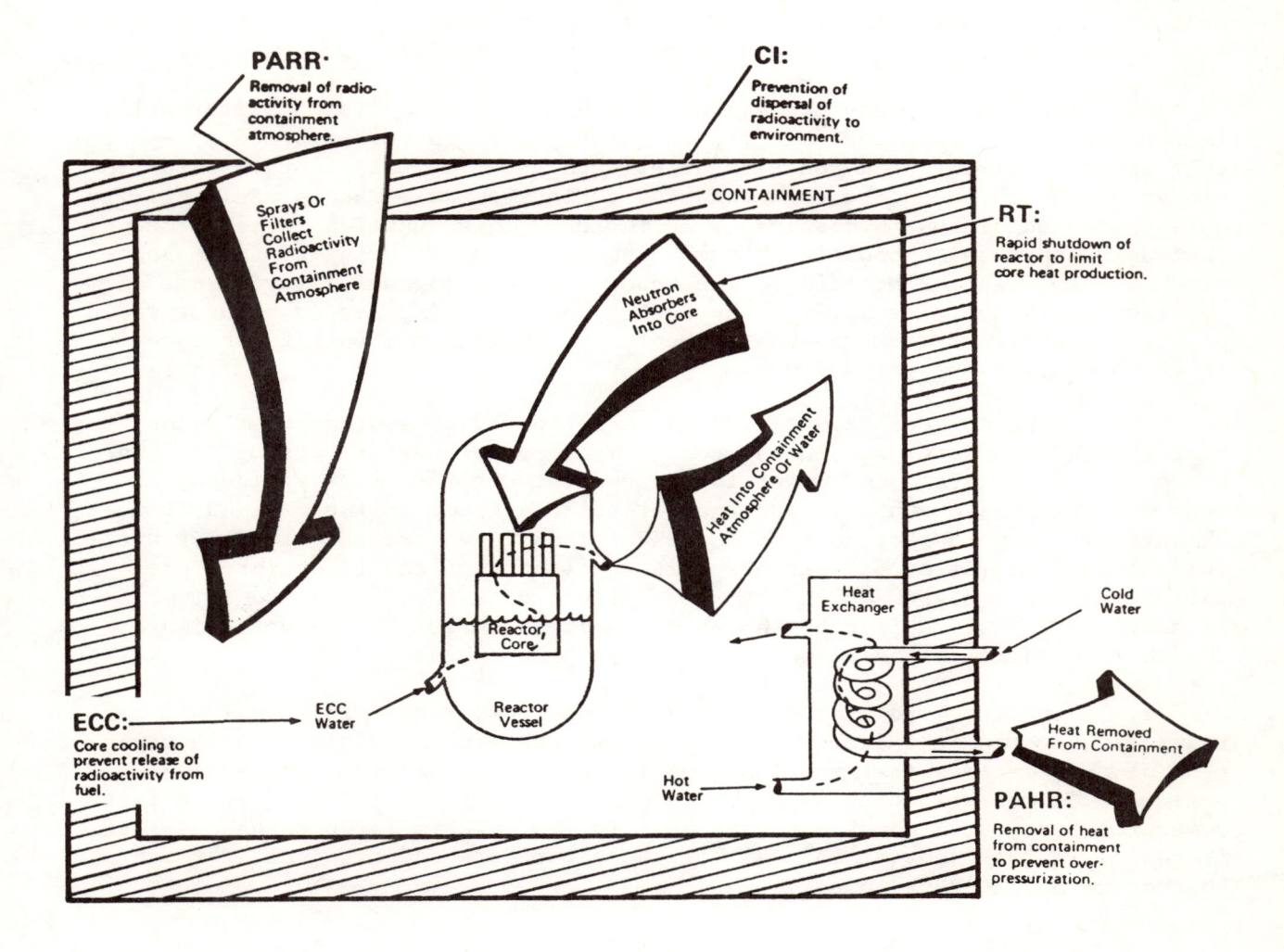

Figure 8. Power Water Reactor Loss of Coolant Accident (LOCA) Engineered Safety Feature (ESF) Functions

The course of events following a LOCA initiating event is strongly influenced by the degree of successful operation of the various ESF's. The ways in which failures of the above functions influence the outcome of LOCA's are discussed briefly below. Reactor trip (RT) is accomplished by rapid insertion of the reactor control rods. The action is initiated automatically by electrical signals generated if any of a number of key operating variables reaches a preset level. The way in which failure of the RT function affects a LOCA is complicated by a number of factors. For example, in the PWR, failure of the RT function in a large LOCA is of no immediate significance since the reactor is rapidly shut down by the loss of core moderator and ECC water contains boron to prevent return to power. However, there are circumstances in which RT is required.

Emergency core cooling (ECC) involves a number of systems that deliver a supply of emergency coolant to the reactor core. Both PWR and BWR plants include high pressure systems primarily for coping with small LOCA's and low pressure systems primarily for large LOCA's. Together, these systems can fulfill the ECC requirement over a wide range of small to large pipe breaks.

Post accident radioactivity removal (PARR) is accomplished differently in the PWR than in the BWR. In the PWR, this function is performed by systems that spray water into the containment atmosphere. The water spray, which includes a chemical additive for enhancing iodine removal, washes radioactivity out of the containment atmosphere. In the BWR, this function is performed by the vapor suppression pool in the containment and a filtering system associated with the reactor building. The vapor suppression pool removes some of the radioactivity released from the core. The filtering system removes radioactivity that leaks from the containment into the reactor building before it is released at an elevated level.

Post accident heat removal (PAHR) is performed by systems that transfer heat from heated water within the containment to cold water outside the containment. The containment water that flows through the primary side of the heat exchanger is taken from the reactor building sump in the PWR and from the pressure suppression pool in the BWR. This is a particularly important function, since failure to perform this function can lead to overpressure failure of the containment and related failure of ECC systems. Containment integrity (CI) is provided by the containment features that serve to isolate the containment atmosphere from the outside environment.

It is evident, from the preceding discussions, that a large release of radioactivity from the reactor core into the containment would require violation of the barriers to the release of radioactivity provided by the fuel pellets, the fuel cladding, and the reactor coolant system. In current large power reactors, the amount of decay heat in the core is large enough so that it could, if not removed, melt all of these barriers and also melt through the bottom of the containment.

In early power reactors the power level was about one tenth that of today's large reactors. It was thought that core melting in those low power reactors would not lead to melt-through of the containment. Further, since the decay heat was low enough to be readily transferred through the steel containment walls to the outside atmosphere, it could not overpressurize and fail the containment. Thus, if a LOCA were to occur, and even if the core were to melt, the low leakage containments that were provided would have permitted the release of only a small amount of radioactivity.

Prior studies have indicated that a core meltdown in a large reactor would likely lead to failure of the containment. Thus, a commonly held opinion regarding core melting is that such an event would result in a very serious accident with large public consequences. This is evidently one of the reasons that major safety efforts have been devoted to the prevention of core meltdown and little attention has been directed toward the examination of the potential relationships between core melting and containment integrity. This study has analyzed such relationships and has found that the containment failure modes, their timing, and the potential radioactive release depend strongly on the operability of the various ESF's. The following paragraphs indicate some general observations based on the containment failure investigations conducted in this study.

Molten Fuel Interactions

Because of the difficulties involved in making precise predictions of the physical processes that accompany core melting in a LOCA, the study has not investigated the potential consequences of partial melting of reactor cores. However, it has been conservatively assumed that if conditions are such that some core melting would result, then essentially complete core melting would occur. It then follows that the core could melt through the bottom of the reactor vessel and through the thick, lower concrete structure of the containment. Melt-through of the containment would be predicted to occur about one-half to one day after the accident, thus providing considerable time for radioactive decay, washout, plateout, etc., to reduce the radioactivity in the containment atmosphere. Furthermore, most of the gaseous and particulate radioactivity that might be released would be discharged into the ground which acts as an efficient filter, thus significantly reducing the radioactivity released to the above-ground environment. Accidents that would follow this path are thus characterized by relatively low releases and consequences. In plants that have relatively large volume containments, the melt-through path described above would represent the most likely course of the accident.

As noted above, the melt-through path would be characterized by low atmospheric releases and consequences. Following this melt-through, there would be the possibility of ground water contamination through a long term process of leaching of radioactivity from the solidifying mass of fuel, soil, etc. The leaching and contamination processes would occur over an extended period of time (several to many years, depending on the particular radioactive species) and the potential contamination levels should not be substantially larger than the maximum permissible concentrations (MPC). The concentrations could potentially be controllable to even lower levels. Accordingly, the potential for ground water contamination therefore has been assessed to have a small contribution to the overall task.

Containments may also fail by overpressure resulting from various non-condensible gases released within the containment as a result of core melting. These gases would arise from a number of sources. At high temperatures the zircaloy cladding of the fuel and the molten iron from support structures would react actively with water to generate large volumes of hydrogen. Also, in penetrating the bottom of the containment, the molten core decomposes the concrete, thus generating large quantities of carbon dioxide. This is true only for concrete which contains limestone. (It is not applicable to basaltic concrete.) For small containments, the pressure due to the combination of these two gases would represent the most likely path to containment failure. Even though such failures would most likely occur in the above ground portion of the containment, this would take several hours from the time of core melt. Thus, there would be considerable time available for reducing the amount of radioactivity released due to decay, plateout, etc. It is not expected that large containments would be failed by this means.

At two key stages in the course of a potential core meltdown there would be conditions which would have the potential to result in a steam explosion* that could rupture the reactor vessel and/or the containment. These conditions may occur when molten fuel would fall from the core region into water at the bottom of the reactor vessel or when it would melt through the bottom of the reactor vessel and fall into water in the bottom of the containment. It is predicted that if such an explosion were to occur in the reactor vessel, it may be energetic enough to change the course of the accident. For reactors enclosed in relatively large volume containments, it is considered improbable that a steam explosion outside the reactor vessel would rupture the containment. If a steam explosion were to occur within the reactor vessel, it is considered possible that both large and small containments could be penetrated by a large missile. Such occurrences might release substantial amounts of radioactivity to the environment. However, these modes of containment failure are predicted to have low probabilities of occurrence.

REACTOR TRANSIENTS

In general, the term reactor transient applies to any significant deviation from the normal operating value of any of the key reactor operating parameters. More specifically, transient events can be assumed to include all those situations (except for LOCA, which is treated separately) which could lead to fuel heat imbalances. When viewed in this way, transients cover the reactor in its shutdown condition as well as in its various operating conditions. The shutdown condition is important in the consideration of transients because many transient conditions result in shutdown of the reactor and decay heat removal systems are needed to prevent fuel heat imbalances due to core decay heat.

Transients may occur as a consequence of an operator error or the malfunction or failure of equipment. Many transients are handled by the reactor control system, which would return the reactor to its normal operating condition. Others would be beyond the capability of the reactor control system and require reactor shutdown by the reactor protection system (RPS) in order to avoid damage to the reactor fuel.

In safety analyses, the principal areas of interest are increases in reactor core power (heat generation), decreases in coolant flow (heat removal) and reactor coolant system (RCS) pressure increases. Any of these could po-

---

*The term steam explosion refers to a phenomenon in which the fuel would have to be in finely divided form and intimately mixed with water so that its thermal energy could be efficiently and rapidly deposited in the water thus creating a large amount of steam.

tentially result from a malfunction or failure, and they represent a potential for damage to the reactor core and/or the pressure boundary of the RCS. In this study the analysis of reactor transients has been directed at identifying those malfunctions or failures that can cause core melting or rupture of the RCS pressure boundary. Regardless of the way in which transients might cause core melting, the consequences are essentially the same; that is, the molten core would be inside an intact containment and would follow the same course of events as a molten core that might result from a LOCA. This fact greatly simplified the determination of the transient contribution to the risk since it permitted the elimination of many transients from the risk determination solely on the basis of their relatively low probabilities compared to those of other transients.

In this study each potential transient is assessed to fall into either one of two general categories, the anticipated (likely) transients and the unanticipated (unlikely) transients. The large majority of potential transients are those that have become commonly known as anticipated transients. There are currently about 10 such occurrences per year at each nuclear power plant, including a few planned shutdowns. Some of the individual types of events, such as loss of offsite power, that contribute to this total number are relatively less likely to occur. All other transients are considered to fall into the unanticipated transient category. As will be shown later, the relatively low probability (unanticipated) transients can be eliminated from the risk determination since their potential contribution to the risk is small compared to that of the more likely (anticipated) transients that produce the same consequence. Similar considerations of the relative probabilities permit elimination of most of the anticipated transients from the risk determination. The transients that were found to be important to the risk assessment are identified later in the text. These are the anticipated (likely) transients that involve the loss of offsite power and loss of plant heat removal systems.

ACCIDENTS INVOLVING THE SPENT FUEL STORAGE POOL

The spent fuel storage pool (SFSP) is identified as having a significant radioactivity inventory, second in amount to the reactor core. Further, the decay heat levels in freshly unloaded fuel assemblies that may be stored in the pool may be sufficiently high to cause fuel melting if the water is completely drained from the SFSP. Because the maximum amount of fuel stored in the pool immediately after refueling is smaller than that in the core and because it has had time (72 hours minimum) for radioactive decay, it is a less intense heat source than a reactor core (about one-sixth) and therefore melt-through of the bottom structure of the pool would occur at a much lower rate and, in fact, may not occur at all. On the average, fuel in the pool will have undergone about 125 days of decay, and it is questionable that such fuel would melt. However, to assure that the risk would not be underestimated, it has been assumed that even this fuel would melt.

Although the pool is not within a containment building, filters in the SFSP building ventilation system and natural deposition of radioactivity within the building both aid in reducing the amount of radioactivity that might be released to the environment in the event of a spent fuel accident.

The most probable ways in which loss of fuel cooling in the SFSP could occur have been determined to be the loss of the pool cooling system or the perforation of the bottom of the pool. The latter could occur, for example, by dropping a shipping cask in the pool or on the top edge of the pool. Both this type of accident and the loss of cooling capability are of low likelihood. The loss of cooling capability, which has been determined to be somewhat more probable, requires that a number of audible alarms be inoperative or ignored and that the visual observation be so lax as to permit the lowering of the pool water level to continue uncorrected for about two weeks --the approximate time required to boil off the SFSP water if cooling capability is lost. It has been shown that the size of such potential accidents is smaller than those that could involve the core.

RISK ASSESSMENT METHODOLOGY

The risk determination was divided into the three major tasks:

Task I includes the identification of potential accidents and the quantification of both the probability and magnitude of the associated radioactive releases to the environment. The major part of the work of the study was devoted to this task.

Task II uses the radioactive source term defined in Task I and calculates how the radioactivity is distributed in the environment and what effects it has on public health and property.

Task III combines the consequences calculated in Task II, weighted by their respective probabilities to produce the overall risk from potential nuclear accidents. To give some perspective to these results, they are compared to a variety of non-nuclear risks.

QUANTIFICATION OF RADIOACTIVE RELEASES

The objective of this task is to generate a histogram, of the form shown in Fig. 9, which shows the probability and magnitude of the various accidental radioactive releases. The isotopic composition, elevation of the release point above ground level and the timing and energy content associated with the release must also be determined to permit the calculation of consequences due to the releases.

This histogram could be determined for a single type of accident (such as a loss of coolant accident). By combining many accidents one can obtain a composite histogram for all important contributors. Since the histogram could be different for the various isotopes released, a full characterization of all accidents could involve a large number of such histograms. A significant effort was devoted to combining all isotopes and accidents into a single histogram for each reactor.

To generate a composite histogram of the type shown in Fig. 9, the methodology employed must in principle be able to identify the accidents that can produce significant releases and determine their probability. To do this for all accidents in a system as complicated as a nuclear power plant is a formidable task because of the very large number of accidents that can be imagined. The problem becomes more manageable, however, when it is realized that, of this large number of potential events, many have trivial releases, many are illogical (i.e., violate known physical conditions) and others have very small probabilities compared to accidents which result in essentially the same release magnitude.. To ensure that unnecessary analyses are not pursued, the methods used must provide a way for logically eliminating accidents that do not significantly contribute to the radioactive source term.

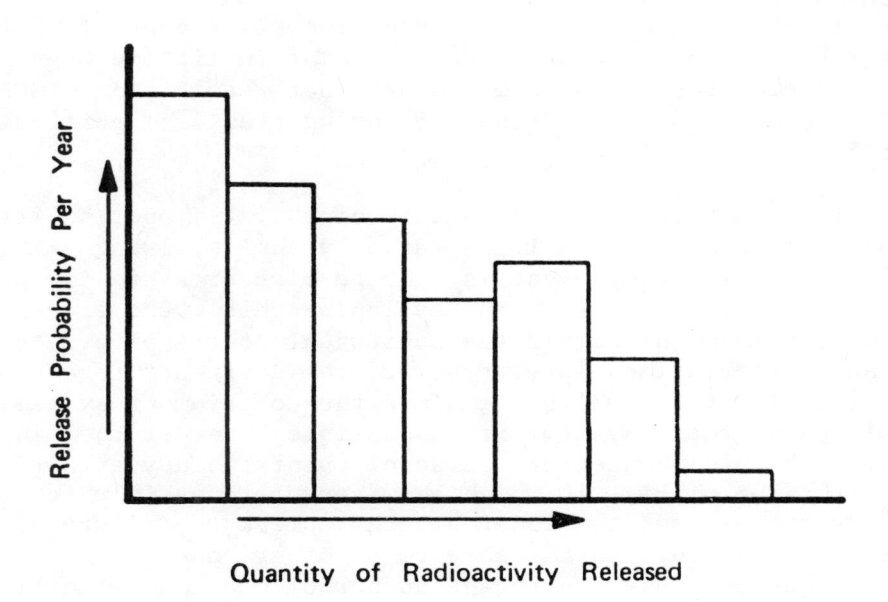

Figure 9. Illustrative Release Probability versus Release Magnitude Histogram

Definition of Accident Sequences--Event Trees

A major element in the characterization of the radioactive releases associated with potential nuclear power plant accidents is the identification of the accident sequences that can potentially influence the public's risk from such accidents. The study employed event tree methodology as the principal means for identification of the significant accident sequences.

An event tree is a logic method for identifying the various possible outcomes of a given event which is called the initiating event. The number of possible final outcomes depends upon the various options that are applicable following the initiating event. In the application to reactor safety studies the initiating event is generally a system failure and the subsequent events are, for the most part, determined by system characteristics and engineering data. In this study the trees are called event trees, and a particular sequence from the initiating event to a final outcome is termed an accident sequence.

In this study the application of event trees was limited to the analysis of potential accidents involving the reactor core. For this purpose it was found convenient to separate the event trees into two types of trees. The first was used to determine how potential accidents were affected by failures in major systems, particularly the engineered safety systems. They cover the significant LOCA and transient initiating events.

These trees were supplemented with a second type of tree, the containment event tree, to provide combined accident sequences from the initiating event to release of radioactivity from the containment. This procedure is described briefly below. It produced a list of systematically defined sequences leading to the release of radioactivity to the environment. The starting point for the development of an event tree is the event (failure) that initiates a potential accident situation. The initiating event is basically either a reactor coolant system rupture that results in a LOCA or any of a number of reactor transients. The initiating events of particular significance will be discussed later.

The application of event trees in determining system operability effects on potential accident sequences is illustrated by the following simplified example in which the initiating event is a large pipe break in the primary system of a reactor. The first step in developing this event tree is to determine which systems might affect the subsequent course of events. In this example these are station electric power, the emergency core cooling system, the radioactivity removal system, and the containment system. Through a knowledge of these systems it is possible to order them in the time sequence in which they influence the course of events. They are ordered in this way across the top of Fig. 10 which shows event trees in which the upper branch represents success and the lower branch represents failure of the system to fulfill its function. In the absence of other constraints there are $2^{(n-1)}$ accident sequences, where n is the number of headings (functions, systems, etc.) included on the tree. However, there are known relationships (constraints) between system functions. For example, if station electric power fails none of the other systems can operate because they depend upon power. In addition to such functional relations there may also be hardware common to more than one system. Once these functional and hardware relationships are incorporated, many of the chains shown in the upper tree of Fig. 10 can be eliminated because they represent illogical sequences. Such sequences are eliminated in the lower tree shown in Fig. 10. Note that elimination of the choices following failure of electric power reduces the number of sequences by about half.

The probability of failure of each system is indicated by the P values noted in Fig. 10. The probability of success is (1-P) since it is assumed that a system operates successfully if it does not fail. If the events (failures, successes) are independent then the probability of occurrence of a given sequence is the product of the probabilities of the individual events in that sequence, as indicated in Fig. 10. Since the failure probabilities are almost always 0.1 or less it is common practice to approximate $(1-P) \approx 1$, as shown in Fig. 10. The probability of occurrence of each system failure is shown to be different in each accident sequence in which it appears. This is done to account for the differences in system failure probabilities that may occur due to the differing dependencies in each accident sequence.

It should be noted that, as indicated by Fig. 10, the study developed event trees in which each branch point provides only two options, system

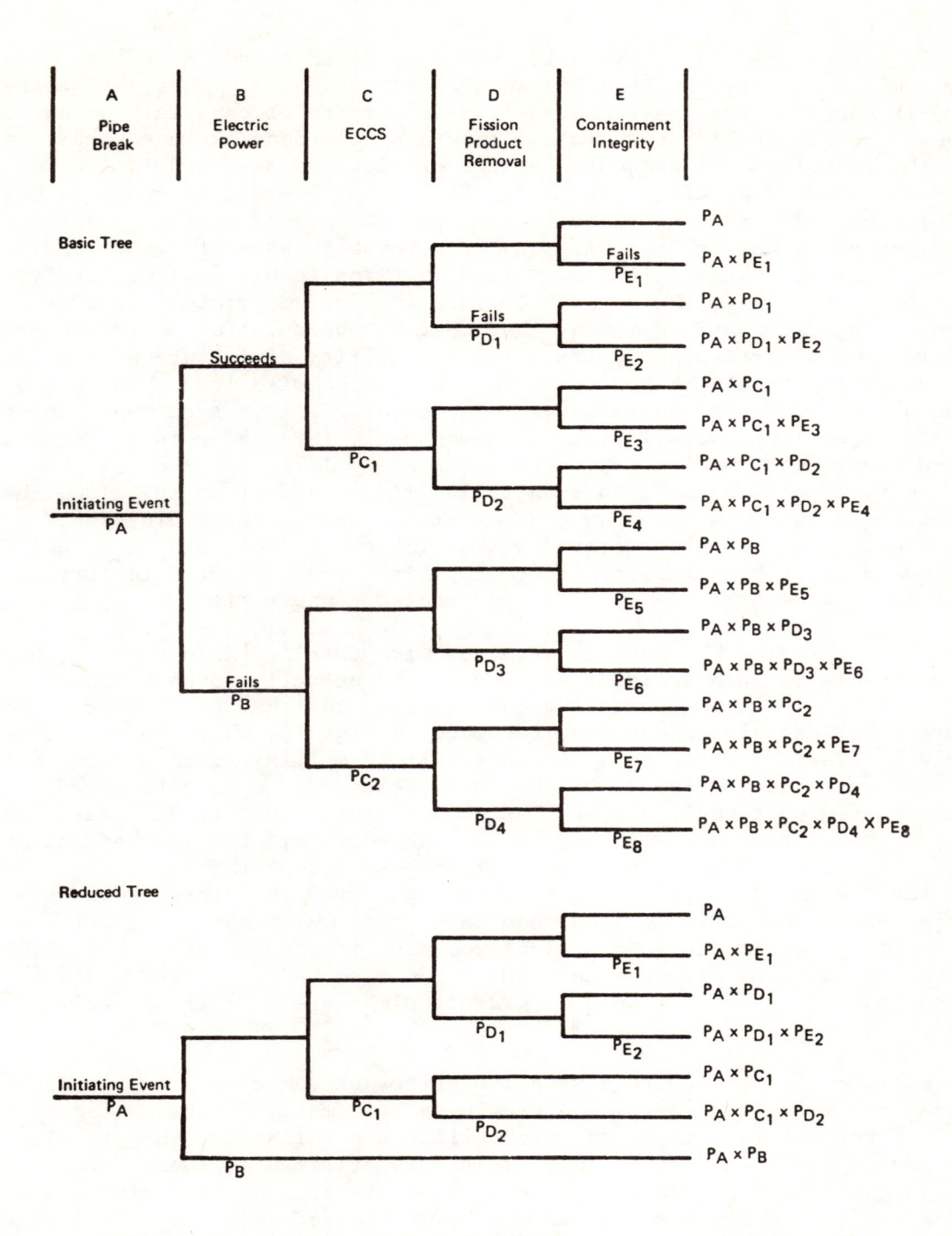

A	B	C	D	E
Pipe Break	Electric Power	ECCS	Fission Product Removal	Containment Integrity

Basic Tree

Initiating Event P_A

Succeeds branch:
- P_A
- Fails P_{E_1}: $P_A \times P_{E_1}$
- Fails P_{D_1}: $P_A \times P_{D_1}$
- P_{E_2}: $P_A \times P_{D_1} \times P_{E_2}$
- P_{C_1}: $P_A \times P_{C_1}$
- P_{E_3}: $P_A \times P_{C_1} \times P_{E_3}$
- P_{D_2}: $P_A \times P_{C_1} \times P_{D_2}$
- P_{E_4}: $P_A \times P_{C_1} \times P_{D_2} \times P_{E_4}$

Fails P_B branch:
- $P_A \times P_B$
- P_{E_5}: $P_A \times P_B \times P_{E_5}$
- P_{D_3}: $P_A \times P_B \times P_{D_3}$
- P_{E_6}: $P_A \times P_B \times P_{D_3} \times P_{E_6}$
- P_{C_2}: $P_A \times P_B \times P_{C_2}$
- P_{E_7}: $P_A \times P_B \times P_{C_2} \times P_{E_7}$
- P_{D_4}: $P_A \times P_B \times P_{C_2} \times P_{D_4}$
- P_{E_8}: $P_A \times P_B \times P_{C_2} \times P_{D_4} \times P_{E_8}$

Reduced Tree

Initiating Event P_A
- P_A
- P_{E_1}: $P_A \times P_{E_1}$
- P_{D_1}: $P_A \times P_{D_1}$
- P_{E_2}: $P_A \times P_{D_1} \times P_{E_2}$
- P_{C_1}: $P_A \times P_{C_1}$
- P_{D_2}: $P_A \times P_{C_1} \times P_{D_2}$
- P_B: $P_A \times P_B$

Note — Since the probability of failure, P, is generally less than 0.1, the probability of success (1-P) is always close to 1. Thus, the probability associated with the upper (success) branches in the tree is assumed to be 1.

Figure 10. Simplified Event Trees for a Large LOCA

success or system failure. No consideration is given to the fact that partial system success may occur within an accident sequence. Thus, an accident sequence is conservatively assumed to lead only to total core melt or no core melt, but never to partial core melt. This has been done because uncertainties in the calculational methods preclude predictions of the detailed conditions that lead to partial core melt. Similarly, because of the difficulty in calculating, with reasonable certainty, the effects of partial system failure, the study has treated all such questionable cases of system operability as complete system failures. Since most applicable systems involve considerable redundancy, the basic procedure involved determination of the fraction of the redundant equipment that must be operable to assure successful function of a particular system. The probability of failure of the system is the probability that the system is in a condition with less than this fraction of the equipment operating. This success/failure treatment can significantly affect the overall risk assessment only through accident sequences which are important contributors to risk. Since a large fraction of the sequences analyzed was found to have an insignificant effect on the risk, the fact that their analysis was done conservatively has a negligible effect on the magnitude of the total estimated risk. Those accident sequences that were found to contribute importantly to the risk were subjected to further analysis in an attempt to remove any unwarranted conservatisms.

If the event tree has been constructed with detailed information, the series of events in each accident sequence would be well enough defined so that it is possible to calculate the consequences for that particular series of events. For example, the bottom sequence in Fig. 10, where no core cooling would be available, can be shown to result in melting of the core and the fraction of core radioactivity that would be released can be calculated. Since the molten core would violate the containment, the accident could produce a release of radioactivity outside of the containment. The mode of containment failure would affect the overall probability of the sequences as well as the magnitude of the release. The event tree therefore provides a definition of the possible accident sequences from which the radioactive releases to the environment can be calculated and, if the failure probabilities are known, the probability of each release can also be calculated. Again, it should be noted that this example is greatly simplified for illustrative purposes.

In summary, the event trees were the principal vehicles, supplemented by additional analyses, utilized for achieving a systematic determination of the radioactive release magnitudes and probabilities associated with potential nuclear power plant accidents. They first were utilized to identify the many possible significant accident sequences. Then, through an iterative process involving successive improvements in the definition of failure probabilities, the incorporation of system interactions and the resolution of physical process descriptions, they provided for the identification of those accident sequences that are important to the achievement of a realistic risk assessment. These selected accident sequences served as the basis for determining the magnitude of applicable radioactive releases. They also served as the vehicle for combining the initiating event probabilities, system failure probabilities, and containment failure probabilities into the composite probabilities applicable to the radioactive releases. With respect to system failure probabilities, the event trees were the principal means of identifying the various system failure definitions needed in the fault trees that were used for determining system failure probabilities.

Probability of Releases

As noted previously, there were a large number of iterations in various parts of the risk assessment cycle. These iterations were necessary in order to determine the dominant accident sequences for use in the final overall risk assessment. The methods described below were utilized in this iterative process and aided in the selection of the dominant accident sequences. However, such iterations and other exploratory analyses are neglected in the following discussion, which is generally concerned with the determination of the radioactive release probabilities for the final overall risk assessment.

The final risk assessment is based on a number of different release categories. Each of these release categories is associated with a specific type and magnitude of release. The final risk assessment requires the probability applicable to each of these release categories. In general a specific release category applies to many accident sequences but, because of the wide range in probability of occurrence of these sequences, it is found that only a few sequences determine the probability of occurrence of a particular category. Thus, the determination of the release probability associated with each release category required the determination and summation of the probability of occurrence of each of the dominant accident sequences in the category.

The probability of occurrence of an accident sequence is composed of the initiating event probability, the failure probabilities of systems included in the sequence, and the containment failure probability. The probabilities for LOCA initiating events such as pipe breaks, vessel ruptures, transients, etc. were determined by deriving appropriate failure rates from available failure rate data. The large majority of the system failure probabilities were determined with the aid of the fault tree technique. This technique, discussed in the next section, is suited for analysis of failures of complex systems. To account for probable dependencies in failures of components and systems involved in the fault trees and event tree accident sequences, many special analyses were performed for the purpose of determining significant common mode failures. These analyses are discussed later. The applicable containment failure modes are largely determined by the accident sequences and the various physical processes that can result from the accident sequences. The basis for the likelihood of containment failure modes was determined by fault trees and the analysis of applicable physical processes.

Fault Trees. As noted above the event trees define certain system failures whose probabilities are needed to determine the risk. In this study the fault tree method has been used to estimate the majority of these failure probabilities. The method uses a logic that is essentially the reverse of that used in event trees. Given a particular failure, the fault tree method is used to identify the various combinations and sequences of other failures that lead to the given failure. The technique is particularly suited to the analysis of the failure of complex systems. The effective utilization of this logic requires that the analyst have a thorough understanding of the system components and their functions. This section gives a general discussion of the fault tree method.

The fault tree method is illustrated in Fig. 11 which shows the first few steps of a fault tree concerning loss of electric power to all engineered safety features (ESF's). In this case it is known that the electric power to ESF's would require both alternating current (AC) power and direct current (DC) power. The AC provides the energy needed but the DC is required by the control systems which turn on the AC. Thus, failure of each of these systems appears in the first level and they are coupled to the top event by an OR gate. This symbol signifies that either one failure or the other (or both) can cause the top event and that the probability of the top event is, to a close approximation, the sum of the probabilities of the two events in the first level. Thus, if $P_{AC} = 0.001$ and $P_{DC} = 0.001$, then $P_{EP} = 0.002$. The EP failure probability can be computed in this way if there are sufficient failure data to determine P_{AC} and P_{DC} directly. However, in general, such failures have not occurred often enough to provide meaningful statistical data and therefore, the analysis must proceed to lower failure levels. The next level is developed only for loss of AC power. In this case it is known that either offsite power (the electrical grid) or onsite power (the station diesel generators) can supply the needed energy. Failures of these systems are therefore coupled by an AND gate, indicating that both would have to fail in order to produce the failure above.

The above basic method is used to develop the trees until they have identified failures for which statistical data exist to determine their probability. In developing the tree, consideration is given to intrinsic component failures, human factors and test and maintenance. The probabilities of the failures are then assigned to the appropriate elements of the tree and the probability of the top event is calculated. For complex trees, such as those involved in this study, the aid of a computer program is utilized for computing the top event probability. In general the individual probabilities are obtained from a limited amount of experience data so they have an appreciable uncertainty associated with them. A computer code used in the study propagated these uncertainties using a standard statistical procedure and determined an uncertainty for the system failure probability.

Fault trees were developed for almost all the major individual systems represented in the event trees. These systems include the various ESF's and some of the normally operating plant systems. In some cases several different versions of a given system fault tree were required, depending upon the accident conditions prevailing at the time the system failure is postulated. For example, the probability of failure of the ECCS may be different depending upon whether the containment spray system operates or fails. Such differences have been accounted for in the study.

There are a number of limitations in applying fault trees to a risk assessment of nuclear power plant accidents. The most important drawback is probably that detailed fault trees for complex systems are very time consuming to develop. Furthermore, there are different ways in which the logic can be developed. Thus two different analysts are likely to produce different trees for the same system. Although both trees may be logically correct and produce the same system failure probability, the fact that they appear to be considerably different can be misleading.

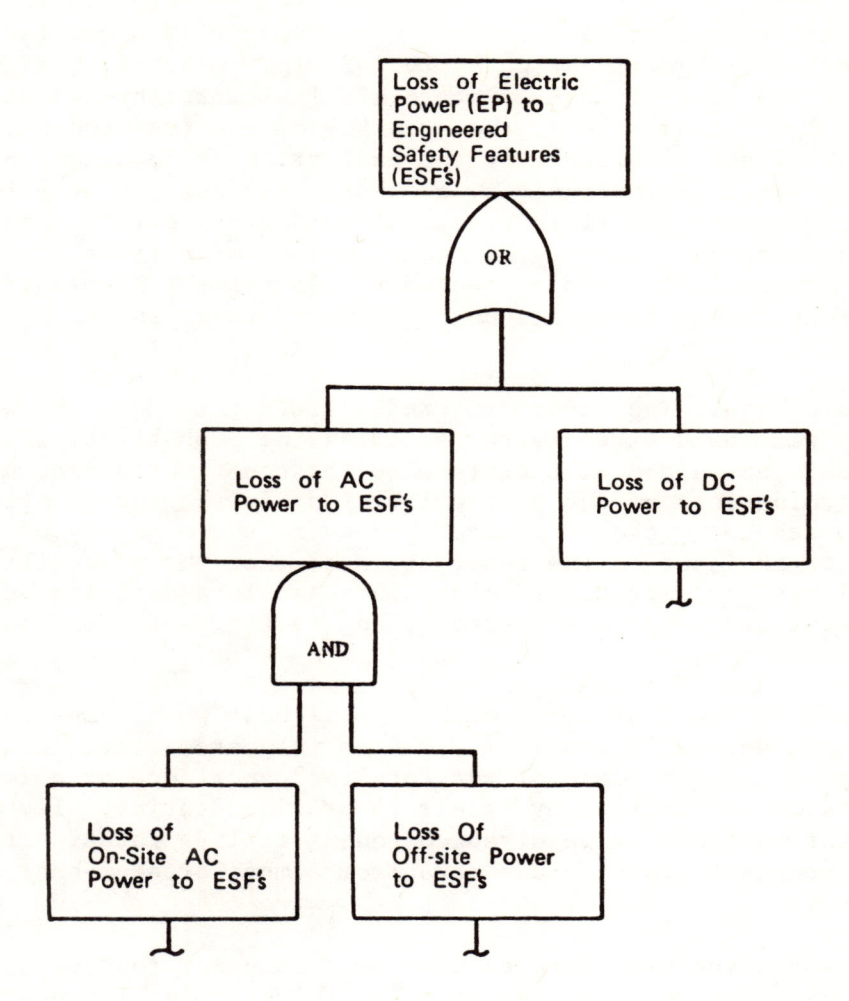

Figure 11. Illustration of Fault Tree Development

As with event trees, serious errors can be made if it is assumed that all failures are independent. A substantial amount of the effort in this study has been expended on the search for common mode failures. The fault trees and event trees have been extremely useful in helping define those common mode failures that can contribute to the overall risk.

As with event trees, there is no way of proving that a complex fault tree includes all the significant paths to failure. Generally, at some point in the analysis, the analyst must truncate his fault tree by assuming certain events are not significant. Thus, the accuracy of the tree depends appreciably upon the skill and experience of the analyst. Any modeling, of course, depends upon the skill of the analyst, however it is particularly important for fault trees where few explicit rules and guidelines exist. However, a good check on the logical adequacy and completeness of a fault tree is obtained when it is quantified and subjected to sensitivity studies. In general all the trees constructed in this study were found to go into more detail than was needed.

Failure Rate Data. The study utilized failure rate data in two principal ways. They were used directly to establish the probabilities of major events (failures) for which fault trees were not constructed. Such uses included the determinations of the probabilities of initiating events such as pipe breaks and reactor vessel ruptures. However, the majority of failure data was utilized as input to the fault trees so that the probabilities of the system failures could be determined. These failure data included estimates of component failures, human errors, and testing and maintenance contributions.

The accuracy of the fault tree method is highest when component failure rates are based on data obtained from failures in systems identical to the one under analysis. In the case of reactors the experience of a few hundred reactor-years is not sufficient by itself to provide statistically meaningful probabilities for most of the required component failure rates. It has therefore been necessary to also use data from a much broader base of industrial experience.

In this study extensive searches have been made for sources of failure rate data. Each source has been investigated to determine its appropriateness for application to nuclear plants. The conditions of service of most of the components in reactors are similar to conditions in many other applications, such as those in fossil-fueled plants and chemical processing plants. The compilations of such industrial experience are the basic source of most of the failure rate data used in the study.

Certain components of nuclear systems may be subjected to rather unique environments, particularly during serious accidents. Foremost among these environmental factors are radiation and high temperature steam. In the process of determining applicable failure rates, the study employed specialists in component reliability to assess the effect of such conditions on system components. Based on their assessments, component failure rates and their uncertainties were increased for the extreme environments.

The design specifications of the components of the ESF's require that they be qualified to operate under a variety of accident conditions. It is, of course, possible that certain components may fail to meet these special design conditions. To ascertain how likely such design errors may be, this study carefully reviewed the components in a selected number of important safety systems to determine how well the design specifications had, in fact, been satisfied. Based on these assessments, component failure rates were modified to account for the deficiencies found.

A common criticism of the fault tree method is that the system failure probabilities are not meaningful because of uncertainties in the knowledge of applicable failure rates. In general, the failure rates used in this study have uncertainties of a factor of 10 or 100 ($\pm$ 3 or $\pm$ 10). In a few cases the uncertainty is a factor of 1000 ($\pm$ 30). Based on these uncertainties, a Monte Carlo technique was used to calculate the uncertainty of the overall system failure probability. The study has used a log-normal distribution for all the uncertainties assigned to component failure rates. The log-normal distributions were combined in a statistical manner to account for the error contributions from different component failure rates. It was found that even with the larger component failure rate uncertainties that were used, the system failure probabilities were sufficiently accurate to obtain meaningful values for risk evaluation.

<u>Common Mode Failures</u>. Common mode failures are multiple failures that result from a single event or failure. Thus, the probabilities associated with the multiple failures become, in reality, dependent probabilities. The single event can be any one of a number of possibilities; a common property, process, environment, or external event. The resulting multiple failures can likewise encompass a spectrum of possibilities including, for example, system failures caused by a common external event, multiple component failures caused by a common defective manufacturing process, and a sequence of failures caused by a common human operator.

Because common mode failures entail a wide spectrum of possibilities and enter into all areas of modeling and analysis, common mode failures cannot be isolated as a separate study, but instead must be considered throughout all the modeling and quantification steps involved in the risk assessment. In the study, common mode and dependency considerations were incorporated in the following stages of analyses:

    Event Tree Construction   Fault Tree Quantification
    Fault Tree Construction   Event Tree Quantification
            Special Engineering Investigations

In the event tree development, common mode failures were first treated in the detailed modeling of system interactions. If failure of one system caused other systems to fail or be ineffective these dependencies were explicitly modeled in the event trees. The systems rendered failed or ineffective by the single system failure were treated in the subsequent analysis as having failed with probability of one and the analysis concerned itself only with the critical single system failure. The changes produced in event trees by these relationships often produced significant increases in predicted accident sequence probabilities since a product of system failure probabilities was replaced by a single system failure probability. The development of the containment event tree that relates various modes of containment failure to system operability states and the physical processes associated with core melt also accounted for dependencies which were due to the initiating event.

The event trees also defined the conditions for which the individual fault trees were constructed. Particular system failures, i.e., the top events of the fault trees, were defined as occurring under specific accident conditions that frequently included the prior failure of other systems. The fault trees were thus coupled to other systems in the accident sequence, as well as to the particular, common, accident environments that existed. The fault trees were drawn to a level such that all relevant common hardware in the systems was identified. This depth of analysis permitted identification of single failures that cause multiple effects or dependencies. These included single failures that cause several systems to fail or be degraded, or cause redundancies to be negated. The failure causes modeled in the fault tree analysis include not only hardware failures, but also include failures caused by human intervention, test and maintenance actions, and environmental effects. Thus, a spectrum of potential dependencies is incorporated in the fault trees. In many cases these additional causes, usually due to human or test and maintenance interfaces, have higher probability contributions to system failure than the hardware causes. In a number of cases these non-hardware actions result in failure probabilities (essentially single failure probabilities) that are high enough to dominate the system failure probability.

The fault tree quantification stage, in which system probabilities were numerically computed, incorporated dependency and common mode considerations throughout the calculations. The failure rate for a particular component included not only contributions from pure hardware failure (sometimes called the random failure rate), but also included applicable contributions from test or maintenance errors, human causes, and environmental causes. Human errors were investigated to identify causes of dependent failures if, for instance, the same operator could perform all the acts. Testing or maintenance activities were examined for causes of dependent failures, for example, when several components would be scheduled for simultaneous testing or maintenance. Components were examined for potential dependent failures that may arise as a result of the environments created by accidents. The quantification formulas treated both hardware and non-hardware contributions with their relevant dependencies. The quantification process included determinations of the maximum possible impacts from common mode failures which might exist but were not previously included in the analyses. These determinations indicated whether additional common mode failures could have significant impact on the computed accident probabilities. The applicable systems and/or components were reexamined to identify the ways, if any, in which such significant common mode failures could occur.

After the fault trees were quantified, the event tree quantification stage combined the individual fault tree probabilities to obtain accident sequence probabilities. Since a sequence in the event trees can be viewed in terms of fault tree logic, the same quantification techniques were used on the individual fault trees. Since multiple systems were analyzed, the couplings now included dependencies across systems.

As a final check on possible dependencies and common mode effects, special engineering investigations were performed to complement the modeling and mathematical techniques which had been used throughout the study. Those event tree accident sequences which dominate the probability of occurrence of the categories were carefully reexamined for any specific dependencies which may have been overlooked.

The probability versus consequence calculations involve several inputs which have no significant common mode failure contributions. The accident probabilities, the weather, and the population distributions used in the calculation of consequences are essentially independent of one another.

Magnitude of Releases

This section discusses the manner in which the magnitude of the radioactivity release from the plant to the environment was determined. The release magnitude is influenced by three major factors; the amount and isotopic composition of radioactivity released from the core, the amount of radioactivity removed within the containment, and the containment failure mode. All of these are time dependent factors which influence the radioactive release magnitude. Thus, the accident sequences defined by the event trees were of particular value in establishing the release magnitudes applicable to each of the release categories noted previously.

As already noted, only those potential reactor accidents that lead to core melting affect the risk significantly. Thus, for the most part, the determination of the release of radioactivity from the reactor core involved the estimation of the fractions of significant radioactive isotopes that are released from cores melting under various conditions. The various conditions and timing of core melting were defined by appropriate analysis of the applicable accident sequences in event trees. A variety of experiments reported in the literature provides information on the radioactivity released from fuel under various conditions. Such information was used in the determination of the applicable release fractions. This work resulted from the deliberations of a group of specialists, who have been conducting work in this area at National Laboratories. In general, the experiments on which these results are based were carried out with relatively small samples of fuel. It is believed that, because of the large surface to volume ratio in such experiments compared to that which would exist in a molten core, the release fractions used in the study tend to overpredict the fraction released from a core. However, since no large scale experiments have been conducted, there is no experimental verification that reductions in the release fractions will in fact be observed. For this reason such potential effects were not taken into account in establishing the release fractions applicable to the various release categories.

Radioactivity released from the core is subjected to a variety of physical processes that reduce the amount of radioactivity available for release to the environment. These processes include wash out by the fission product removal systems, natural plate out and deposition processes on surfaces within the containment, radioactive decay, and the effects of filters. These processes, coupled with the fuel release and the mode and timing of containment failure, are the major determinants of the magnitude of radioactive release to the environment. To account for all these effects a computer code called CORRAL was developed. The various parameters used in the code are based on recent investigations, conducted at National Laboratories, of the transport of radioactive materials within containments. The CORRAL code output is the quantity of each of 54 biologically significant isotopes released to the environment as a result of a given accident sequence.

Many of the accident sequences involve similarity in core melting, similarity in radioactivity removal processes, and the same containment failure mode. This permitted classification of accident sequences into a number of different types called release categories. Thus, the releases produced by core melt are characterized by several different categories, each involving a particular composition, timing and release point.

The work outlined above provided the information for composite histograms of the type shown in Fig. 9, that represent the probability and magnitude of the radioactive releases associated with each consequence category.

CONSEQUENCES OF RADIOACTIVE RELEASES

The objective of this task was the prediction of the public consequences that result from the radioactive releases defined by Task I. The consequence predictions serve as the primary input to Task III, the overall risk assessment. The consequences of a given radioactive release depend upon how the radioactivity is dispersed in the environment, upon the number of people and amount of property exposed, and upon the effects of radiation exposure on people and contamination of property. These major elements of the consequence predictions are indicated in Fig. 12, which shows the principal subtasks involved in this task (Task II).

The dispersion of the radioactivity is determined principally by the weather conditions at the time of release and the release conditions, i.e., ground level, elevated, hot, cold, extended or puff. The population distribution as a function of distance is known for each of 68 reactor sites either in use or planned. The probable effects of radiation on people and property are based on available information on such effects.

The probability associated with a specific consequence is determined by combining the probabilities of the individual input parameters, i.e., by multiplying $P_{release} \times P_{weather} \times P_{population}$. In determining the consequence probability in this way, it is necessary to assure the probabilities are reasonably independent. It is difficult to visualize that the accident and the population densities can significantly affect one another. It seems equally reasonable to assume that the population and the weather have no strong dependency, since the frequency distributions have been obtained by combining actual meteorological and demographic data applicable to a large number of sites. It could also be argued that violent weather might cause an accident. Although this is highly unlikely because of reactor design requirements, it is not impossible. However, violent weather is characterized by extremely high turbulence which would cause very rapid dilution of the radioactivity and this would drastically reduce the consequences. The reduced consequences would likely counteract any probability increase associated with such a dependency, resulting in a negligible effect on the overall risk. Thus, it is

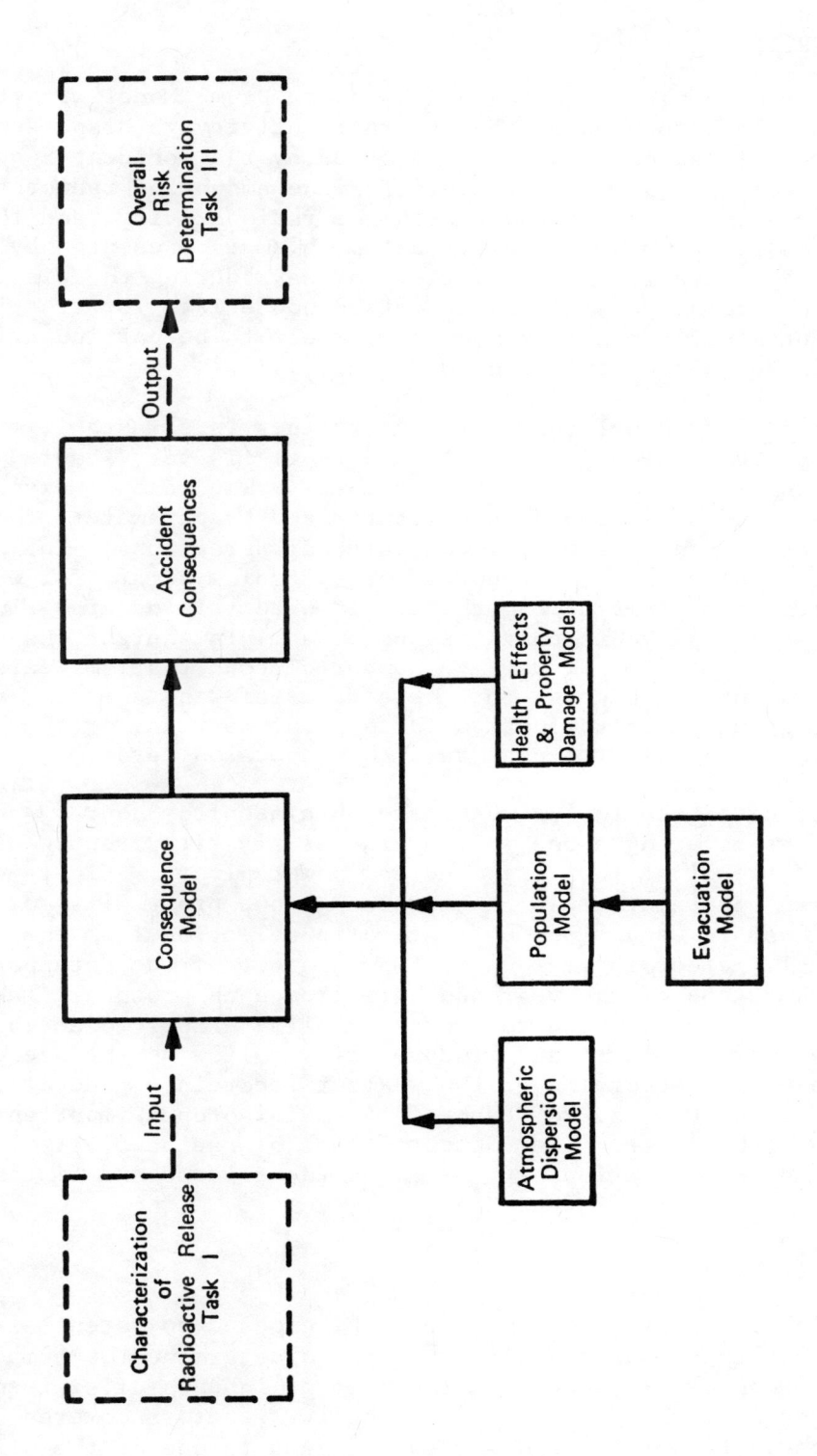

Figure 12. Subtasks in the Determination of the Consequences of Radioactive Releases Task II

reasonable to assume that the three probabilities are independent and a straightforward multiplication is justified.

Atmospheric Dispersion Model

The computer code uses the standard Gaussian Plume Model with the modifications noted below, to predict the way radioactivity is dispersed in the atmosphere. The release conditions are defined by the accident sequences. Each one of the release categories identifies the amount of radioactivity released, the amount of heat released with the radioactivity, and the elevation of the release. A much more difficult problem is presented by the weather which can change in a large variety of ways during the dispersion of radioactivity. No completely realistic method now exists for treating dispersion of pollutants and existing models are felt to be particularly uncertain at long distances from the point of release.

The Gaussian plume model characterizes weather in six stability classes, A through F. Weather type A is unstable and type F is very stable. Wind speed is also required as an input to this model. The rate of dispersion for various types and wind speeds is characterized by parameters that give the spreading rate in the horizontal and vertical directions. In the study's model, the effects of rain are accounted for by adding a rate for washout of radioactivity from the plume during the period when rain occurs. In all cases the radioactive plume is contained below a mixing height characteristic of the site, season and time of day. A number of conservatisms exist in the model in that it does not account for the effects of wind shear, changes in wind direction, ground decontamination factors due to rain, or the potential for strongly heated releases to penetrate the inversion layer.

The weather data used in the model are obtained from hour by hour meteorological records covering a one year period at six sites that would typify the locations of the first 100 large nuclear power plants. Ninety weather samples are taken and each sample is thus assigned a probability of 1/90. The starting times are determined by systematic selection from the various sets of applicable meteorological data. One quarter of the data points are chosen from each season of the year and half from each group are taken in the daytime. This procedure is used to reduce sampling errors to acceptable levels. The weather stability and wind velocity following the accident are then assumed to change according to the weather recordings made at the site. The weather model calculates, for each of 54 radioisotopes important to the prediction of health effects, the concentrations of radioactivity in the air and on the ground as a function of time after the release and distance from the reactor.

Population Model

To determine the population that could be exposed to potential releases of radioactivity, census bureau data are used to determine the number of people as a function of distance from the reactor in each of sixteen 22 1/2° sectors around each of the sites at which the 100 reactors covered by the study are located. Each reactor has been assigned to one of the six typical meteorological data sets and a 16 sector composite population has been developed for each set. Three of these sectors are those which have the largest cumulative population (within 50 miles of the reactor) of all the sectors associated with reactors assigned to that set. The probability of exposing people in these sectors is

$$P_{exposure_{1,2,3}} = \frac{1}{16 \times N}$$

where N is the number of reactors assigned to the set. The other 13 sectors of a typical set are obtained in the same way, except that groups of sectors with approximately the same population density are combined to obtain the population as a function of distance. These sectors were given a probability of exposure of:

$$P_{exposure_{4-16}} = \frac{n}{16 \times N}$$

where n is the number of sectors combined. In this combination process, the n radial sectors are averaged at each mesh point distance to give the value used.

In the case of a potentially serious accidental release, it is assumed that people living within about 25 miles of the plant, and located in the direction of the wind, would be evacuated in order to reduce their exposure to radioactivity. An evacuation model to represent this process has been developed. This model is based on the study's analysis of data collected on a substantial number of actual evacuations that have taken place in the United States.

Health Effects and Property Damage Model

The consequence model calculates the doses from five potential exposure modes; the external dose from the passing cloud, the dose from internally deposited radionuclides which are inhaled from the passing cloud, the external dose from the radioactive material which is deposited on the ground, and the doses from internally deposited radionuclides which are either inhaled after resuspension or ingested from ground contamination.

The potential health effects calculated are early fatalities (i.e., fatalities that occur within one year of the accident), early illnesses (i.e., people needing medical treatment), and late health effects that are estimated from the total man-rem dose to the population. Late health effects may include lethalities from cancers, thyroid illnesses, and genetic effects that can potentially occur at long times after the time of the accident. Radiation exposure information does not provide clear distinctions between probable deaths, injuries and long term effects. The probability of early fatalities and illnesses is computed by using a dose effect relationship. For whole body exposures, the probability of early fatality varies from 0.01 to 99.99% for doses of 320 and 750 rads, respectively, with a median value of 510 rads. The principal early illness involves respiratory impairment whose probability of occurrence varies from 5 to 100% for doses to the lung in the range of 3000 to 6000 rads, respectively. The incidence of latent cancer fatalities and genetic effects is based on the BEIR report with some modification of the former to account for dose rate and dose magnitude dependencies. In addition to whole body effects and doses to the lung, thyroid gland and GI tract doses are also calculated. The effect of thyroid doses is to increase the occurrence rate of thyroid nodules, a portion of which is expected to become cancerous. Since thyroid nodules can be treated very effectively, it is expected that few, if any, deaths will result from thyroid irradiation.

The consequence model also provides for prediction of property damage due to radioactive contamination. It also includes costs associated with relocating people for the time needed to decontaminate their property. Property damage costs are calculated on a per capita basis relative to the total value of property and land in the United States including appropriate values for the loss of agricultural crops.

OVERALL RISK ASSESSMENT

The analysis of accident consequences described in the preceding section yielded the probability/magnitude relationships for each of the specific consequences--early fatalities, early illness, thyroid illness, latent cancer fatalities, genetic effects, property damage, and land contamination. Together, these seven distributions represent the overall public risk from potential nuclear power plant accidents involving nuclear power plants of current design. For reasons discussed earlier, no attempt has been made to consolidate the various consequence types into a single probability/magnitude distribution in which the various types of consequence are represented by a single common unit.

RESULTS

This section, taken from the Executive Summary of WASH-1400, presents a condensed summary of the results of the study. The reader is referred to WASH-1400 for a more detailed breakdown.

The results from this study suggest that the risks to the public from potential accidents in nuclear power plants are comparatively small. This is based on the following considerations:

a. The possible consequences of potential reactor accidents are predicted to be no larger, and in many cases much smaller, than those of non-nuclear accidents. The consequences are predicted to be smaller than people have been led to believe by previous studies which deliberately maximized estimates of these consequences.

b. The likelihood of reactor accidents is much smaller than that of many non-nuclear accidents having similar consequences. All non-nuclear accidents examined in this study, including fires, explosions, toxic chemical releases, dam failures, airplane crashes, earthquakes, hurricanes and tornadoes, are much more likely to occur and can have consequences comparable to, or larger than, those of nuclear accidents.

Figures 13, 14, and 15 compare the nuclear reactor accident risks predicted for the 100 plants expected to be operating by about 1980 with risks from other man-caused and natural events to which society is generally already exposed. The following information is contained in the figures:

a. Figures 13 and 14 show the likelihood and number of fatalities from both nuclear and a variety of non-nuclear accidents. These figures indicate that non-nuclear events are about 10,000 times more likely to produce large numbers of fatalities than nuclear plants. The fatalities shown in Figs. 13 and 14 for the 100 nuclear plants are those that would be predicted to occur within a short period of time after the potential reactor accident. This was done to provide a consistent comparison to the non-nuclear events which also cause fatalities in the same time frame. As in potential nuclear accidents, there also exist possibilities for injuries and longer term health effects from non-nuclear accidents. Data or predictions of this type are not available for non-nuclear events and so comparisons cannot be made easily.

b. Figure 15 shows the likelihood and dollar value of property damage associated with nuclear and non-nuclear accidents. Nuclear plants are about 1000 times less likely to cause comparable large dollar value accidents than other sources. Property damage is associated with three effects:

  1. the cost of relocating people away from contaminated areas,
  2. the decontamination of land to avoid overexposing people to radioactivity,
  3. the cost of ensuring that people are not exposed to potential sources of radioactivity in food and water supplies.

In addition to the overall risk information in Figs. 13-15, it is useful to consider the risk to individuals of being fatally injured by various types of accidents. The bulk of the information shown in Table IV is taken from the 1973 Statistical Abstracts of the U.S. and applies to the year 1969, the latest year for which these data were tabulated when this study was performed. The predicted nuclear accident risks are very small compared to other possible causes of fatal injuries.

In addition to fatalities and property damage, a number of other health effects could be caused by nuclear accidents. These include injuries and long-term health effects such as cancers, genetic effects, and thyroid gland illness. The early illness expected in potential accidents would be about 10 times as large as the fatalities shown in Figs. 13 and 14; for comparison there are 8 million injuries caused annually by other accidents. The number of cases of genetic effects and long-term cancer fatalities is predicted to be smaller than the normal incidence rate of these diseases. Even for a large accident, the small increases in these diseases would be difficult to detect from the normal incidence rate.

Thyroid illnesses that might result from a large accident are mainly the formation of nodules on the thyroid gland; these can be treated by medical procedures and rarely lead to serious consequences. For most accidents, the number of nodules caused would be small compared to their normal incidence rate. The number that might be produced in very unlikely accidents would be about equal to their normal occurrence in the exposed population. These would be observed during a period of 10 to 40 years following the accident.

While the study has presented the estimated risks from nuclear power plant accidents and compared them with other risks that exist in our society, it has made no judgment on the acceptability of nuclear risks. The judgment as to what level of risk is acceptable should be made by a broader segment of society than that involved in this study.

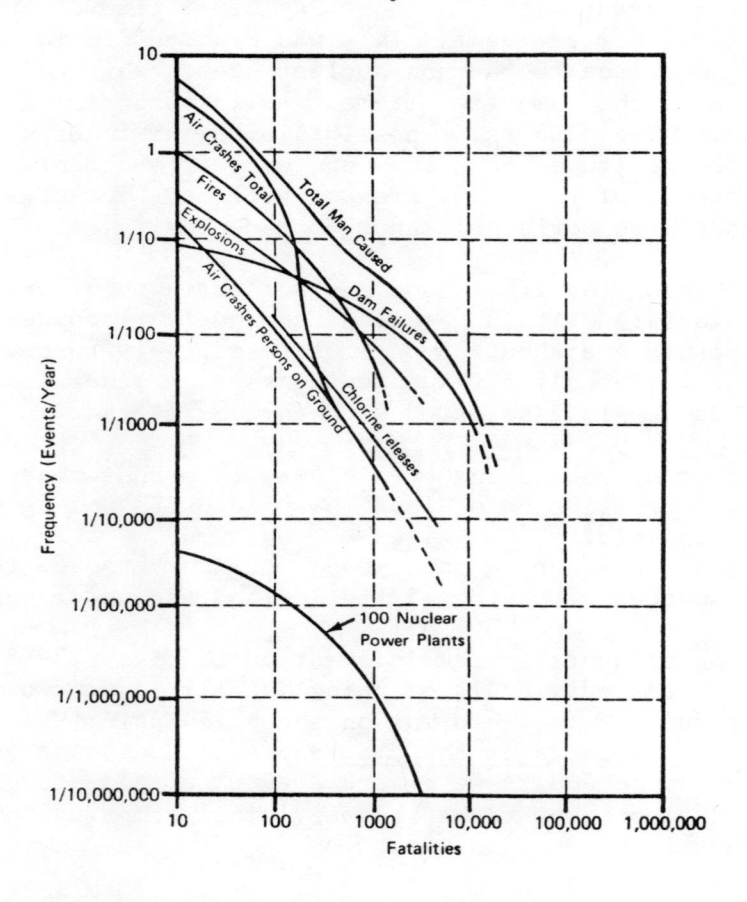

Figure 13. Frequency of Fatalities due to Man-Caused Events

Notes: 1. Fatalities due to auto accidents are not shown because data are not available. Auto accidents cause about 50,000 fatalities per year.

2. Approximate uncertainties for nuclear events are estimated to be represented by factors of 1/4 and 4 on consequence magnitudes and by factors of 1/5 and 5 on probabilities.

3. For natural and man caused occurrences the uncertainty in probability of largest recorded consequence magnitude is estimated to be represented by factors of 1/20 and 5. Smaller magnitudes have less uncertainty.

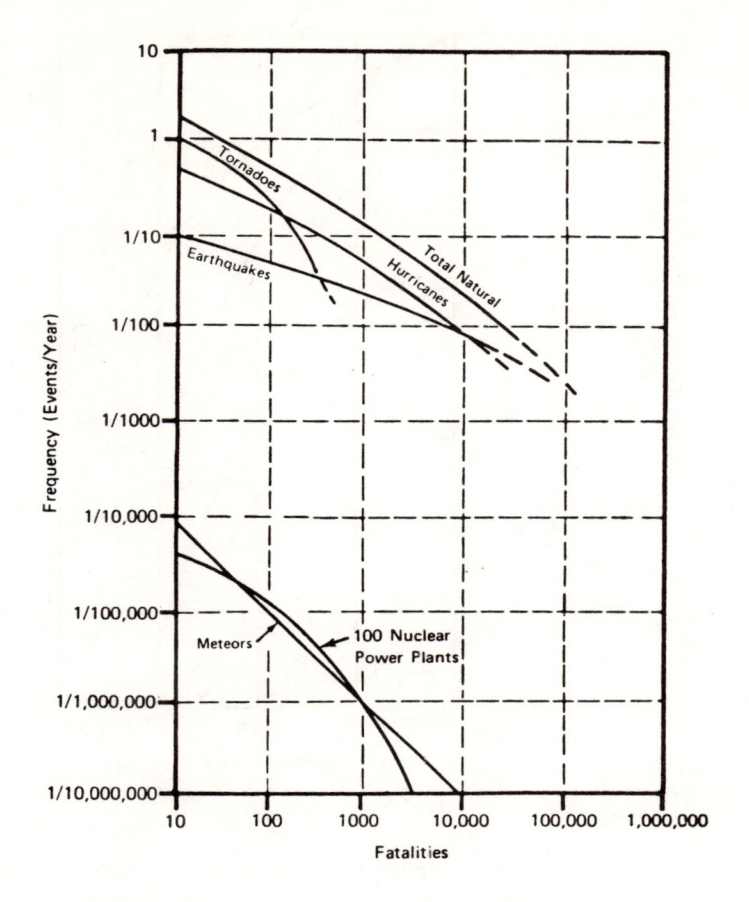

Figure 14. Frequency of Fatalities due to Natural Events

Notes: 1. For natural and man-caused occurrences the uncertainty in probability of largest recorded consequence magnitude is estimated to be represented by factors of 1/20 and 5. Smaller magnitudes have less uncertainty.

2. Approximate uncertainties for nuclear events are estimated to be represented by factors of 1/4 and 4 on consequence magnitudes and by factors of 1/5 and 5 on probabilities.

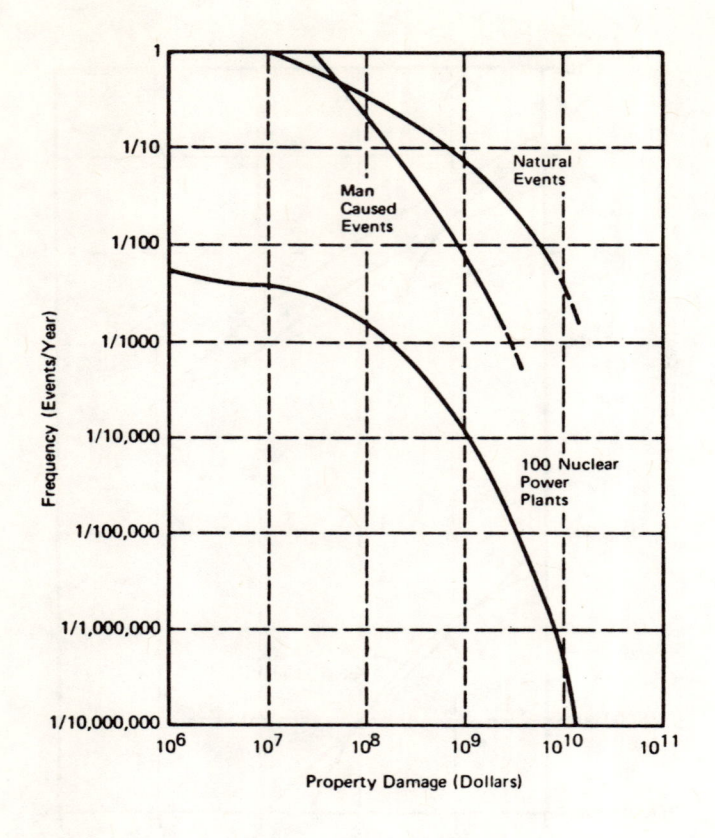

Figure 15. Frequency of Property Damage due to Natural and Man-Caused Events

Notes: 1. Property damage due to auto accidents is not included because data are not available for low probability events. Auto accidents cause about $15 billion damage each year.

2. Approximate uncertainties for nuclear events are estimated to be represented by factors of 1/5 and 2 on consequence magnitudes and by factors of 1/5 and 5 on probabilities.

3. For natural and man-caused occurrences the uncertainty in probability of largest recorded consequence magnitude is estimated to be represented by factors of 1/20 and 5. Smaller magnitudes have less uncertainty.

Table IV. Average Risk of Fatality by Various Causes

Accident Type	Total Number	Individual Chance per Year
Motor Vehicle	55,791	1 in 4,000
Falls	17,827	1 in 10,000
Fires and Hot Substances	7,451	1 in 25,000
Drowning	6,181	1 in 30,000
Firearms	2,309	1 in 100,000
Air Travel	1,778	1 in 100,000
Falling Objects	1,271	1 in 160,000
Electrocution	1,148	1 in 160,000
Lightning	160	1 in 2,000,000
Tornadoes	91	1 in 2,500,000
Hurricanes	93	1 in 2,500,000
All Accidents	111,992	1 in 1,600
Nuclear Reactor Accidents (100 Plants)	-	1 in 5,000,000,000

CURRENT PROBLEMS IN REACTOR SAFETY

John T. Telford
Office of Nuclear Reactor Regulation
Nuclear Regulatory Commission
Washington, D.C. 20555

INTRODUCTION

The principal aim of the Nuclear Regulatory Commission in regulating the nuclear power industry is to provide for public safety. Nuclear power plants are allowed to be built and operated only after careful and detailed reviews at the design and construction stages have been conducted that show the plant and methods of operation meet prescribed safety standards. The Commission's standards and regulations ensure that the plant design is conservative, that there are ample safety margins, and that redundant components and systems are provided.

In spite of the measures taken to ensure safety and the application of a defense-in-depth philosophy, the operation of nuclear power plants is not completely risk-free. Reactor safety problems do arise and it is the intent of this paper to discuss some of the issues that are currently in review by the nuclear industry and the NRC.

NRC REVIEW PHILOSOPHY

The aforementioned safety objective for nuclear plants is being achieved through NRC requirements for a defense-in-depth safety concept. This concept includes conservative design practices, a comprehensive quality assurance program, the provision of multiple safety systems, and physical barriers between the radioactivity contained in the reactor fuel and the plant environs. This safety concept acknowledges the fact that no single step can be made error-free and relies instead upon multiple lines of defense to provide the necessary level of safety. Thus, this concept assumes that all defects will not be eliminated and that men will err and material will fail, despite our best efforts to the contrary. Further assurance that this safety objective has been achieved is provided by thorough and comprehensive safety reviews and inspections by the NRC staff during plant design, construction, testing and operation.

The technical adequacy of the NRC staff review is further assured by the independent review of the ACRS. The validity of the total process is subjected to a public hearing where members of the public may challenge the technical and judgmental basis for the staff's conclusion. This hearing process allows a high degree of public participation in the safety review process before an independent panel. In the event of unresolvable differences between the parties, the differences may be appealed to the Atomic Safety and Licensing Appeal Board and if necessary the Commissioners, and in turn, their decisions may be appealed to the Federal judicial system.

In addition, operating experience provides a practical test of whether our safety objective is being achieved. Although the operating experience at U.S. commercial reactors to date includes about 250 reactor-years, it does not provide a sufficient statistical basis for a quantitative determination of the actual probability of occurrence of low probability accidents; however, the fact that no member of the public has been injured by radiation as a result of an accident during operation of a licensed nuclear power reactor provides demonstrable evidence of successful results to date. The lessons of operating experience and a continuing research and development and analysis program as applied to nuclear power plants, help assure us of an even greater degree of safety.

SAFETY ISSUES

It is apparent that technical issues will be identified as a result of operating experiences and as a result of continuing NRC staff and ACRS reviews of plants prior to operation. These issues often relate to a particular type or size of nuclear facility and to both operating nuclear facilities and facilities still under design or construction.

Because of the broad applicability of certain technical issues and because of the varying degrees of significance, the resolution of the issues for various facilities is quite different. This variation in resolution of the issue sometimes results in different plant modifications, different administrative procedures, or different schedules for implementation of the solution to a generic technical issue. After first assuring that a sufficient level of safety is present and compliance with the Commission's regulations is provided, the appropriate resolution for a particular nuclear facility is determined by balancing the risk and the benefit of various courses of action to achieve our overall safety objective. In general, these technical issues result from the fact that uncertainties associated with highly unlikely events cannot be completely eliminated. The large safety margins incorporated in each plant provide allowance for such uncertainties. The likelihood of significant consequences associated with such occurrences over the period it takes to resolve an issue must be acceptably small to permit construction or operation to continue while the issue is under review.

A better understanding of the treatment of generic technical issues can be gained by examination of examples of such issues. Recent operating experience has resulted in generic evaluations of the cable fire at Browns Ferry, BWR channel box wear, waterhammer in feedwater piping, and fuel pellet-clad interactions. The significance of each of these issues has resulted in a reevaluation of the potentially affected nuclear facilities. When safety margins are thought to be significantly reduced as in the case of channel box wear, immediate action, in this case derating, was taken to restore safety margins. Longer term modifications are being made to both increase safety margins and allow less restrictive operation.

Other technical issues, such as anticipated transients without scram

(ATWS), analysis of reactor vessel supports, and pool dynamic loads in BWR containments, have been identified in the regulatory review process. These issues are the subject of ongoing reviews to quantify existing safety margins and investigate the appropriateness of increased margins. Each of these issues must be appropriately treated in the evaluation for each specific nuclear reactor to determine whether the incremental increase in safety associated with a potential design solution warrants it being required. A discussion of the significances of specific reactor safety issues follows.

Reactor Vessel Supports

In May 1975, the NRC was informed that a reanalysis of reactor vessel support stresses by one licensee revealed that margins of safety for lateral loads on the reactor vessel supports may be less than originally thought. These same analyses indicated that the supports will retain essential structural integrity even with these assumed accident loadings. This means that the vessel remains fully supported while undergoing some translation and that the ultimate consequences of this postulated accident are no more severe than those accommodated by the original design. This condition could come about only as a result of the very unlikely event of a large LOCA that occurs and only at a few very specific locations in the piping. The likelihood of this occurring is very small. The NRC has underway an active review effort to reassess the adequacy of the design of certain reactor vessel support structures to withstand the asymmetric loads that could result from a highly unlikely combined seismic event and loss-of-coolant accident.

Pool Dynamic Loads

This technical issue pertains to the ongoing evaluations of the structural adequacy of BWR suppression containments to sustain pool dynamic loads following a LOCA, i.e., downward and upward loads due to water movement during a LOCA. The significance of these loads was first ascertained during the NRC review of the most recent General Electric pressure-suppression containment design, i.e., the Mark III. Since these types of loads were not specifically considered in the design of Mark I and Mark II containments, it was necessary to initiate re-evaluations to these designs.

Accordingly, all utilities with facilities having either Mark I or Mark II containments were advised by the NRC in April 1975 of this deficiency and were requested to provide information on the potential magnitude of these loads and their effect, in combination with other design loads, on the integrity of containment structures.

There has been a succession of responses from each licensee to this request. This was necessary because all the information needed to perform these evaluations was not immediately available and the time required to generate the information was considerable.

In May 1975, an owner's group was formed of all utilities owning plants with Mark I type containments. A similar group was formed for the Mark II owners. The purpose of the groups was to determine the magnitude and significance of these loads as quickly as possible and identify courses of action needed to resolve any outstanding concerns. The General Electric Company was contracted as the lead technical organization. With regard to the Mark I containments, the group established a short-term and a long-term program. The short-term program is intended to make an assessment of each Mark I plant using current information to determine the significant loads and structural capability of the containments. The long-term program to be completed in 1977 would include large scale tests and additional refined analyses to evaluate any outstanding concerns which were identified during the short-term program to confirm the data base for that program. The short-term program is presently continuing and reports were issued in September of 1975 with subsequent amendments to be submitted through April of 1976, including responses to staff's questions.

The approach taken by the licensees has been to provide evaluations of plant designs based on the best available load information and to continuously update these evaluations as new information was generated. The basis for the evaluations has been to assure that there is no loss of containment function for the LOCA event.

The results of calculations performed by the General Electric Company and the Bechtel Corporation show that the structural margins of the Mark I plants would be substantially reduced in certain structural members of the torus support structure in the unlikely event of a LOCA. In this regard the margins would be restored to acceptable levels by operating the current Mark I plants with a differential pressure between the drywell and wetwell of about 1 psi. On February 27, 1976 all Mark I owners were directed to operate their plants in this manner.

In defining the LOCA event, consideration has been given to a realistic assessment of the magnitude and combination of pool dynamic and other design basis loads. For example, the postulated actuation of a relief valve coincident with the LOCA has not been assumed because one does not necessarily result from the other. Also, as another example, the torus upward loads specified in the short-term program represent, in the opinion of the staff, "best estimates" due to the manner in which the loads were derived from test data as discussed in the staff Safety Evaluation Report on Vermont Yankee, February 13, 1976. In keeping with the staff approach, the maintenance of containment function is not predicated on stress levels being within code allowable limits, but rather the margins are evaluated; i.e., the additional capability of structures to sustain loads even with limited yielding which is short of failure.

It is clear that these assumptions regarding the course of the event and the response of the structure are not consistent with the design of current plants that are not in late stages of construction. The question which arises deals with the adequacy of this alternate basis in assessing the safety of operating plants. The answer lies in the context in which these criteria are being applied.

The purpose of the short-term program was to provide a quick estimate of the potential loads and an assessment of the acceptability of the continued operation of each plant during the course of the long-term program. Its purpose was not to make final evaluations or establish acceptance criteria to be used in the ultimate resolution of the problem. Therefore, the staff has recognized other factors in applying the short-term program criteria to operating plants and in determining that there is reasonable assurance that there is no undue risk to public health and safety. These include the low probability of a sudden and large loss-of-coolant accident occurring in the interim period and the amount of margin in the plant structural capability which is necessary to provide reasonable assurance. The first factor is significant in that the likelihood of excessive dynamic loads being imposed on the containment structure during a LOCA is directly related to the failure probability of high energy piping within the containment. Although not all containment load combinations have been evaluated, and in some local areas the calculated code stress levels were exceeded in the short-term program, loss of containment function is not predicted based on these calculations. Further, the probability of the event occurring during the period of evaluation is small.

The second factor is directly related to the amount of conservatism which should be present in the design. The staff believes that the conclusion for Mark I plants that there would be no loss of containment function is adequate for the short-term. The results of the ongoing review program are expected to confirm the adequacy of these containments. However, some plants may have to make some modifications to restore original margins. In addition, all operating Mark I plants will use the differential pressurization method currently in use at Vermont Yankee to provide further assurance of increased margin.

Plants currently under review for construction permits are being evaluated on a more conservative basis, to provide additional margin. Similarly, there will be provision in the long-term Mark I program to incorporate features that would restore margins in Mark I plants.

ATWS

This issue is primarily related to the common mode failure of reactor shutdown systems during anticipated plant transients. The significance of this concern goes back many years to discussions that the industry, the ACRS, and the staff had concerning interactions between the normal control system and the protection system circuitry. Since it was decided that adequate separation and isolation were being achieved, emphasis shifted to equipment failures. Various studies showed that wear out or random failures would not cause a significant reduction in reliability because of the redundancy provided in the plant. It was not clear that systems were adequately protected against common mode failures.

The consequences of an ATWS event were provided by the nuclear power plant vendors. The NRC review of the vendors' analyses that the great majority of postulated ATWS events did not lead to serious consequences, but that design changes to improve protection against ATWS would be appropriate in anticipation of the large number of plants expected in the future. Anticipated transients are those events that might occur one or more times during the life of the plant. Although the designs include consideration of these items, that is not to say that they are likely to occur. The Reactor Safety Study, WASH-1400, has estimated that the reactor protection system has a failure probability of about 10^{-5} or 10^{-6} per demand. Therefore, an anticipated transient without scram event is a very unlikely occurrence.

Staff evaluations, conclusions, and positions on ATWS were published in a September 1973 Technical Report on Anticipated Transients Without Scram (ATWS), WASH-1270. The staff's position expressed in WASH-1270, is that recent vintage plants would be required to incorporate any design changes necessary to assure that the consequences of a postulated ATWS event would be acceptable, and that those portions of the reactor shutdown system that might be particularly vulnerable to common mode failures would be identified and corrected. Certain older operating reactors would be required to analyze the consequences of an ATWS event. Based on these analyses, decisions as to whether or not any plant changes are required by these older plants would be made on an individual plant basis.

In WASH-1270, the staff concluded that in view of the very low likelihood of an ATWS event for the limited number of plants now in operation, limitations on operation due to this issue were neither necessary nor appropriate. In response to the staff position in WASH-1270, all of the reactor manufacturers have submitted reports which analyze the consequences of ATWS events and which review the reactor shutdown system design. These reports are under review by the NRC staff to determine what, if any, modifications should be required.

Fire in Electrical Cable Trays

On March 22, 1975, an event occurred in the Tennessee Valley Authority Browns Ferry Nuclear Plant, Decatur, Alabama, which shutdown two of the three nuclear units at the station for an extended period. Each unit is powered by a boiling water reactor (BWR), and has a net electrical capacity of over 1,000 megawatts.

A fire at the plant caused both Units 1 and 2, which were in full power operation, to be shut down shortly after the fire started. Unit 3 was under construction and was not affected by this event. The fire originated in an electrical cable penetration between the cable spreading room located beneath the common control room for Units 1 and 2, and the reactor building. The fire burned for approximately seven hours, and spread horizontally and vertically from its point of origin to all ten cable trays within the penetration, into the cable spreading room for several feet, and along the cables via the cable insulation through the penetration for about 40 feet into the reactor building. About 2,000 cables were damaged.

Fire damage was primarily to electrical power and control cables, and was localized in an area roughly 40 feet by 20 feet in an equipment room within the Unit 1 secondary containment building.

Both units were shut down safely. However, because of the fire, all normally used shutdown cooling systems and other components which comprise the emergency core cooling system (ECCS) for Unit 1 were inoperable for several hours. In this unusual circumstance, TVA used other installed equipment and maintained sufficient cooling capability to protect the nuclear fuel from overheating. There were no significant problems associated with the shutdown cooling of the Unit 2 reactor.

Even though the normal and emergency core cooling systems were initially unavailable in Unit 1, at least <u>five methods</u> were available to provide core cooling within the required time frame. During the event, time was always available to take the required actions to maintain the plant in a safe shutdown condition.

There was no adverse radiological impact on the public, plant personnel, or the environment as a result of the fire. Sampling indicated that airborne release rates were less than 10 percent of the technical specification limit. Minor injuries, including ten cases of smoke inhalation and a fractured wrist, were sustained by personnel engaged in the fire fighting activities.

A basic cause of the fire was failure to recognize the significance of the flammability of the materials involved. The immediate cause of the fire was the ignition of polyurethane used for cable penetration sealing material. Construction workers checking for air leaks in a penetration connecting the cable spreading room with the reactor building used a candle flame to detect air flow. The candle flame ignited the polyurethane.

The NRC sent a team of inspectors to the site immediately upon notification of the occurrence to review the safety of the units under the post-incident conditions and assessed license changes necessary to maintain the units in a safe shutdown condition during repair activities and safely return the units to operation.

The NRC is evaluating the licensee's proposed modifications to equipment and systems, administrative procedures and controls. A special review group was established within the NRC to study corrective measures to prevent or mitigate the consequences of similar events as a generic issue.

Waterhammer in Feedwater Piping

In pressurized water reactors, an essential part of the secondary water system is the feedwater system. This system returns the condensed steam from the turbine condenser to the steam generators and maintains the water inventory in the secondary system. Each pressurized water reactor has at least two steam generators.

Loss of the normal feedwater system as a result of either a pipe or a valve failure could affect the ability of the plant to cool down after a reactor shutdown. Auxiliary feedwater systems are provided as a backup to the normal feedwater system to provide redundancy of feedwater supply.

The feedwater system piping in some pressurized water reactor plants has experienced a phenomenon described as feedwater flow instability (characterized as "water-hammer"). The significance of the water hammer varies from plant to plant. Large sized feedwater system piping and associated components have been damaged by water hammers in certain events, e.g., excessive pipe movement and valve operator damage. Most water hammers have occurred subsequent to restarting feed flow after an operational transient such as a loss of steam generator water level.

Because significant forces are being generated during some events, there is a concern for the unlikely event that both the normal and the auxiliary feedwater systems might be lost to several steam generators and plant cooldown, a safety related function, affected. The potential consequences are considered serious enough to warrant study and the development of design and operational modifications to reduce the incidence and the severity of the phenomenon.

The NRC has tentatively concluded that the redundancy of steam generators and of feedwater systems, the high quality of components used, and conservative designs provide satisfactory safety margins. The complete termination of feedwater flow to several steam generators has not occurred and no radioactive material releases have been associated with these events.

TRANSPORTATION OF IRRADIATED FUEL

Jack Rollins
Nuclear Assurance Corporation
Atlanta, Georgia 30329

1.0 INTRODUCTION

This chapter discusses the safety aspects of the transportation portion of the nuclear fuel cycle; and, in particular, the transportation of irradiated fuel which has been discharged from nuclear reactors. The subject has been limited to irradiated fuel transportation, first of all, because a discussion of other types of nuclear transportation, i.e. including UF_6 and other recovered products, fresh fuel, low-, medium-, and high-level wastes would add a significant volume to the text, and second, irradiated fuel transportation alone provides an adequate illustration of the major nuclear safety considerations in this area of the fuel cycle.

Previous discussions of irradiated fuel transportation (1,2) have included projections of rapidly increasing cask shipping requirements as a function of time based on a proliferation of nuclear reactors and their associated fuel discharges. However, due to a significant delay in reprocessing plant startup, particularly in the U.S., and corresponding plans by Utilities to increase reactor site fuel storage capability, these projections have had to be modified considerably. Nevertheless, irradiated fuel transportation on a worldwide basis presently totals several hundred shipments per year and will continue to increase albeit at a lesser rate than previously anticipated to an expected level of thousands of (truck and rail) shipments per year during the 1980's. This translates into a need for dozens of irradiated fuel casks during this period. Suffice it to say, then, that the magnitude of irradiated fuel movement predicted for coming years warrants an intensive consideration of the nuclear safety aspects of this segment of the fuel cycle. Certainly, the precautionary measures to be maintained during cask development and subsequent irradiated fuel shipments should be consistent with those associated with the nuclear power plant per se.

Already public attention has focused on this heretofore unrecognized area of the fuel cycle as articles, for example, on the hazards of nuclear transportation on the freeways in one U.S. state (3), and the ban on nuclear shipments through a major U.S. city in another state (4) have appeared. In addition, the continued reluctance of U.S. railroads to ship radioactive material (2) has contributed to the public's ever-increasing awareness of questions related to the safety of nuclear transportation. Accordingly, the reader is encouraged to not only review the limited content of material presented in this chapter, but to puruse also the other documents referenced herein, to obtain a thorough knowledge of the broad safety aspects of irradiated fuel transportation.

2.0 FEDERAL REGULATIONS

The first step in maintaining nuclear safety for irradiated fuel transportation is implemented through regulatory requirements for the associated cask design. Specifically, the design of a cask planned for use in shipping irradiated fuel must be reviewed with respect to compliance with the Nuclear Regulatory Commission (NRC) criteria contained in the Code of Federal Regulations, Title 10, Part 71. The more limiting of these criteria along with corresponding cask design requirements are summarized in Table I (5-8). These are, obviously, performance criteria and, as such, serve in assessing the adequacy of the shipping cask design. In essence, the license applicant who seeks NRC approval of a given cask design must prepare and submit to NRC a detailed safety analysis report which demonstrates that the cask design complies with the above criteria.

After a detailed safety review of the license application and possible amendments thereto (which, depending on the complexity of the design, may take 1 to 5 years), NRC typically approves the design and issues a certificate of compliance which outlines the licensed and operational characteristics of the cask along with a designation of specific fuel authorized for shipment therein.

Once the certificate of compliance is received, the licensee may proceed with cask fabrication. The latter, however, must be conducted under a rigorous Quality Assurance (QA) Program, the scope of which must have been approved during the aforementioned licensing process. The standard QA references emphasized by NRC are Appendix B of 10 CFR 50 and Proposed Appendix E of 10 CFR 71 (9). To ensure compliance with approved Quality Assurance procedures, periodic audits entailing site visits are made by NRC representatives.

Not only is Quality Assurance maintained during the fabrication phase of the cask, but also into post-fabrication use and maintenance of the cask in the form of periodic procedural and maintenance checks/tests. Again, as during cask fabrication, periodic audits are made by NRC representatives to make sure QA requirements are being enforced.

In addition to NRC criteria, safety standards developed by both NRC and the Department of Transportation (DOT), in the form of DOT regulations (7-8) which are imposed on shippers and carriers, influence the design and analysis of irradiated fuel casks. Such standards relate primarily to radiation dose levels and temperatures on the exterior of the cask and/or transporting vehicle. Also the DOT regulations cover radioactive material classification, labeling and marking.

Design and safety analysis of future U.S. casks (within the next few years) will most probably be reviewed under the International Atomic Energy Agency (IAEA) regulations (10) which are now under consideration for adoption by the U.S. These regulations are already in use by most European countries. The most significant impact of the change in regulations as far as U.S. cask reviews are concerned would be in two areas: 1) transport groups (6) would

Table I. Summary of Limiting Regulatory Criteria (5-8)

Design Requirement	Normal Conditions	Accident Conditions
Structural Criteria		
External pressure	25 psig	25 psig
Internal pressure	25 psig--Must not exceed 50% of design gage pressure	Must not release more than 10 curies of Group III or IV radionuclides or 1000 curies of noble gas
Load resistance	Must support 5 times weight of container	
Penetration	Must resist penetration by 13 lb-1¼ in. diameter bar dropped 4 feet	Must withstand 40" drop onto 6" diameter x 8" long bar (position of max. damage)
Drop	For containers weighing 30,000 lb or more, all exterior components must operate after drop of 1 ft	30 ft drop onto a flat unyielding surface followed by loss of coolant and fire
Lifting devices	Must support 3 times cask weight	Failure must not impair shielding
Tiedown devices	Must sustain forces of 10 g fore and aft, 5 g side, 2 g vertical combined	Failure must not impair shielding
Thermal Criteria		
Ambient conditions	Must function in range of -40F to 130F + solar heat load in stagnant air	Must function following fire at 1475F for 30 min with fire emissivity of 0.9 and container absorptivity of 0.8*
Fuel clad temperature	Must be held below clad failure temperature	Must not release more than 10 curies of Group III or IV

Table I. Continued

Design Requirement	Normal Conditions	Accident Conditions
		radionuclides or 1000 curies of noble gas
	Shielding Criteria	
Dose rate	Restricted to 10 mr/hr at 3 ft from cask surface and 200 mr/hr at cask surface	Restricted to 1 r/hr at 3 ft from cask surface
	Restricted to 10 mr/hr at 6 ft from "sole use" transport vehicle	
	Criticality Criteria	
Multiplication constant	Must be subcritical ($k_{eff} < 1.0$) when fully moderated	Must be subcritical ($k_{eff} < 1.0$) when fully moderated (immersed in 3 ft of water for period of not less than 8 hours)
	Contamination Criteria	
Primary coolant activity	Must not exceed 300 μc/ml of Group III and IV radionuclides	Must not release more than 10 curies of Group III and IV radionuclides
Surface limits	Must not exceed 10^{-8} curies of β- and γ-emitting material per 100 cm^2 nor 10^{-9} curies of α-emitting materials per 100 cm^2	

*The package shall not be cooled artificially until 3 hours after the test period unless it can be shown that the temperature on the inside of the package (cask) has begun to fall in less than 3 hours.

be classified into specific radionuclides, and 2) additional assessment of cask designs to determine allowable leakage rates after normal and accident tests. Reviews of the effects of these changes are continuing in the U.S.

3.0 CASK DESIGN REQUIREMENTS

An appreciation of the nuclear safety considerations that go into the design of a cask to ship irradiated fuel can best be obtained by reviewing the respective areas of the cask design and safety analysis that must be documented for purposes of NRC licensing approval. Accordingly, the following sections highlight the important technical aspects of a typical cask design and its associated safety evaluation. These sections have been purposely abbreviated for this chapter, and the reader is referred to References 11-18 for a more extensive coverage.

3.1 Structural Integrity

To assure containment of the irradiated fuel under normal as well as accident conditions (see Table I), a high degree of structural integrity must be incorporated into the cask design. The more stringent performance requirements are those of the 30-foot free fall onto an unyielding surface and the 40-inch drop onto a 6-inch diameter steel pin. To meet these criteria the cask design typically includes a stainless steel framework, the outer geometry and thickness of which are determined on the basis of the anticipated cask weight and the above drop conditions. Present day features include an outer stainless steel shell on the order of 1 to 2 inches thick to survive the 40-inch drop (puncture) test, and external impact limiters (shock absorbers) in the form of steel fins or steel encased wood to provide low "g" loadings and rupture-free impact under the 30-foot drop test. Vulnerable areas of the cask to be considered from the standpoint of possible release of radioactive contents include the shielded lid closure (normally secured by a number of bolts), and the drain and pressure relief valves. In addition, in the case of lead-shielded casks, lead slump during impact and resultant radiation streaming are of concern.

Adequacy of the structural design is demonstrated by structural analysis computer programs and/or scale model drop tests. Numerous data from the latter, which have ranged from small to large scale models, have been documented (19-20) and serve as a guide for future test programs. In addition to scale model testing, full-scale tests on a number of previously used casks are now being planned by the U.S. Energy Research and Development Administration for the fall of 1976 (21).

3.2 Criticality Safety

One of the more challenging areas in the safety analysis and design of shipping casks for irradiated fuel is that of the nuclear criticality safety evaluation. Simply stated, the cask together with fuel contents must remain subcritical under a prescribed set of accident conditions. Furthermore, it must be demonstrated that the loaded cask has a k_{eff} value of well below 1.0 under normal conditions of transport. Design features to meet these criteria typically include a fuel basket which, in addition to providing structural support for the fuel assemblies, contains neutron poison compositions such as

B$_4$C in SS, Cd-Cu, and Ag-In-Cd in SS and/or aluminum. Material selection is essentially up to the cask designer since any one of the foregoing neutron poisons can effectively control reactivity within a multi-element cask. However, supplemental use of highly conducting materials such as copper or aluminum may be dictated by secondary heat transfer considerations (5,22) particularly in the case of dry (cavity) cask designs where fuel cladding temperatures are of primary concern.

Since proof of subcriticality of a cask design is not practical from the direct experimental approach, procedure usually entails a detailed analytical approach whereby highly sophisticated computer programs are utilized (23). Results from the latter are qualified by means of benchmark calculations based on actual data from selected critical experiments.

For conservatism, the above calculations are based on a cold, clean fuel assumption. In addition, positive reactivity contributions are considered such as the greater reflectivity (as compared to water) of lead or uranium shielding. Also, increased moderation (fuel assemblies are typically under-moderated) and corresponding increased reactivity effects due to fuel assembly (rod) bowing or fuel rod drop out under accident conditions are taken into account (24).

3.3 Shielding

One of the more salient features of a shipping cask designed to accommodate highly radioactive fuel is its heavily shielded walls. Historically, such shielding has been provided to attenuate the highly-penetrating fission product gamma radiation (see Table II) and has consisted of either lead, steel, uranium or a combination thereof.

Table II. Radioactivity of Irradiated Fuel* (15,16)
(curies per metric ton of uranium)

	Cooling Period (in days)			
	90	150	365	3650
Fission Products	6.19 x 10^6	4.39 x 10^6	2.22 x 10^6	3.17 x 10^5

*Typical PWR fuel irradiated to 33,000 MWD/MT.

In recent years, however, with the attainment of higher fuel burnups (> 20,000 MWD/MTU) and the corresponding production of neutron emitting transuranium elements (primarily Cm-242 and Cm-244--see Tables III and IV) supplementary neutron shielding in the form of hydrogenous materials (liquid and solid) has been added. Not only must the fixed source neutrons (from spontaneous fission of the curium isotopes) and α,n neutrons (alpha particles from curium decay interacting with the light nuclei of oxygen in the oxide fuel) be considered, but also the uranium fission neutrons resulting from subcritical multiplication in the fuel. The latter source of neutrons adds to the complexity in designing the neutron shielding for casks of the wet cavity type. That is, the internal cavity water at the periphery of the fuel attenuates neutrons emanating from the fuel thereby minimizing the need for external neutron shielding; conversely, this water also moderates neutrons thereby increasing the neutron source due to uranium fission which, in turn, increases the need for external neutron shielding. The obvious solution of borating the internal cavity water is not feasible due to restriction on receipt of borated water at fuel unloading sites.

Table III. Major Neutron Sources in Irradiated Fuel (25)

Isotope	Half Life (years)	Source Strength (neutrons/sec-gram)
^{238}Pu	89 (α)	5.1×10^4
	4.9×10^{10} (SF)	3.4×10^3
^{240}Pu	6580 (α)	6.8×10^2
	1.2×10^{11} (SF)	1.4×10^3
^{242}Cm	0.44 (α)	2.0×10^7
	7.2×10^6 (SF)	2.3×10^7
^{244}Cm	18.1 (α)	4.2×10^5
	1.4×10^7 (SF)	1.2×10^7

Table IV. Curium Inventory and Associated Neutron Source Strength for Irradiated Fuel (25)

Fuel Burnup (MWD/MT)	Isotope Production (grams/metric ton) ^{242}Cm	^{244}Cm	Neutron Source Strength (neutrons/sec per metric ton $\times 10^{-7}$) ^{242}Cm	^{244}Cm	Total
14,700	2	1	8.6	1.2	9.8
20,400	5	4	21.5	4.8	26.3
25,200	8	8	36.4	9.6	46.0
30,500	12	16	51.6	19.2	70.8

From a nuclear safety standpoint, the adequacy of the cask shield design is assessed based on an analysis of post-accident dose rates external to the cask. Specifically, the above dose rate (neutron and gamma radiation combined) must not exceed 1 rem/hr at 3 feet from the external surface of the cask. The crux of the safety analysis, therefore, is to determine the loss of shielding loss factor (or, stated differently, the dose increase factor) due to cask damage under accident conditions. This factor along with the allowable post-accident dose rate (1 rem/hr) is used to establish the allowable dose level at 3 feet from the cask surface prior to shipment. If the measured dose rate level is above this value, which would indicate that shielding loss under accident conditions would result in a dose rate greater than 1 rem/hr (at 3 feet from the cask surface), then shipment is not allowed.

3.4 Heat Removal

The most limiting aspect of present day cask designs is that of efficient heat removal. That is, the quantity of irradiated fuel that can be transported in casks from a shielding or criticality standpoint is most often limited by heat load. This refers to the problem of effectively dissipating the decay heat from the irradiated fuel (see Table V) in order to minimize fuel cladding temperatures, excessive values of which might result in unacceptable releases of radioactivity from the cask.

Table V. Thermal Energy in Irradiated Fuel* (15,16)
(watts per metric ton of uranium)

	Cooling Period (in days)			
	90	150	365	3650
Thermal Energy	2.71×10^4	2.01×10^4	1.04×10^4	1.06×10^3

*Typical PWR fuel irradiated to 33,000 MWD/MT.

As part of the safety evaluation for the thermal design of the cask, detailed heat transfer calculations are performed including radiation and convection heat transfer from the cask surface, conduction and radiation through the cask walls, and convection and radiation within the cask cavity. In addition, stress-rupture calculations are performed for the fuel cladding to establish temperatures below which clad perforation either does not occur or else is minimized. Aside from fuel clad temperature determinations, the extent and consequences of lead (shielding) melt must be analyzed for certain cask designs. The restriction here, as noted in the above discussion on shielding, is that the post-accident dose rate cannot exceed a specified level. Also, in cask designs of the wet cavity type internal pressure due to the water is an important (thermal) consideration.

Typical features of present day casks approved from a thermal safety standpoint include fins on the outer cask surface to increase the heat transfer area (corrugated surfaces are also utilized), internal peripheral fins to minimize temperature drops across shield-shell interfaces, asbestos layers to minimize burning of balsa impact limiters, surge tanks to accommodate thermal expansion of neutron shielding fluids, relief valves and rupture discs to release internal pressure under thermal accident conditions, and high temperature cask lid seals.

3.5 Decontamination

During fuel loading and unloading operations the shipping cask is submerged in the spent fuel storage pool. As a result, the cask surface becomes coated with radioactive contaminants present in the pool water. Accordingly, decontamination operations are required to clean the cask surface to an acceptable level (see Table I) prior to release of the (empty or loaded) cask for over the road/rail transport. Proposed NRC regulations (9) require that the external surfaces of the cask, as far as practicable, be designed, fabricated and finished to facilitate decontamination.

Experience has shown that after extended usage cask surfaces exhibit what is known as a "weeping" effect whereby contaminants within the pores of the cask surface material (usually stainless steel) appear on the surface of the cask at unacceptable levels. This phenomenon occurs after the cask has been decontaminated and is enroute. Accordingly, design emphasis is presently directed toward preventing the cask surface from becoming grossly contaminated (which, essentially, is a function of temperature, exposure time and pool contamination activity). This is achieved by using a cover or shroud around the cask during loading/unloading operations.

4.0 RELATED TOPICS ON TRANSPORTATION SAFETY

In addition to the nuclear safety aspects of irradiated cask design discussed in the foregoing paragraphs, there are other safety related topics that have been extensively covered in the open literature. These are the related topics of 1) _risk_ based on the probability of transportation accidents and 2) _environmental effects_ of transportation under normal and accident conditions (13,16,26). Although pertinent to the general subject of transportation safety, these topics are considered beyond the scope of this chapter and are noted here for supplemental study by the reader.

4.1 Emergency Response Plan

Also in the interest of safety it is pertinent to note that follow up in the unlikely event of an accident during transportation is provided through what is known as an Emergency Response Plan. Such a plan is generally provided by the appropriate supplier of the transportation service, and is intended to initiate immediate assistance to the carrier of the cask as well as local civil authorities by providing radiological surveillance of the shipment. In addition, the plan intends to establish the extent, if any, of the physical damage to the cask and associated transport vehicle. Federal assistance is also provided through the Interagency Radiological Assistance Plan (IRAP), which is a voluntary agreement entered into by federal agencies to establish an organization and operating agreements to be mobilized in the event of a major accidental release of radioactive material which might seriously endanger public health and safety. The USERDA is the designated agency under IRAP responsible for administration, coordination, and implementation of emergency measures in cooperation with other participating federal agencies (27).

4.2 In-Plant Cask Drop Protection

Another safety topic related to irradiated fuel transportation, but falling under the more general heading of cask handling or cask-plant interfaces is that of in-plant cask drop protection. This topic relates to an NRC requirement for nuclear reactor owners to evaluate an in-plant accident involving the dropping of an irradiated fuel cask (28). Such a requirement in no way affects the licensing (safety) of the cask per se, but does influence the cask design in the area of cask (trunnions) and ancillary (yoke) lifting components due to handling considerations.

The cask drop protection approach being followed in nuclear plants varies from plant to plant and includes: use of impact pads, restricted travel paths, pool dashpots, air effects machines, and redundant cranes (28-29). It is only in the latter approach that the cask design is influenced. That is, redundant trunnions (two sets) and a redundant lifting yoke (part of the standard cask transportation equipment) may be desirable.

5.0 PAST, PRESENT AND FUTURE CASKS

Discussion thus far in this chapter has dealt with the general design and safety analysis requirements for irradiated fuel casks. To illustrate the effect that these requirements have had on actual cask designs, a review is provided in this section of the more important casks of past, present and future generations.

5.1 Earlier Casks

Table VI lists some of the major first generation casks for various types of nuclear fuel along with their more important parameters. These casks are, in general, devoid of neutron shielding and impact limiters and have a relatively low heat load capacity. Additional information on various earlier government-owned casks may be found in Ref. 30.

5.2 Present Generation Casks

Tables VII and VIII list the major second-generation casks that are presently in various stages of licensing, fabrication and operation. These cask models represent state-of-the-art in irradiated LWR fuel cask design. The reader will note the alternative approaches to design based on, for example, shield material, method of cooling, capacity and mode of transport. This simply indicates that a given level of safety can be achieved through widely different cask designs, the approach to which is strongly influenced by economic considerations. Figures 1a and 1b illustrate a typical second-generation cask of the truck-cask variety which has made an extensive number of successful shipments.

5.3 Future Irradiated Fuel Casks--Safety Features

Future cask designs are to an extent being influenced by the licensing activities related to present generation casks. For example, the number of safety questions related to lead slump and lead melting under respective impact and fire conditions is leading cask designers to reconsider steel as a cask shielding material. This material was rejected in earlier cask designs on the basis of overall transportation economics. In addition, the trend is to slightly reduce the cask heat load capability in large, multi-element casks in balance with a more defensible position regarding safety under accident conditions. Also, the design concept of "zero-release", wherein no radioactive contents are released from the cask under accidents, will most likely continue.

Whether wet or dry casks will predominate is a moot question since the operating experience to date has not decidedly favored either approach. Impact limiters of the wood (as opposed to steel fin) variety will be frequently seen in cask structural designs. This is due to the copious quantity of test impact data documented for this material.

Comprehensive test programs for irradiated fuel casks are now underway and continued efforts are planned for the future. As the results of these tests become available, appropriate use will be made by private industry in the case of cask design and by the regulatory agencies in cask safety review. Efforts such as these will ensure cask designs that will result in the continued safety of irradiated fuel transportation.

Table VI. Irradiated Fuel Shipping Cask Characteristics

Designation	NFS-1/S-1/Stanray	Model No. 100 (NFS-2)	IF-100	IF-200
Owner	Nuclear Fuel Services, Inc.	Nuclear Fuel Services, Inc.	General Electric Co.	General Electric Co.
Designer	Stearns Roger	Battelle Memorial Institute	National Lead	National Lead
Fabricator	Stearns Roger		National Lead/ Stearns Roger	National Lead
Docket No.	70-666	70-1104	70-729, 70-754, 70-1053	70-951, 70-1053, 70-534
Date of Design Report	August 1964	Feb. 1, 1968	1962	1965
Design Report Title		The Shipment of Big Rock Point and Humboldt Bay Irradiated Fuel Subassemblies in NFS Shipping Cask		
Transport Mode	Rail	Rail	Truck	Truck
Loaded Weight, lbs	140,000 +	120,000	44,000/45,000	56,000
Gamma Shield	9-1/2 in. Lead + 1-1/2 in. Steel	8-3/4 in. Lead + 2-3/8 in. Steel	8-3/8 in. Lead	8-3/8 in. Lead + 1-3/4 in. Steel
Neutron Shield	none	none	none	Wood Crate
Internal Coolant	Water	Water	Water	Water
Cavity Dimensions length/diameter, in.	114-1/2 - 140-3/4/41 across hex flats	103/40	129-137/13	140/15
Contents	12 Big Rock Point/18 Humboldt Bay Fuel Assemblies		4 Dresden 1/7 CVTR Assemblies	2 Indian Point Assemblies
Decay Heat, Btu/hr	100,000	68,900	20,000	45,000
External Heat Transfer	Natural convection - finned surface	Natural convection - finned surface	Natural convection - finned surface	Natural convection - finned surface
Maximum Normal Operating Pressure, psig	120	100	100	

Designation	Yankee - Rail ELC-1002	Yankee - Truck	M-100	Piqua/Elk River	NLC-6502
Owner	Westinghouse	Westinghouse	U.S. Navy	Atcor	National Lead
Designer	Westinghouse/ Battelle	Edlow Lead	National Lead	National Lead	National Lead
Fabricator	Edlow Lead	Edlow Lead	National Lead	National Lead	National Lead
Docket No.	70-596				
Date of Design Report	October 1961	February 1965		February 1962	
Design Report Title	Shipment of Irradiated Yankee Fuel Elements WCAP-1859	Experience in the Shipment of Yankee Fuel Assemblies for Post-Irradiation Examination WCAP-6062		Piqua/Elk River Reactors Spent Fuel Shipping Cask Design Report TID-21246	
Transport Mode	Rail	Truck	Truck	Truck	Truck
Loaded Weight, lbs	150,000	44,000	220,000	60,000	44,000
Gamma Shield	10 in. Lead + 1-3/4 in. Steel	9-3/8 in. Lead 1 in. Steel	10 1/2 in. Lead + 2 in. Steel	8 in. Lead + 1-3/4 in. Steel	Lead, Uranium & Steel - 10 in. Lead equiv.
Neutron Shield	none	none	none	none	none
Internal Coolant	Water	Water	Water	Air	
Cavity Dimensions length/diameter, in.	119/38	115/12	132/55	83/30	12 x 12 x 130
Contents	9 Yankee Rowe Assemblies	1 Yankee Rowe Assembly		19 Piqua and 19 Elk River Fuel Assemblies	20 NRU/20 NRX elements
Decay Heat, Btu/hr	273,000	29,000	300,000	7000	25,000
External Heat Transfer	Forced internal convection - external fins	Natural convection - external fins	Air cooled heat exchanger	Natural convection - external fins	Natural convection - unfinned surface
Maximum Normal Operating Pressure, psig	75	50	300	100	

Table VI. Continued

Designation	JMTR	Argonne TREAT Cask	FSV-1	PB-1
Owner	Japanese Atomic Energy Research Institute	USAEC/Argonne	Gulf General Atomic	Philadelphia Electric Company
Designer	National Lead	Argonne	GGA/NL	Battelle Memorial Institute
Fabricator	National Lead	Argonne	National Lead	Whitehead and Kales
Docket No.			70-72, 70-1165	70-1234
Date of Design Report	May 1970	June 1970	April 17, 1969	January 15, 1970
Design Report Title	Design Report JMTR Spent Fuel Transport Cask W-5200	Evaluation of the Uranium Shielded Shipping Cask for Radioactive Materials/Spec. ETD-PD-0029	Final Design Report for Ft. St. Vrain Fuel Shipping Cask	The Shipment of Peach Bottom No. 1 Irradiated Fuel in Whitehead & Kales Shipping Cask Model PB-1
Transport Mode	Truck	Truck	Truck	Truck
Loaded Weight, lbs	41,061	19,360	46,500	62,800
Gamma Shield	9-3/8 in. Lead + 1-3/8 in. Steel	3 in. Depleted Uranium + 2.5 in. Steel	3-1/2 in. Depleted Uranium + 1-5/8 in. Steel	6-1/2 in. Lead + 1-3/4 in. Steel
Neutron Shield	none	none	none	none
Internal Coolant	Air	Air	Helium	Helium
Cavity Dimensions length/diameter, in.	32.35/30.0	113-5/8/ 11-3/8	208-11/16/28	173.1/40.5
Contents	41 MTR Fuel Assemblies	1 Special Form Container containing TREAT experimental 19-37 U-Pu fuel pins	6 spent fuel elements (in a container) Ft. St. Vrain HTGR station	19 spent fuel elements (in a container) Peach Bottom 1 HTGR station
Decay Heat, Btu/hr	28,295	109	2,322	14,250
External Heat Transfer	Natural convection - finned surface	Natural convection - unfinned surface	Natural convection - unfinned surface	Natural convection - unfinned surface
Maximum Normal Operating Pressure, psig	70	7.5	50	100

Table VI. Continued

Designation	BMI-1	PRDC	MTR	Trino	HNPF Cask
Owner	Battelle Memorial Institute	Battelle Memorial Institute	National Lead	Transnucleaire	USAEC
Designer	Battelle Memorial Institute	Battelle Memorial Institute	National Lead	Transnucleaire	Atomics International
Fabricator	Battelle/Edward Lead	Central Ohio Welding	National Lead		
Docket No.	70-813				
Date of Design Report	November 14, 1963	September 30, 1966		March 1968	November 1967
Design Report Title	Battelle Research Reactor Spent Fuel Shipping Cask	The Shipment of Power Reactor Development Company's Irradiated Fermi Fuel Assemblies		TN.2/TRI/00	HNPF Six Element Irradiated Fuel Shipping Cask NAA-SR-12547
Transport Mode	Truck	Truck	Truck	Truck	Rail
Loaded Weight, lbs	23,200	34,250	24,000	70,548	85,000
Gamma Shield	8 in. Lead + 3/4 in. Steel	8-1/4 in. Lead + 1 in. Steel	8 1/2 in. Lead + 1 in. Steel	8.07 in. Lead + 1.18 in. Steel	7 in. Lead + 2-1/2 in. Steel
Neutron Shield	none	none	none	none	none
Internal Coolant	Water	Water	Air	Air	Air
Cavity Dimensions length/diameter, in.	54/15-1/2	121/28.5	29 5/16 - 34/22 1/2	152.6/17.3 x 17.3	197-1/2 - 204/ 18-1/2
Contents	24 MTR Assemblies	1 Fermi Subassembly	28 MTR elements	51,200	6 HNPF Assemblies
Decay Heat, Btu/hr	7682	4100	18,460	51,200	922
External Heat Transfer	Natural convection - finned surface	Natural convection - unfinned surface	Natural convection - finned surface	Natural convection - pin type fins	Natural convection - unfinned surface
Maximum Normal Operating Pressure, psig	100	75	20	0	5

Table VII. Technical Data on Licensed Second-Generation Irradiated Fuel Casks (31)

CASK IDENTIFICATION	NAC-1	NFS-4	NLI-1/2	TN-8	TN-9	IF-300	EXCELLOX-3	NLI-10/24
Owner	NAC	NFS	NLI	Transnu-cleaire	Transnu-cleaire	GE	BNFL	NLI
Date Licensed	5/74	11/72	2/74	5/74	5/74	9/73	--	6/76
Cask Type	LWT	LWT	LWT	OWT	OWT	OWT/RAIL	MULTI-RAIL	RAIL
Capacity (PWR/BWR)	1/2	1/2	1/2	3/0	0/7	7/18	7/15	10/24
Loaded Cask Wt. (Tons)	25	25	23.8	40	38	68-70	70	98.9
Overall Cask Length (In.)	214	214	227.3	217.5	226.5	209.9	210	254.5
Cask Diameter (In.)	50	50	64	79.6	79.6	75.8	69	96
Cavity Length (In.)	178	178	178	168.5	178	180.3	184	179.5
Cavity Diameter (In.)	13.5	13.5	12.6	Complex	Complex	37.5	34	45
Cavity Condition	Wet	Wet	Dry	Dry	Dry	Wet	Wet	Dry
Decay Heat (kw)	11.5	11.5	10.6	35.5	24.5	61.5	35	97.2
Cooling Method	Natural Convection	Natural Convection	Natural Convection	Natural Convection	Natural Convection	Forced Convection	Natural Convection	Natural Convection
Shielding Material (Gamma/Neutron)	Pb/H$_2$O	Pb/H$_2$O	Pb-U/H$_2$O	Pb/Resin	Pb/Resin	U/H$_2$O	Pb/H$_2$O	Pb-U/H$_2$O

Table VIII. Technical Data on Unlicensed Second-Generation Irradiated Fuel Casks (31)

CASK IDENTIFICATION	BCL-6	ENL-1300	NFS-5	HZ-75-T	Genas 1-2	Genas 10/24	TN-12
Owner	BCL	ENL	NFS	Hitachi	Robatel	Robatel	Transnucleaire
Cask Type	LWT	LWT	LWT	MULTI-	LWT	RAIL	RAIL
Capacity (PWR/BWR)	1/1	1/3	2/3 or 2-4	7/17	1/2	10/24	12/32
Loaded Cask Wt. (Tons)	24.2	22.8	25	81.9	28.7	105	110
Overall Cask Length (In.)	211.9	199	214.5	224.9	--	224	272
Cask Diameter (In.)	49.6	32	52	94.5	--	103.5	100
Cavity Length (In.)	178.8	178.5	178	179.3	--	180	177.6
Cavity Diameter (In.)	9.5 x 9.5	16.5	9.5 x 18.2	37.5	--	43.5	46.4
Cavity Condition	Wet	Wet	Wet	Wet	Wet	Wet	Dry
Decay Heat (kw)	12	--	25	84	15	110	120-150
Cooling Method	Natural Convection	Natural Convection	Natural Convection	Forced Convection	Natural Convection	Natural Convection	Natural Convection
Shielding Material (Gamma/Neutron)	Pb/H$_2$O	U/H$_2$O	U/H$_2$O	Pb/H$_2$O	Pb/ -	Pb/Solid	SS/Resin-H$_2$O

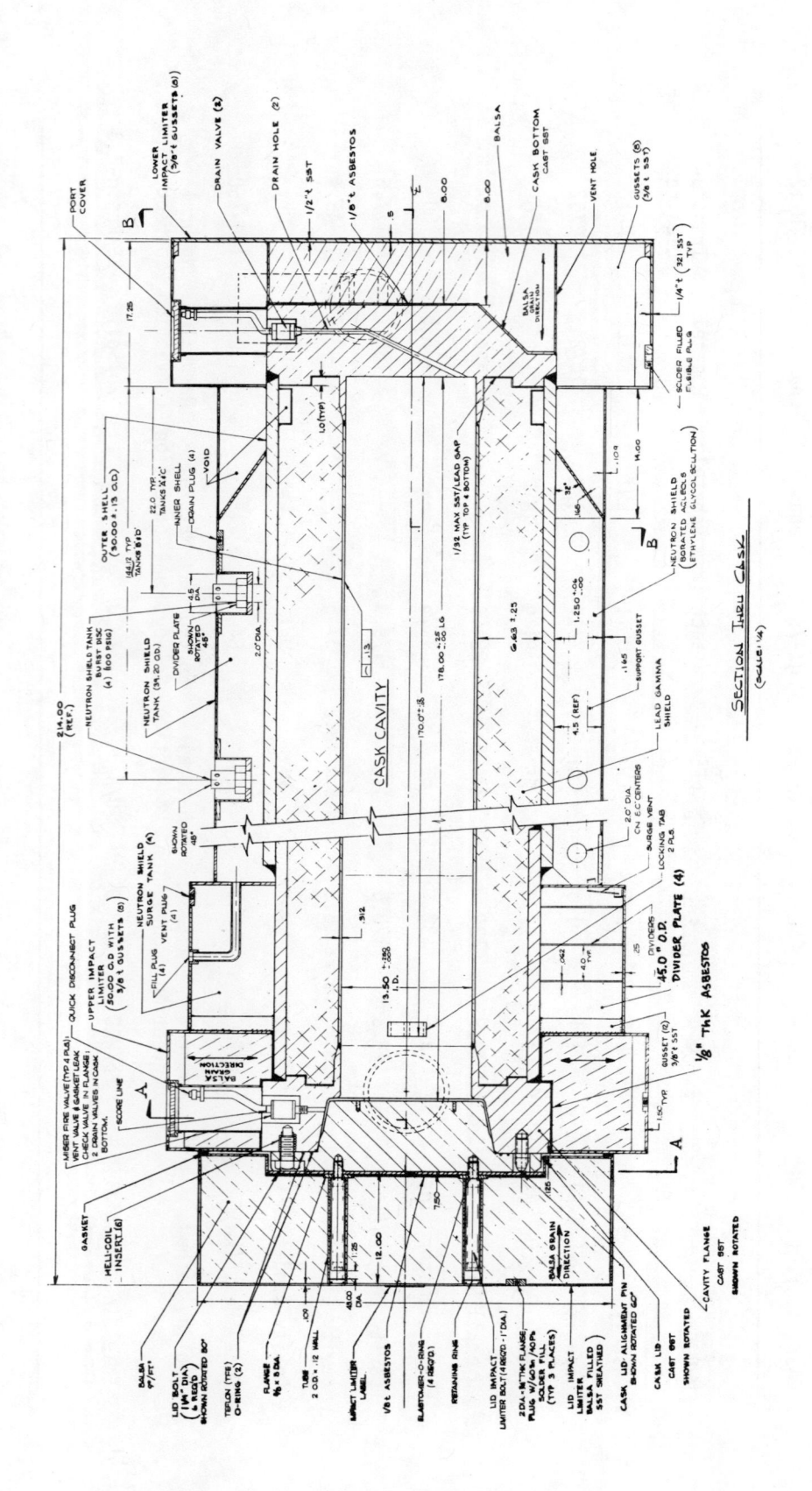

Figure 1a. NAC-1 Irradiated Fuel Cask (1 PWR/2 BWR Capacity)

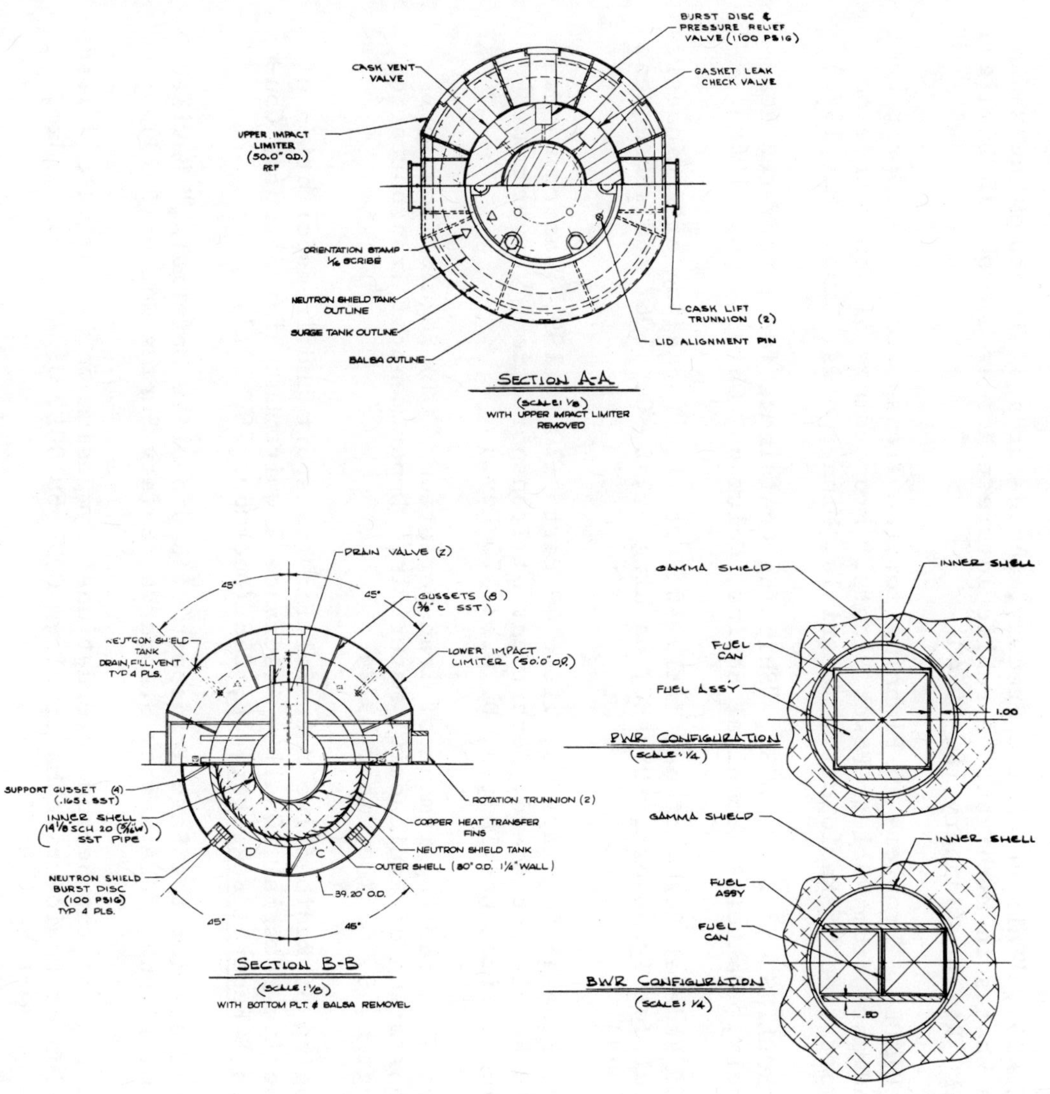

Figure 1b. NAC-1 Irradiated Fuel Cask (1 PWR/2 BWR Capacity)

REFERENCES

1. R. SALMON, J. O. BLOMEKE, J. P. NICHOLS, "Trends and Projected Shipments in the Nuclear Fuel Cycle Industry to the Year 2000," Proceedings of the Fourth International Symposium on Packaging and Transportation of Radioactive Materials, CONF-740901-P1, Miami Beach, Florida (September 1974)

2. C. W. SMITH, "Railroad Transportation of Radioactive Materials--Can We Count on IT?," Proceedings of Conference on Uranium/Fuel Cycle 74, New Orleans, Louisiana (March 1974)

3. M. ANDERSON, "Fallout on the Freeway, the Hazards of Transporting Radioactive Wastes in Michigan," A Public Interest Research Group in Michigan (PIRGIM) Report (January 18, 1974)

4. D. J. BINDER, "Problems in Planning for the Transportation of Spent Fuel and Radwaste from Nuclear Power Plants," Proceedings of Conference on Transportation of Radioactive Materials, Minneapolis, Minn. (May 1976)

5. J. D. ROLLINS and E. C. LUSK, "Shipping Containers for Irradiated Reactor Fuels," Battelle Technical Review, Columbus, Ohio (August 1968)

6. Code of Federal Regulations, Title 10 Part 71, "Packaging of Radioactive Material for Transport," US AEC, Washington, D. C. (1967 and subsequent revisions)

7. Code of Federal Regulations, Title 49 Parts 170 to 199, "Hazardous Materials Regulating Board, Department of Transportation," Government Printing Office, Washington, D. C. (1970 Rev.)

8. "A Review of the Department of Transportation (DOT) Regulations for Transportation of Radioactive Materials," Department of Transportation, Washington, D. C. (December 1972)

9. "Packaging of Radioactive Material for Transport and Transportation of Radioactive Material, Quality Assurance Requirements for Shipping Containers," Federal Register, 38, 248 (December 28, 1973)

10. "Regulations for the Safe Transport of Radioactive Materials," Revised Edition, Vienna, IAEA Safety Standards, Safety Series No. 6 (1973)

11. Proceedings of the Second International Symposium on Packaging and Transportation of Radioactive Materials, CONF-681001, US AEC, Oak Ridge, Tennessee (October 1968)

12. Proceedings of the Third International Symposium on Packaging and Transportation of Radioactive Materials, CONF-710801, Richland, Washington (August 1971)

13. Proceedings of the Fourth International Symposium on Packaging and Transportation of Radioactive Materials, CONF-740901, Miami Beach, Florida (September 1974)

14. C. E. MacDONALD, "Radioactive Material Package Design Reviews," Proceedings of the Fourth International Symposium on Packaging and Transportation of Radioactive Materials, CONF-740901-P3, Miami Beach, Florida (September 1974)

15. "The Safety of Nuclear Power Reactors (Light Water-Cooled) and Related Facilities," WASH-1250, US AEC, Washington, D. C. (July 1973)

16. "Environmental Survey of Transportation of Radioactive Materials to and from Nuclear Power Plants," WASH-1238, US AEC, Washington, D. C. (December 1972)

17. L. B. SHAPPERT, "A Guide for the Design, and Operation of Shipping Casks for Nuclear Applications," Cask Designers Guide ORNL-NSIC-68, Oak Ridge, Tennessee (February 1970)

18. J. W. LANGHAAR, "Casks for Irradiated Fuel: A Look at the Cask Designers' Guide," Nuclear Safety, 12, 6 (Nov.-Dec. 1971)

19. The Proceedings of a Seminar on Test Requirements for Packaging for the Transport of Radioactive Materials, Vienna (February 1971)

20. L. B. SHAPPERT and J. M. EVANS, "The Testing of Large, Obsolete Casks," Trans. Am. Nucl. Soc., 20, 624 (1975)

21. W. S. HOLMAN, "ERDA: Energy Materials - Transportation - Research - Development," Proceedings of Conference on Transportation of Radioactive Materials, Minneapolis, Minnesota (May 1976)

22. J. D. ROLLINS and J. L. RIDIHALGH, "Evaluation of the Criticality Problem in Irradiated Fuel Shipment," Trans. Am. Nucl. Soc., 12, 892 (1969)

23. R. H. ODEGAARDEN and C. R. MAROTTA, "Criticality Calculation Method for Multi-Element LWR Fuel Shipping Packages," Proceedings of Fourth International Symposium on Packaging and Transportation of Radioactive Materials, CONF-740901-P2, Miami Beach, Florida (September 1974)

24. R. E. BEST and J. D. ROLLINS, "An Engineering Approach to Criticality Safety in an Irradiated Fuel Cask," Trans. Am. Nucl. Soc., 17, 277 (1973)

25. J. D. ROLLINS and R. E. BEST, "Neutron Shielding Problems Associated with the Shipment of High-Burnup LWR Fuel," Trans. Am. Nucl. Soc., 12, 970 (1969)

26. "Draft Environmental Statement on the Transportation of Radioactive Material by Air and Other Modes," NUREG-0034, U.S. Nuclear Regulatory Commission, Washington, D. C. (March 1976)

27. W. E. POLLOCK and W. M. ROGERS, JR., Editors, "Transportation of Radioactive Materials in the Western States," Western Interstate Nuclear Board (March 1974)

28. R. H. JONES, "What's a Nice Cask Like You Doing in a Place Like This: A Discussion of USAEC Requirements for In-Plant Cask Drop Analysis and Prevention," Proceedings of the Fourth International Symposium on Packaging and Transportation of Radioactive Materials, CONF-740901-P3, Miami Beach, Florida (September 1974)

29. D. M. COLLIER, "Spent Fuel Handling and Shipping Capability," Nuclear Assurance Corporation Report NAC-19, Atlanta, Georgia (January 1973)

30. "Directory of Packagings for Transportation of Radioactive Materials," WASH-1279, US AEC, Washington, D. C. (October 1973)

31. J. D. ROLLINS and D. M. COLLIER, "Spent Fuel Casks--Worldwide Status," Proceedings of the Fourth International Symposium on Packaging and Transportation of Radioactive Materials, CONF-740901-P3, Miami Beach, Florida (September 1974)

APPENDIX A

ACCIDENT ANALYSIS

taken from Chapter 15 of

"Standard Format and Content of Safety Analysis Reports
for Nuclear Power Plants," Regulatory Guide 1.70
(Revision 2), Nuclear Regulatory Commission
(September 1975)

15. ACCIDENT ANALYSES

The evaluation of the safety of a nuclear power plant should include analyses of the response of the plant to postulated disturbances in process variables and to postulated malfunctions or failures of equipment. Such safety analyses provide a significant contribution to the selection of the design specifications for components and systems from the standpoint of public health and safety. These analyses are a focal point of the Commission's construction permit and operating license reviews of plants.

In previous chapters of the SAR, the structures, systems, and components important to safety should have been evaluated for their susceptibility to malfunctions and failures. In this chapter, the effects of anticipated process disturbances and postulated component failures should be examined to determine their consequences and to evaluate the capability built into the plant to control or accommodate such failures and situations (or to identify the limitations of expected performance).

The situations analyzed should include anticipated operational occurrences, (e.g., a loss of electrical load resulting from a line fault), off-design transients that induce fuel failures above those expected from normal operational occurrences, and postulated accidents of low probability (e.g., the sudden loss of integrity of a major component). The analyses should include an accident whose consequences are not exceeded by any other accident considered credible so that the site evaluation required by 10 CFR Part 100 may be conducted.

Transient and Accident Classification

The approach outlined below is intended to organize the transients and accidents considered by the applicant and presented in the SAR in a manner that will:

1. Ensure that a sufficiently broad spectrum of initiating events has been considered,

2. Categorize the initiating events by type and expected frequency of occurrence so that only the limiting cases in each group need to be quantitatively analyzed,

3. Permit the consistent application of specific acceptance criteria for each postulated initiating event.

To accomplish these goals, a number of disturbances of process variables and malfunctions or failures of equipment should be postulated. Each postulated initiating event should be assigned to one of the following categories:

1. Increase in heat removal by the secondary system (turbine plant),

2. Decrease in heat removal by the secondary system (turbine plant),

3. Decrease in reactor coolant system flow rate,

4. Reactivity and power distribution anomalies,

5. Increase in reactor coolant inventory,

6. Decrease in reactor coolant inventory,

7. Radioactive release from a subsystem or component, or

8. Anticipated transients without scram.

Typical initiating events that are representative of those that should be considered by the applicant in this chapter of the SAR are presented in Table 15-1. The evaluation of each initiating event should be presented in a separate subsection corresponding to the eight categories defined above. The information to be presented in these subsections is outlined in Section 15.X.X.

One of the items of information that should be discussed for each initiating event relates to its expected frequency of occurrence. Each initiating event within the eight major groups should be assigned to one of the following frequency groups:

1. Incidents of moderate frequency,

2. Infrequent incidents, or

3. Limiting faults.

The initiating events for each combination of category and frequency group should be evaluated to identify the events that would be limiting. The intent is to reduce the number of initiating events that need to be quantitatively analyzed. That is, not every postulated initiating event needs to be completely analyzed by the applicant. In some cases a qualitative comparison of similar initiating events may be sufficient to identify the specific initiating event that leads to the most limiting consequences. Only that initiating event should then be analyzed in detail.

It should be noted, however, that different initiating events in the same category/frequency group may be limiting when the multiplicity of consequences are considered. For example, within a given category/frequency group combination, one initiating event might result in the highest reactor coolant pressure boundary (RCPB) pressure while another initiating event might lead to minimum core thermal-hydraulic margins or maximum offsite doses.

Plant Characteristics Considered in the Safety Evaluation

A summary of plant parameters considered in the safety evaluation should be given; e.g., core power, core inlet temperature, reactor system pressure, core flow, axial and radial power distribution, fuel and moderator temperature coefficient, void coefficient, reactor kinetics parameters, available shutdown rod worth and control rod insertion characteristics. A range of values should be specified for plant parameters that vary with fuel exposure or core reload. The range should be sufficiently broad to cover all expected changes predicted for the entire life of the plant. The permitted operating band (permitted fluctuations in a given parameter and associated uncertainties) on reactor system parameters should be specified. The most adverse conditions within the operating band should be used as initial conditions for transient analysis.

Assumed Protection System Actions

Settings of all protection system functions that are used in the safety evaluation should be listed. Typical protection system functions are reactor trips, isolation valve closures, ECCS initiation, etc. The uncertainty (combined effect of calibration error, drift, instrument error, etc.) associated with each function should also be listed together with the expected and maximum delay times.

15.X Evaluation of Individual Initiating Events

The applicant should provide an evaluation of each initiating event using the format of Section 15.X.X (e.g., 15.2.7 for a loss of normal feedwater flow initiating event). As shown in Table 15-1, a particular initiating event may be applicable to more than one category. The SAR sections should be appropriately referenced to indicate this.

The detailed information listed in Section 15.X.X, paragraphs 1 and 2, should be given for each initiating event. However, the extent of the quantitiative information in Section 15.X.X, paragraphs 3 through 5, that should be included will differ for the various initiating events. For those situations where a particular initiating event is not limiting, only the qualitative reasoning that led to that conclusion need be presented, along with a reference to the section that presents the evaluation of the more limiting initiating event. Further, for those initiating events that require a quantitative analysis, such an analysis may not be necessary for each of Section 15.X.X, paragraphs 3 through 5. For example, there are a number of plant transient initiating events that result in minimal radiological consequences. The applicant should merely present a qualitative evaluation to show this to be the case. A detailed evaluation of the radiological consequences need not be performed for each such initiating event.

3. Core and system performance.

    a. Mathematical model. The mathematical model employed, including any simplifications or approximations introduced to perform the analyses, should be discussed. Any digital computer programs or analog simulations used in the analyses should be identified. If a set of codes is used, the method combining these codes should be described. Important output of each code should be presented and discussed under "results." Principal emphasis should be placed on the input data and the extent or range of variables investigated. This information should include figures showing the analytical model, flow path identification, actual computer listing, and complete listing of input data. The detailed description of mathematical models and digital computer programs or listings are preferably included by reference to documents available to the NRC with only summaries provided in the SAR text.

    b. Input parameters and initial conditions. The input parameters and initial conditions used in the analyses should be clearly identified. Table 15-2 provides a representative list of these items. However, the initial values of other variables and additional parameters should be included in the SAR if they are used in the analyses of the particular event being analyzed.

    The parameters and initial conditions used in the analyses should be suitably conservative for the event being evaluated. The bases used to select the numerical values that are input parameters to the analysis, including the degree of conservatism, should be discussed in the SAR.

    c. Results. The results of the analyses should be presented and described in detail in the SAR. As a minimum, the following information should be presented as a function of time during the course of the transient or accident:

    (1) Neutron power,

    (2) Heat fluxes, average and maximum,

    (3) Reactor coolant system pressure,

    (4) Minimum CHFR, DNBR, or CPR, as applicable,

    (5) Core and recirculation loop coolant flow rates (BWRs),

    (6) Coolant conditions - inlet temperature, core average temperature (PWR), core average steam volume fraction (BWR), average exit and hot channel exit temperatures, and steam volume fractions,

    (7) Temperatures - maximum fuel centerline temperature, maximum clad temperature, or maximum fuel enthalpy,

15.X.X Event Evaluation

1. Identification of causes and frequency classification. For each event evaluated, include a description of the occurrences that lead to the initiating event under consideration. The probability of the initiating event should be estimated and the initiating event should be assigned to one of the following groups:

    a. Incidents of moderate frequency - these are incidents, any one of which may occur during a calendar year for a particular plant.

    b. Infrequent incidents - these are incidents, any one of which may occur during the lifetime of a particular plant.

    c. Limiting faults - these are occurrences that are not expected to occur but are postulated because their consequences would include the potential for the release of significant amounts of radioactive material.

2. Sequence of events and systems operation. The following should be discussed for each initiating event:

    a. The step-by-step sequence of events from event initiation to the final stabilized condition. This listing should identify each significant occurrence on a time scale, e.g., flux monitor trip, insertion of control rods begin, primary coolant pressure reaches safety valve set point, safety valves open, safety valves close, containment isolation signal initiated, and containment isolated. All required operator actions should also be identified.

    b. The extent to which normally operating plant instrumentation and controls are assumed to function.

    c. The extent to which plant and reactor protection systems are required to function.

    d. The credit taken for the functioning of normally operating plant systems.

    e. The operation of engineered safety systems that is required.

The effect of single failures in each of the above areas and the effect of operator errors should be discussed and evaluated. The discussion should provide enough detail to permit an independent evaluation of the adequacy of the system as related to the event under study. One method of systematically investigating single failures is the use of a plant operational analysis or a failure mode and effects analysis. The results of these types of analyses can be used to determine which functions, systems, interlocks, and controls are safety related and what readouts are required by the operator under anticipated operational occurrence and accident conditions.

(8) Reactor coolant inventory - total inventory and coolant level in various locations in the reactor coolant system,

(9) Secondary (power conversion) system parameters - steam flow rate, steam pressure and temperature, feedwater flow rate, feedwater temperature, steam generator inventory, and

(10) ECCS flow rates and pressure differentials across the core, as applicable.

The discussion of results should emphasize the margins between the predicted values of various core parameters and the values of these parameters that would represent minimum acceptable conditions.

4. Barrier performance. This section of the SAR should discuss the evaluation of the parameters that may affect the performance of the barriers, other than fuel cladding, that restrict or limit the transport of radioactive material from the fuel to the public.

a. Mathematical model. The mathematical model employed, including any simplifications or approximations introduced to perform the analyses, should be discussed. If the model is identical, or nearly identical, with that used to evaluate core performance, this should be stated in the SAR. In that case, only the differences, if any, between the models need be described.

A detailed description of the model used to evaluate barrier performance should be presented if it is significantly different from the core performance model. The information that should be included is indicated in paragraph 3 of Section 15.X.X, item a.

b. Input parameters and initial conditions. Any input parameters and initial conditions of variables relevant to the evaluation of barrier performance that were not presented and discussed in paragraph 3 of Section 15.X.X., item b, should be discussed in this section. The discussion should present the numerical values of the input to the analyses and should discuss the degree of conservatism of the selected values.

c. Results. The results of the analyses should be presented and described in detail in the SAR. As a minimum, the following information should be presented as a function of time during the course of the transient or accident:

(1) Reactor coolant system pressure,

(2) Steam line pressure,

(3) Containment pressure,

(4) Relief and/or safety valve flow rate,

(5) Flow rate from the reactor coolant system to the containment system, if applicable.

5. Radiological consequences. This section of the SAR should summarize the assumptions, parameters, and calculational methods used to determine the doses that result from limiting faults and infrequent incidents. Sufficient information should be given in this section to fully substantiate the results and to allow an independent analysis to be performed by the NRC staff. Thus, this section should include all of the pertinent plant parameters that are required to calculate doses for the exclusion boundary and the low population zone as well as those locations within the exclusion boundary where significant site-related activities may occur (e.g., the control room).

The elements of the dose analysis that are applicable to several accident types or that are used many times throughout Chapter 15 can be summarized in this section (or cross-referred) with the bulk of the information appearing in appendices.

If there are no radiological consequences associated with a given initiating event, this section for the event should simply contain a statement indicating that containment of the activity was maintained and by what margin.

Two separate analyses should be provided for each limiting fault. The first analysis should be based on design basis assumptions acceptable to the NRC for purposes of determining adequacy of the plant design to meet 10 CFR Part 100 criteria. These design basis assumptions can, for the most part, be found in regulatory guides that deal with radiological releases. For instance, when calculating the radiological consequences of a loss-of-coolant accident (LOCA), it is suggested that the assumptions given in Regulatory Guide 1.3, "Assumptions Used for Evaluating the Potential Radiological Consequences of a Loss-of-Coolant Accident for Boiling Water Reactors," and Regulatory Guide 1.4, "Assumptions Used for Evaluating the Potential Radiological Consequences of a Loss-of-Coolant Accident for Pressurized Water Reactors," be used. This analysis should be referred to as the "design basis analysis." There may be instances in which the applicant will not agree with the conservative margins inherent in the design basis approach approved by the NRC staff. If this is the case, the applicant may provide an indication of the assumptions he believes to be adequately conservative, but the known NRC assumptions should nevertheless be used in the design basis analysis.

The second analysis should be based on what the applicant believes to be realistic assumptions. This analysis will help quantify the margins that are inherent in the design basis approach. The analysis will also be useful in determining the expected environmental consequences of accidents. This second analysis should be referred to as the "realistic analysis."

The parameters and assumptions used for these analyses, as well as the results, should be presented in tabular form. Table 15-3 provides a representative list of these items. Table 15-4 summarizes additional items that should be provided when dealing with specific types of accidents. When possible, the summary tabulation should provide the necessary quantitative information. If, however, a particular assumption cannot be simply or clearly stated in the table, the table should reference a section or an appendix that adequately discusses the information.

Judgment should be used in eliminating unnecessary parameters from the summary table or in adding parameters of significance that do not appear in Table 15-3 or 15-4. The summary table should have two columns. One column should indicate the assumptions used in the design basis analysis, while the other should indicate assumptions used in the realistic analysis.

A diagram of the dose computation model, labeled "Containment Leakage Dose Model," should be appended to Chapter 15. An explanation of the model should accompany the diagram. The purpose of the appendix is to clearly illustrate the containment modeling, the leakage or transport of radioactivity from one compartment to another or to the environment, and the presence of engineered safety features (ESF) such as filters or sprays that are called on to mitigate the consequences of the LOCA. The diagram should employ easily identifiable symbols, e.g., squares to represent the containment or various portions of it, lines with arrowheads drawn from one compartment to another or to the environment to indicate leakage or transport of radioactivity, and other suitably labeled or defined symbols to indicate the presence of ESF filters or sprays. Individual sketches (or equivalent) may be used for each significant time interval in the containment leakage history (e.g., separate sketches showing the pulldown of a dual containment annulus and the exhaust and recirculation phases once negative pressure in the annulus is achieved, with the appropriate time intervals given).

In presenting the assumptions and methodology used in determining the radiological consequences, care should be taken to ensure that analyses are adequately supported with backup information, either by reporting the information where appropriate, by referencing other sections within the SAR, or by referencing documents readily available to the NRC staff. Such information should include the following:

    a. A description of the mathematical or physical model employed, including any simplifications or approximations introduced to perform the analyses.

    b. An identification and description of any digital computer program or analog simulation used in the analysis. The detailed description of mathematical models and programs are preferably included by reference with only summaries provided in the SAR text.

c. An identification of the time-dependent characteristics, activity, and release rate of the fission products or other transmissible radioactive materials within the containment system that could escape to the environment via leakages in the containment boundaries and leakage through lines that could exhaust to the environment.

d. The considerations of uncertainties in calculational methods, equipment performance, instrumentation response characteristics, or other indeterminate effects taken into account in the evaluation of the results.

e. A discussion of the extent of system interdependency (containment system and other engineered safety features) contributing directly or indirectly to controlling or limiting leakages from the containment system or other sources (e.g., from spent fuel handling areas), such as the contribution of (1) containment water spray systems, (2) containment air cooling systems, (3) air purification and cleanup systems, (4) reactor core spray or safety injection systems, (5) postaccident heat removal systems, and (6) main steam line isolation valve leakage control systems (BWR).

This section should present the results of the dose calculations giving the potential 2-hour integrated whole body and thyroid doses for the exclusion boundary. Similarly, it should provide the doses for the course of the accident at the closest boundary of the low population zone (LPZ) and, when significant, the doses to the control room operators during the course of the accident. Other organ doses should be presented for those cases where a release of solid fission products or transuranic elements are postulated to be released to the containment atmosphere.

TABLE 15-1

REPRESENTATIVE INITIATING EVENTS
TO BE ANALYZED IN SECTIONS 15.X.X OF THE SAR

1. Increase in Heat Removal by the Secondary System

   1.1 Feedwater system malfunctions that result in a decrease in feedwater temperature

   1.2 Feedwater system malfunctions that result in an increase in feedwater flow

   1.3 Steam pressure regulator malfunction or failure that results in increasing steam flow

   1.4 Inadvertent opening of a steam generator relief or safety valve

   1.5 Spectrum of steam system piping failures inside and outside of containment in a PWR

2. Decrease in Heat Removal by the Secondary System

   2.1 Steam pressure regulator malfunction or failure that results in decreasing steam flow

   2.2 Loss of external electric load

   2.3 Turbine trip (stop valve closure)

   2.4 Inadvertent closure of main steam isolation valves

   2.5 Loss of condenser vacuum

   2.6 Coincident loss of onsite and external (offsite) a.c. power to the station

   2.7 Loss of normal feedwater flow

   2.8 Feedwater piping break

3. Decrease in Reactor Coolant System Flow Rate

   3.1 Single and multiple reactor coolant pump trips

   3.2 BWR recirculation loop controller malfunctions that result in decreasing flow rate

   3.3 Reactor coolant pump shaft seizure

   3.4 Reactor coolant pump shaft break

TABLE 15-1 (Continued)

4. **Reactivity and Power Distribution Anomalies**

    4.1 Uncontrolled control rod assembly withdrawal from a subcritical or low power startup condition (assuming the most unfavorable reactivity conditions of the core and reactor coolant system), including control rod or temporary control device removal error during refueling

    4.2 Uncontrolled control rod assembly withdrawal at the particular power level (assuming the most unfavorable reactivity conditions of the core and reactor coolant system) that yields the most severe results (low power to full power)

    4.3 Control rod maloperation (system malfunction or operator error), including maloperation of part length control rods

    4.4 Startup of an inactive reactor coolant loop or recirculating loop at an incorrect temperature

    4.5 A malfunction or failure of the flow controller in a BWR loop that results in an increased reactor coolant flow rate

    4.6 Chemical and volume control system malfunction that results in a decrease in the boron concentration in the reactor coolant of a PWR

    4.7 Inadvertent loading and operation of a fuel assembly in an improper position

    4.8 Spectrum of rod ejection accidents in a PWR

    4.9 Spectrum of rod drop accidents in a BWR

5. **Increase in Reactor Coolant Inventory**

    5.1 Inadvertent operation of ECCS during power operation

    5.2 Chemical and volume control system malfunction (or operator error) that increases reactor coolant inventory

    5.3 A number of BWR transients, including items 2.1 through 2.6 and item 1.2.

6. **Decrease in Reactor Coolant Inventory**

    6.1 Inadvertent opening of a pressurizer safety or relief valve in a PWR or a safety or relief valve in a BWR

TABLE 15-1 (Continued)

    6.2   Break in instrument line or other lines from reactor coolant pressure boundary that penetrate containment

    6.3   Steam generator tube failure

    6.4   Spectrum of BWR steam system piping failures outside of containment

    6.5   Loss-of-coolant accidents resulting from the spectrum of postulated piping breaks within the reactor coolant pressure boundary, including steam line breaks inside of containment in a BWR

    6.6   A number of BWR transients, including items 2.7, 2.8, and 1.3

7.   <u>Radioactive Release from a Subsystem or Component</u>

    7.1   Radioactive gas waste system leak or failure

    7.2   Radioactive liquid waste system leak or failure

    7.3   Postulated radioactive releases due to liquid tank failures

    7.4   Design basis fuel handling accidents

    7.5   Spent fuel cask drop accidents

8.   <u>Anticipated Transients Without Scram</u>

    8.1   Inadvertent control rod withdrawal

    8.2   Loss of feedwater

    8.3   Loss of a.c. power

    8.4   Loss of electrical load

    8.5   Loss of condenser vacuum

    8.6   Turbine trip

    8.7   Closure of main steam line isolation valves

TABLE 15-2

INPUT PARAMETERS AND
INITIAL CONDITIONS FOR TRANSIENTS AND ACCIDENTS

Neutron Power

Moderator Temperature Coefficient of Reactivity

Moderator Void Coefficient of Reactivity

Doppler Coefficient of Reactivity

Effective Neutron Lifetime

Delayed Neutron Fraction

Average Heat Flux

Maximum Heat Flux

Minimum DNBR, CHFR, or CPR

Axial Power Distribution

Radial Power Distribution

Core Coolant Flow Rate

Recirculation Loop Flow Rate (BWR)

Core Coolant Inlet Temperature

Core Average Coolant Temperature (PWR)

Core Average Steam Volume Fraction (BWR)

Core Coolant Average Exit Temperature, Steam Quality, and Steam Void Fraction

Hot Channel Coolant Exit Temperature, Steam Quality, and Steam Void Fraction

Maximum Fuel Centerline Temperature

Reactor Coolant System Inventory (lb)

Coolant Level in Reactor Vessel (BWR)

Coolant Level in Pressurizer (PWR)

Reactor Coolant Pressure

TABLE 15-2 (Continued)

Steam Flow Rate

Steam Pressure

Steam Quality (temperature if superheated)

Feedwater Flow Rate

Feedwater Temperature

CVCS Flow and Boron Concentration (if these vary during the course of the transient or accident being analyzed)

Control Rod Worth, Differential, and Total

TABLE 15-3

REPRESENTATIVE PARAMETERS TO BE TABULATED*
FOR POSTULATED ACCIDENT ANALYSES

	Design Basis Assumptions	Realistic Assumptions
1. Data and assumptions used to estimate radioactive source from postulated accidents		
a. Stretch power level		
b. Burnup		
c. Percent of fuel perforated		
d. Release of activity by nuclide		
e. Iodine fractions (organic, elemental, and particulate)		
f. Reactor coolant activity before the accident (and secondary coolant activity for PWR)		
2. Data and assumptions used to estimate activity released		
a. Primary containment volume and leak rate		
b. Secondary containment volume and leak rate		
c. Valve movement times		
d. Adsorption and filtration efficiencies		
e. Recirculation system parameters (flow rates versus time, mixing factor, etc.)		
f. Containment spray first order removal lambdas as determined in Section 6.2.3		
g. Containment volumes		
h. All other pertinent data and assumptions		
3. Dispersion Data		
a. Location of points of release		
b. Distances to applicable receptors (e.g., control room, exclusion boundary, and LPZ)		
c. χ/Qs at control room, exclusion boundary, and LPZ (for time intervals of 2 hours, 8 hours, 24 hours, 4 days, 30 days)		

*As applicable to the event being described

TABLE 15-3 (Continued)

			Design Basis Assumptions	Realistic Assumptions
4.	Dose Data			
	a.	Method of dose calculation		
	b.	Dose conversion assumptions		
	c.	Peak [or f(t)] concentrations in containment		
	d.	Doses (whole body and thyroid doses for LPZ and exclusion boundary; beta, gamma, and thyroid doses for the control room)		

TABLE 15-4

ADDITIONAL PARAMETERS AND INFORMATION TO BE PROVIDED
OR REFERENCED IN THE SUMMARY TABULATION FOR SPECIFIC DESIGN
BASIS ACCIDENTS

1. <u>Loss-of-Coolant Accident</u> (Section 15.6.5)

   a. <u>Hydrogen Purge Analysis</u>

      (1) Holdup time prior to purge initiation (assuming recombiners are inoperative)

      (2) Iodine reduction factor

      (3) χ/Q values at appropriate time of release

      (4) Purge rates for at least 30 days after initiation of purge

      (5) LOCA plus purge dose at LPZ

   b. <u>Equipment Leakage Contribution to LOCA Dose</u>

      (1) Iodine concentration in sump water after LOCA

      (2) Maximum operational leak rate through pump seals, flanges, valves, etc.

      (3) Maximum leakage assuming failure and subsequent isolation of a component seal

      (4) Total leakage quantities for (2) and (3)

      (5) Temperature of sump water vs time

      (6) Time intervals for automatic and operator action

      (7) Leak paths from point of seal or valve leakage to the environment

      (8) Iodine partition factor for sump water vs temperature of water

      (9) Charcoal adsorber efficiency assumed for iodine removal

   c. <u>Main Steam Line Isolation Valve Leakage Control System Contribution to LOCA Dose (BWR)</u>

      (1) Time of system actuation

      (2) Fraction of isolation valve leakage from each release point

      (3) Flow rates vs time for each release path

TABLE 15-4 (Continued)

    (4) Location of each release point

    (5) Transport time to each release point

2. <u>Waste Gas System Failure</u> (Section 15.7.1)

   a. Activity transfer times to waste gas system components

   b. Number of tanks or other holdup components

   c. Tank volumes

   d. Charcoal bed delay times for Xe and Kr

   e. Seismic classification of tank and associated piping

   f. Decontamination factors of components

   g. Primary coolant volume

   h. Isotopic activity in each system component including daughter products

   i. Time to isolate air ejector

   j. Delay time in delay pipe

   k. Design basis activity measured at air ejector (Ci/sec) including contribution due to activity spiking in coolant

3. <u>Main Steam Line and Steam Generator Tube Failures</u> (Sections 15.1.5, 15.6.3, 15.6.4)

   a. Characterization of primary and secondary (PWR) system (e.g., temperatures, pressures, steam generator water capacity, steaming rates, feedwater rates, and blowdown rates); parameter values should be given for periods prior to, during, and following the accident

   b. Iodine spiking produced by the shutdown and depressurization (ratio of concentration of all iodine isotopes before and after depressurization)

   c. Chronological list of system response times, operator actions, valve closure times, etc.

   d. Steam and water release quantities, and all assumptions made in their computation

   e. Iodine partition factors and their bases

TABLE 15-4 (Continued)

    f.   Fuel damage resulting from single rod stuck in combination with the accident

4. **Fuel Handling Accident** (Section 15.7.4)

    a.   Number of fuel rods in core

    b.   Number, burnup, and decay time of fuel rods assumed to be damaged in the accident

    c.   Radial peaking factor for the rods assumed to be damaged

    d.   Earliest time after shutdown that fuel handling begins

    e.   Amounts of iodines and noble gases released into pool

    f.   Pool decontamination factors

The following items should be provided to determine if the calculational methods of Regulatory Guide 1.25 (Safety Guide 25), "Assumptions Used for Evaluating the Potential Radiological Consequences of a Fuel Handling Accident in the Fuel Handling and Storage Facility for Boiling and Pressurized Water Reactors," apply:

    g.   Maximum fuel rod pressurization

    h.   Minimum water depth between top of fuel rods and fuel pool surface

    i.   Peak linear power density for the highest power assembly discharged

    j.   Maximum centerline operating fuel temperature for the fuel assembly in item i above

    k.   Average burnup for the peak assembly in item i above

5. **Control Rod Ejection and Control Rod Drop Accidents** (Sections 15.4.8 and 15.4.9)

    a.   Percent of fuel rods undergoing clad failure

    b.   Radial peaking factors for rods undergoing clad failure

    c.   Percent of fuel reaching or exceeding melting temperature

    d.   Peaking factors for fuel reaching or exceeding melting temperature

    e.   Percent of core fission products assumed released into reactor coolant

TABLE 15-4 (Continued)

    f.   Summary of primary and secondary system parameters used to determine activity release terms (see 3 above) from steam line path (PWR)

    g.   Summary of containment system parameters used to determine activity release terms from containment leak paths

    h.   Summary of system parameters and decontamination factors used to determine activity release from condenser leak paths (BWR)

6.   <u>Spent Fuel Cask Drop</u> (Section 15.7.5)

    a.   Number of fuel elements in largest capacity cask

    b.   Number, burnup, and decay time of fuel elements in cask assumed to be damaged

    c.   Number, burnup, and decay time of fuel elements in pool assumed to be damaged as a consequence of a cask drop (if any)

    d.   Average radial peaking factor for the rods assumed to be damaged

    e.   Earliest time after reactor fueling that cask loading operations begin

    f.   Amounts of iodines and noble gases released into air and into pool

    g.   Pool decontamination factors, if applicable.

APPENDIX B

ABBREVIATIONS

AC	alternating current
ACI	American Concrete Institute
ACRS	Advisory Committee on Reactor Safeguards
AEC	Atomic Energy Commission
AGAP	as great as practicable
ALAP	as low as practicable
ALARA	as low as reasonably achievable
ANS	American Nuclear Society
ANSI	American National Standards Institute
ASLAB	Atomic Safeguard Licensing Appeal Board
ASLB	Atomic Safeguard Licensing Board
ASME	American Society of Mechanical Engineers
ASTM	American Society for Testing and Materials
ATWS	anticipated transients without scram
BEIR	biological effects of ionizing radiation
BOP	balance of plant
BSAR	Babcock & Wilcox standard safety analysis report
BWR	boiling water reactor
CANDU	Canadian heavy water-moderated reactor
CESSAR	Combustion Engineering standard safety analysis report
CFR	Code of Federal Regulations
Ci	Curie
CI	containment integrity
CIS	containment isolation system
CSS	containment spray system
DBA	design-base accident
DC	direct current
DNB	departure from nucleate boiling
DNBR	departure from nucleate boiling ratio
ECCS	emergency core cooling system
EPRI	Electric Power Research Institute
ERDA	Energy Research and Development Administration
ESF	engineered safety features
FLECHT	full length emergency cooling heat transfer
FNP	floating nuclear plant
F-P SSAR	Fluor Pioneers standard safety analysis report
FRC	Federal Radiation Council
FSAR	final safety analysis report
GCR	gas-cooled reactors
GE	General Electric
GESSAR	General Electric standard safety analysis report
GPM	gallons per minute

GW	gigawatt
HEPA	high-efficiency particulate air
HPCS	high pressure core spray
HPIS	high pressure injection system
HTGR	high temperature gas-cooled reactors
HWR	heavy water-moderated reactor
ICRP	International Commission on Radiological Protection
I&E	inspection and enforcement
IEEE	Institute of Electrical and Electronic Engineers
INEL	Idaho National Engineering Laboratory
INMM	Institute of Nuclear Materials Management
LCVIP	Licensee Contractor Vender Inspection Program
LET	linear energy transfer
LMFBR	liquid-metal-cooled fast breeder reactor
LOCA	loss-of-coolant-accident
LOFT	loss-of-fluid-test
LPCI	low pressure coolant injection
LPCS	low pressure core spray
LPIS	low pressure injection system
LWA	limited work authorization
LWR	light water reactor
LWST	liquid waste storage tanks
MCi	megacurie
MOX	mixed oxide fuels
MPC	maximum permissible concentration
MPE	maximum permissible exposure
MREM	milli-Roentgen equivalent man
MWD/MTU	megawatt days per metric ton uranium
MWD/T	megawatt days per tonne
MWe	megawatt electrical
MWt	megawatt thermal
NAS	National Academy of Science
NCRP	National Council on Radiation Protection
NEPA	National Environmental Policy Act
NMSS	Nuclear Materials Safety and Safeguards
NRC	Nuclear Regulatory Commission
NRR	Office of Nuclear Regulatory Research
NSSS	nuclear steam supply system
OBE	operating basis earthquake
PAHR	post accident heat removal
PARR	post accident radioactivity removal
PI	proportional plus integral
PMH	probable maximum hurricane
PNL	Pacific Northwest Laboratories
PSAR	preliminary safety analysis report
PSIA	pounds per square inch absolute
PSID	pounds per square inch difference
PWR	pressurized water reactor
QA	quality assurance
RCG	radioactive concentration guide
RCS	reactor coolant system
REM	Roentgen equivalent man
RESAR	Westinghouse standard safety analysis report

RHR	residual heat removal
RRRC	regulatory requirements review committee
RT	reactor trip
SAR	safety analysis report
SER	safety evaluation report
SFSP	spent fuel storage pool
SINB	Southern Interstate Nuclear Board
SIS	safety injection system
SIW	submarine prototype reactor
SNUPPS	standard nuclear power plant system
SRP	standard review plan
SSAR	standard safety analysis report
SSE	safe shutdown earthquake
SWESSAR	Stone & Webster standard safety analysis report
TVA	Tennessee Valley Authority
WGST	waste gas storage tanks
WUP	Wisconsin Utility Project
y	year

INDEX

A

abbreviations   403
accidents
  analysis   381
  calculational methods   233-250, 291-294, 298-300
  classification   209-210, 310-312
  containment, confinement   1-2
  design basis   39
  ECCS acceptance criteria   237-238, 281-282
  heat removal   19, 21-27
  loss-of-coolant (LOCA)
    See loss-of-coolant accidents
  probabilities of   42-44, 304-305, 308-309, 343-349
Advisory Committee on Reactor Safeguards (ACRS)   60, 65-66
air filters   9, 16
American National Standards Institute (ANSI)   72, 77, 90
anticipated transients without scram (ATWS)   352-353, 355-356
ASME Boiler and Pressure Vessel Code   30, 74, 77
Atomic Safety and Licensing Appeal Board (ASLAB)   60
Atomic Safety and Licensing Board (ASLB)   68

B

boiling water reactors (BWR)
  See reactor types
Browns Ferry fire   352, 356-357
burnout
  See DNB

C

charcoal adsorbers   9, 16
chemical reactions
  hydrogen generation   21, 28
  metal-water   28, 236-238, 282

chemical reactions (continued)
  radiolysis of water   2
COBBA   228, 232-233
Code of Federal Regulations
  10CFR 2   97-98
  10CFR 50   30, 60, 62, 65, 67, 74, 77, 78, 89, 93, 240, 248-249
  10CFR 55   82, 85
  10CFR 100   1, 9, 39, 41, 42, 60, 65, 74
common mode failures   337-339, 355
computer codes   233-234, 248-250
  See also specific code names
construction permits   60, 68, 93
containment
  concepts   2
  design   29-31
  double   16, 18-20
  dry   2, 3
  environmental effects   29-31
  heat removal systems   19, 21-27, 324
  HTGR   29
  ice condenser   2, 4
  Mark I   2, 5, 28, 353
  Mark II   2, 6, 28, 353
  Mark III   2, 7, 28
  penetrations   8-15
  philosophy   1
  pressure-suppression   2, 324
  seismic effects   30, 36
  sprays   16, 324
control systems
  BWRs   175, 197-203
  general   172-177, 205-207
  PWRs   175, 177-197, 206
core melting   325-326
critical heat flux
  See DNB

D

defense-in-depth philosophy   310, 351
departure from nucleate boiling (DNB)   204-205

design basis accidents
   See accidents
Doppler coefficient   172-173

E

ECCS acceptance criteria
   See accidents
effluents
   See radiation
emergency core cooling systems (ECCS)
   boiling water reactors   295-296
   pressurized water reactors   283-285
enforcement   97-98
engineered safeguards   35, 311, 322-325
environmental effects   29-31, 35, 36
environmental report   66
environmental statement   67, 68
event tree   310, 329-333

F

fault tree analysis   310, 333-339
final safety analysis reports (FSAR)   69
fires in cable trays
   See Browns Ferry fire
fission product decay heat   197, 249
fission product release   116-134
fission product removal systems   9-19
fission products   101-104, 107-108
FLECHT (full length emergency cooling heat transfer   249, 289-291
fluid flow
   computer codes   233-234, 291
   critical flow   242-247
   ECC bypass   237, 287-289
   flow distribution   228, 232-233

G

gas cooled reactors
   See reactor types
gaseous radwastes
   See radioactive wastes

H

heat transfer
   burnout   See critical heat flux or DNB
   contact (gap)   218-220
   critical heat flux   220, 226-228
   DNB   204-205, 220, 224-226
   film boiling   237, 239-240
   fuel temperatures   211-218
   heat-transfer coefficients   218-220, 240
   radiation heat transfer   240-242
high pressure core spray (HPCS)
   See emergency core cooling systems
high pressure injection systems (HPIS)
   See emergency core cooling systems

I

ice-condenser containment
   See containment
inspection and enforcement   70-71, 87-89
inspection and enforcement bulletin   98-99
inspection programs   92-98
instrumentation   205-206
iodine hazard   101, 133, 137, 141, 144
iodine removal   16

J

K

kinetics
   See neutron kinetics and reactivity effects

L

LASER   215
leakage control   16, 21
licensing process   60-65, 360-363
limited work authorization (LWA)   68, 69, 93
loss-of-coolant accidents (LOCA)
   causes   321-322
   for boiling water reactors   297-298
   for pressurized water reactors   285-291

loss-of-coolant accidents (continued)
　general　1-2, 233, 236-237, 248-250, 311, 321-326
loss-of-fluid test (LOFT)　282
low pressure injection system (LPIS)
　See emergency core cooling systems

M

maximum permissible concentrations (MPC)　101, 325
meltdown　325-326
metal-water reaction
　See chemical reactions
meteorology　342
missiles　30, 31

N

neutron kinetics
　basic equation　169-172
　delayed neutrons　170
　multiplication factor　170
　neutron generation time　171
　neutron lifetime　170
　reactivity　170-171
　reactor period　171
noble gases　104, 109-111, 129-130, 134
nuclear power growth　255-258
Nuclear Regulatory Commission (NRC)　57-60

O

operating license　69, 70
operator licensing　82, 85

P

penetrations, containment
　See containment
plutonium hazards　101, 147
pool dynamic loads (for BWR)　353-355
preliminary safety analysis report (PSAR)　62, 64-65, 69
pressure-suppression systems
　See containment
pressurized water reactors (PWR)
　See reactor types
pressurizer control　188-191
public hearings　67-69

Q

quality assurance　70, 77, 90-91

R

radiation
　hazards　157, 159-162
　main isotopic hazards　101-106, 141-150
　nuclear plant releases　39, 48, 49, 105, 109-111, 116-141, 328-329, 333, 339-341
　reduction of hazards　162-165
radioactive wastes
　actinides　101, 127, 137, 264-266
　categories　258, 260
　characteristics　261-262
　disposal　270, 272-277
　fission products　101
　gaseous　127-131
　liquid　118-125
　neutron activation　101
　power plant inventory　312, 319-321
　projected　258-260
　shipping　267, 270-271
　solids　112, 266-267
　storing　267-270
radiological effects　127, 132-137, 157
Rassmussen report
　See WASH-1400
reactivity coefficients
　See reactivity effects
reactivity effects
　burnup　174
　control changes　174-176
　feedback　172
　fission product buildup　174
　moderator coefficient　179
　temperature　172-174
　void coefficients　198
reactivity feedback
　See reactivity effects
reactor dynamics
　See neutron kinetics
reactor licensing
　See licensing process
reactor period
　See neutron kinetics
reactor safety study
　See WASH-1400

reactor siting
  exclusion area 39, 41
  low population zone 39, 41, 42
  off-shore 48, 50-52
  philosophy 33-43
  population center distance 39, 41, 42
  selection considerations 33-43
  underground 50, 53-55
reactor types
  boiling water reactors (BWR) 197-200, 295-296, 312-313, 315-317
  gas-cooled reactors 29
  pressurized water reactors (PWR) 177-178, 282-283, 312-314
reactor vessel supports 353
regulatory guides 74, 76, 90
RELAP 291
risk concept 301-304, 328
risk criteria 43-48

S

safety analysis report (SAR) 62-69
safety evaluation report (SER) 65, 66, 69
safety features
  See accidents, containment, emergency core cooling systems, engineered safeguards
safety injection systems
  See emergency core cooling systems
scram systems 205-207
seismic effects
  containment *See* containment
  sloshing 2
shipping containers 360-363
siting
  See reactor siting
standardization 78-82
standard review plan (SRP) 64, 74, 77
standards 71-78, 90
steam binding 237
steam generators 194-197

T

technical specifications 70
temperature coefficients
  See reactivity effects
testing
  operational 96-97

testing (continued)
  startup 96-97
  preoperational 95-96
transients 311-312, 326-327, 382
transportation of fuel 34, 359
tritium 104, 122-128, 137

U

uranium oxides 211-218
U.S. Atomic Energy Commission
  See Nuclear Regulatory Commission

V

valves
  containment isolation 9-15
  pressure relief 296
  vent 289, 301
violations
  See enforcement
void coefficient
  See reactivity coefficients

W

WASH-1400 39, 42-44, 303-349, 355
waste
  See radioactive wastes
waste management 261-277
waste storage 48
water-hammer in feedwater piping 352, 357-358
water reactors
  See reactor types
winds 30

X

Y

Z

zirconium-water reaction
  See chemical reactions